ADVANCES IN WAVELETS

Springer
*Singapore
Berlin
Heidelberg
New York
Barcelona
Budapest
Hong Kong
London
Milan
Paris
Tokyo*

ADVANCES IN WAVELETS

Ka-Sing Lau (ed.)

Springer

Professor Ka-Sing Lau
Department of Mathematics
The Chinese University of Hong Kong
Shatin
New Territories
Hong Kong (SAR)
China

Library of Congress Cataloging-in-Publication Data

Advances in wavelets / edited by Ka-Sing Lau
    p.   cm.
Includes bibliographical references.
ISBN 9814021083
1. Wavelets (Mathematics)   2. Lau, Ka-Sing.
QA403.3.A38   1998
515'.2433--dc21   98-29266
                                      CIP

ISBN 981-4021-08-3

Department of Mathematics This work is subject to copyright. All rights are reserved, whether the whole or part of the material is concerned, specifically the rights of translation, reprinting, reuse of illustrations, recitation, broadcasting, reproduction on micro-films or in any other way, and storage in databanks or in any system now known or to be invented. Permission for use must always be obtained from the publisher in writing.

© Springer-Verlag Singapore Pte. Ltd. 1999
Printed in Singapore

The publisher makes no representation, express or implied, with regard to the accuracy of the information contained in this book and cannot accept any legal responsibility or liability for any errors or omissions that may be made.

Typesetting: Camera-ready by editor
SPIN 10691536   5 4 3 2 1 0

Preface

Wavelets is a new area that stands at the intersection of the frontiers of mathematics, scientific computing and signal and image processing. It has been one of the major research directions in science in the last decade and is still undergoing rapid growth. The Far East shares with the world the great enthusiasm and development on this new and exciting topic. In the year of 1996 - 97, the Institute of Mathematical Sciences at the Chinese University of Hong Kong(CUHK) sponsored a year-long programme on "Wavelets and their Applications". The programme was intended to bring together interested researchers from local and abroad to discuss and initiate wavelet study in Hong Kong. There were seminars and lectures throughout the year and a four-day workshop on May 5 - 8, 1997. More than 70 participants attended the events, including faculty members, scientists and graduate students from local and visitors from China, Singapore, North and South America and Europe.

The present monograph contains eleven polished versions of the lectures selected from the programme. It includes expository articles and current research findings and covers the following topics

- Theory of frames and applications - Wavelet coefficients - Refinable functions and approximation - Multiwavelets - Tiling

We apologize that we are not able to include all the interesting lectures in the monograph because of the limit of the size.

The theory of frame was introduced in the 50's and is now one of the fundamental parts of the wavelet theory. We have two chapters on this topic: Benedetto first outlines the theory and its application to signal processing; Ron and Shen discuss their new work of shift-invariant tight frame versus the non shift-invariant wavelet basis.

There are two articles concerning the important topic of wavelet coefficients. Mallat and Falcon study the distortion rate for high bit rate and low bit rate compression. To some extend they justify the folklore: If one could approximate an image with a small number of wavelet coefficients, then the basis should be well-adapted to performing bit compression. In another direction Perrier and Wickerhauser introduce the "connection coefficients" to facilitate the wavelet expansion of the multiplication of two functions.

Refinable functions, in particular those related to the conjugate quadrature filters, play a central role in wavelet theory. In an article, Lawton makes use of the Hardy space theory and the Lie group of paraunitary matrices to investigate an approximation question of such filters. For the approximation to polynomials, Cabrelli, Heil and Molter present a detailed exposition of the "accuracy" of the refinable function, and J. Wang considers the vector-valued case (multi-wavelet) in relation to the independence of the translations.

The theory of multiwavelets is a more recent topic. It is well-known that a scalar-valued wavelet cannot be continuous, compactly supported, orthogonal and symmetric simultaneously. In 1994, Donovan, Geromino, Hardin and Massopust first constructed a vector-valued wavelet with all these properties and initiated the interest on multiwavelets. In this monograph they present a brief account on their construction based on fractal interpolation. Goh, Lee and Tang give a historical exposition from the point of view of cardinal Hermite splines and summarize some of their recent contributions. There are two other important topics of multiwavelet:one by Jia on the biorthogonality and the other by J. Wang on the accuracy as mentioned above.

The Haar wavelet is closely connected with tiling. There are exciting progress in this geometric and algebraic direction. In the last chapter, Y. Wang gives a survey article on his recent joint work with Lagarias.

The idea of this year-long programme on wavelet was first initiated with Professor Ruilin Long of the Institute of Mathematics, Academia Sinica. Unfortunately Professor Long passed away before the programme started. We are very sad to lose such a distinguished mathematician. The co-organizational task was taken up by his colleague, Professor Han Lin Chen, to whom we express our sincere thanks.

We are also grateful to Professors Raymond Chan, Chi Wei Leung, Kung Fu Ng, Dr. Sze-Man Ngai and Ms. Pauline Chan for their help in carrying out the programme and to the Department of Mathematics of CUHK for providing the facilities and secretarial support throughout the year. Special thanks are due to Professor Kam Wing Leung for managing the computer files of the manuscripts, without his help the monograph will not be able to materialize.

Finally, we like to acknowledge our thanks to the following organizations for their generous financial support to the programme: Lee Hysan Foundation, Hong Kong Pei Hua Education Foundation, K.C. Wong Education Foundation, U.S. Army Research Office(Far East). Hong Kong Mathematical Society and the Institute of Mathematical Sciences, CUHK.

The editor

Contents

1   Frames, Sampling and Seizure Prediction   1
     J. J. Benedetto

2   Construction of Compactly Supported Affine Frames in $L_2(\mathbb{R}^d)$ 27
     A. Ron and Z. Shen

3   Understanding Wavelet Image Coding   51
     S. Mallat and F. Falzon

4   Multiplication of Short Wavelet Series using Connection Coefficients   77
     V. Perrier and M.V. Wickerhauser

5   Conjugate Quadrature Filters   103
     W. M. Lawton

6   Polynomial Reproduction by Refinable Functions   121
     C. A. Cabrelli, C. Heil and U. M. Molter

7   From Cardinal Hermite Splines to Multiwavelets   163
     S.S. Goh, S.L. Lee and W.S. Tang

8   Orthogonal Multiwavelet Constructions: 101 Things to do with a Hat Function   187
     G.C. Donovan, J.S. Geronimo and D.P. Hardin

9   Convergence of Vector Subdivision Schemes and Construction of Biorthogonal Multiple Wavelets   199
     R. Q. Jia

10   Study of Linear Independence and Accuracy of Scaling Vectors via Two-scale Factors   229
     J. Z. Wang

11   Self-Affine Tiles   261
     Y. Wang

1
Frames, Sampling, and Seizure Prediction

John J. Benedetto[1]

ABSTRACT The theory of frames is presented from the point of view of sampling theory. A proof is given of an iterative implementable algorithm designed to reconstruct a large class of signals in terms of their sampled values. The sampling point of view is expanded historically and theoretically. The former outlines the contributions of Cauchy and Whittaker, and notes applications to number theory and interpolation theory. The latter recalls the sampling theorem for locally compact abelian groups, and explains its role in optimizing bandwidth. A relevant application to epileptic seizure prediction is given.

1 Introduction

The theory of frames was introduced by Duffin and Schaeffer [DS52] in 1952 to deal with expansions of functions in terms of nonharmonic Fourier series. Their work originated in problems of completeness for sets of exponential functions, spearheaded by the fundamental work of Paley and Wiener [PW34] in 1934 and culminating in the profound papers of Beurling and Malliavin [BM62], [BM67] in the 1960s.

In 1986, Daubechies, Grossmann, and Meyer [DGM86] reintroduced the theory of frames in the context of signal processing and the then new theory of wavelets. Their point of view was thoroughly and quantitatively developed by Daubechies in the late 1980s, e.g., [Dau92]. In the process, the theory of frames has emerged as a tool for dealing with noise reduction; and the interaction of frames and signal processing has established the relationship between sampling theory and the theory of frames.

---

[1]Department of Mathematics, University of Maryland, College Park, MD 20742
The author is grateful for informative and stimulating conversations with Stéphane Mallat, Linwood Vincent, Gregory Warhola, and Victor Wickerhauser. He also wants to express his appreciation to the AFOSR (F496209610193) for their generous support of the work.

Independent of [DGM86] and the emergence of wavelet theory by Daubechies, Grossmann, Mallat, Meyer, and Morlet, there were related, important, and fundamental results due to Frazier and Jawerth [FJ85], and Feichtinger and Gröchenig [FG86].

There are many natural avenues of investigation relating frames, sampling, and applications to signal processing. For example, the theory of frames arises in a natural way in auditory modelling, and the associated irregular sampling theory leads to an effective noise reduction and speech compression algorithm called WAM [BT93](U.S. Patent 5,388,182 (1995)). Another avenue involves uniform oversampling used in quantization noise reduction problems associated with Sigma-Delta ($\Sigma\Delta$) Analogue to Digital Converters (ADCs), e.g., [CT92]. The theory of frames, sampling, and noise reduction combine in this case into our theory of frame multiresolution analysis (FMRA) [BL97].

In this exposition, we shall travel a path involving frames, sampling, and applications in the following way. *Section 2* is devoted to the theory of frames, recording a useful algorithm and relating frames to sampling. In *Section 3*, we shall discuss uniform sampling, beginning with Cauchy's original sampling theorem (1841) and ending with Kluvánek's beautiful generalization (1965), whose basic idea is also used in tiling and wavelet problems. We have chosen to treat uniform sampling in *Section 3* because our application to seizure prediction in *Section 4* makes essential use of uniform oversampling. Specifically, we are led to a wavelet based uniform sampling algorithm, which makes best possible use of preseizure epileptic data in order to determine the fundamental periodicities that characterize certain types of epileptic seizures. This detection of periodicities has many other applications including problems associated with ocean wave prediction and cockpit motion sickness.

Our notation is described in *Section 5*. To begin the paper, however, we do point out that integration over Euclidean space, $\mathbb{R}^d$, resp., summation over the d-dimensional integer lattice $\mathbb{Z}^d$, is designated by "$\int$" instead of "$\int_{\mathbb{R}^d}$", resp., "Σ" instead of "$\Sigma_{n\in\mathbb{Z}^d}$". We shall usually deal with the case $d=1$.

2 The theory of frames

Let H be a separable Hilbert space with inner product $\langle x,y\rangle$ and norm $\|x\| = \langle x,y\rangle^{1/2}$.

Definition 2.1 (Bases)
 a. A sequence $\{x_n : n \in \mathbb{Z}^d\} \subseteq H$ is a *Schauder basis* or *basis* for H if each $y \in H$ has a unique decomposition $y = \sum c_n(y)x_n$ in H, where $\{c_n(y)\} \subseteq \mathbb{C}$.

b. A basis $\{x_n\}$ for H is an *unconditional basis* for H if

$$\exists C > 0 \text{ such that } \forall F \subseteq \mathbb{Z}^d, \text{ finite, and } \forall \{b_j, c_j : j \in F\} \subseteq \mathbb{C},$$

where $|b_j| \leq |c_j|$ for each $j \in F$, we have

$$\|\sum_{n \in F} b_n x_n\| \leq C \|\sum_{n \in F} c_n x_n\|.$$

An unconditional basis $\{x_n\} \subseteq H$ is *bounded* if

$$\exists A, B > 0 \text{ such that } \forall n \in \mathbb{Z}^d, \ A \leq \|x_n\| \leq B.$$

c. It is well-known that separable Hilbert spaces have orthonormal bases (ONBs) [GG81]; and it is elementary to see that ONBs are bounded unconditional bases.

Definition 2.2 (Frames)
 a. A sequence $\{x_n : n \in \mathbb{Z}^d\} \subseteq H$ is a *frame* for H if there exist $A, B > 0$ such that

$$\forall y \in H, \quad A\|y\|^2 \leq \sum |\langle y, x_n \rangle|^2 \leq B\|y\|^2. \qquad (2.1)$$

A and B are *frame bounds*, and a frame is *tight* if $A = B$. A frame is *exact* if it is no longer a frame whenever any one of its elements is removed.
 b. The *frame operator* of the frame $\{x_n\}$ is the function $S : H \longrightarrow H$ defined as $Sy = \sum \langle y, x_n \rangle x_n$ for all $y \in H$.
 c. The theory of frames is due to Duffin and Schaeffer [DS52], cf., [You80], [DGM86], [Dau92], [BF94, Chapter 3]. An exact frame is a bounded unconditional basis and vice-versa, e.g., [You80].

Theorem 2.3 (Frame Decomposition Theorem)
 Let $\{x_n : n \in \mathbb{Z}^d\} \subseteq H$ be a frame for H with frame bounds A and B.
 a. The frame operator S is a topological isomorphism with inverse S^{-1} : $H \longrightarrow H$. $\{S^{-1}x_n\}$ is a frame with frame bounds B^{-1} and A^{-1}, and

$$\forall y \in H, \quad y = \sum \langle y, S^{-1}x_n \rangle x_n = \sum \langle y, x_n \rangle S^{-1}x_n \text{ in } H. \qquad (2.2)$$

 b. If $\{x_n\}$ is a tight frame for H, if $\|x_n\| = 1$ for all n, and if $A = B = 1$, then $\{x_n\}$ is an orthonormal basis for H.
 c. If $\{x_n\}$ is an exact frame for H, then $\{x_n\}$ and $\{S^{-1}x_n\}$ are biorthonormal, i.e.,

$$\forall m, n, \quad \langle x_m, S^{-1}x_n \rangle = \delta(m, n),$$

and $\{S^{-1}x_n\}$ is the unique sequence in H which is biorthonormal to $\{x_n\}$.
 d. If $\{x_n\}$ is an exact frame for H, then the sequence resulting from the removal of any one element is not complete in H, i.e., the linear span of the resulting sequence is not dense in H.

Remark 2.4 (Frames, Coherent Light, and Vitali's Theorem)

a. If $\{x_n : n \in Z^d\}$ is a frame for H with frame bounds A and B and frame operator S, then it is easy to see that

$$\|I - \frac{2}{A+B}S\| \leq \frac{B-A}{A+B} < 1, \tag{2.3}$$

where $I : H \longrightarrow H$ is the identity operator. The inequality (2.3) allows us to prove that

$$S^{-1} = \frac{2}{A+B} \sum_{k=0}^{\infty} \left(I - \frac{2}{A+B}S\right)^k, \tag{2.4}$$

which, in turn, can be used to prove part a of *Theorem 2.3*.

We also mention (2.3) because of the notion of *visibility* V, which is defined as

$$V = \frac{I_{\max} - I_{\min}}{I_{\max} + I_{\min}},$$

where $I_{\max}$ and $I_{\min}$ are maximum and minimum light intensities, e.g., [Kla96]. In the case of full interference of light waves (for the classical two-slit experiment), one has $I_{\min} = 0$; and, hence, the visibility is 1, a value associated with *coherent light*. In the context of frames, the analogy is that $A = 0$, so that a frame is not obtained.

b. We shall verify *Theorem 2.3b* because of a stronger result proved by Vitali which deserves comment.

Since $\{x_n\}$ is tight and $A = 1$ we can write

$$\|x_m\|^2 = \|x_m\|^4 + \sum_{n \neq m} |\langle x_m, x_n \rangle|^2,$$

thereby obtaining the orthonormality of $\{x_n\}$ since each $\|x_n\| = 1$. *Theorem 2.3b* then follows by the following well-known result: if $\{x_n\} \subseteq H$ is orthonormal then it is an orthonormal basis for H if and only if

$$\forall y \in H, \quad \|y\|^2 = \sum |\langle y, x_n \rangle|^2.$$

In 1921, Vitali proved that an orthonormal sequence $\{x_n\} \subseteq L^2[a, b]$ is complete, and so $\{x_n\}$ is an orthonormal basis, if and only if

$$\forall t \in [a, b], \quad \sum |\int_a^t x_n(u) du|^2 = t - a. \tag{2.5}$$

For the case $H = L^2[a, b]$, Vitali's result is stronger than *Theorem 2.3b* since (2.5) is tightness with $A = 1$ for functions $y = \mathbf{1}_{[a,t]}$. Other remarkable contributions by Vitali are highlighted in [Ben76].

Theorem 2.5 (Characterization of Frames)

a. A sequence $\{x_n : n \in \mathbb{Z}^d\}$ in H is a frame for H with frame bounds A and B if and only if the mapping

$$L : H \longrightarrow \ell^2(\mathbb{Z}^d)$$
$$y \longmapsto \{\langle y, x_n \rangle\}$$

is a topological isomorphism of H onto a closed subspace of $\ell^2(\mathbb{Z}^d)$. In this case,

$$\|L\| \leq B^{1/2} \text{ and } \|L^{-1}\| \leq A^{-1/2},$$

where L^{-1} is defined on the range $L(H)$.

b. A sequence $\{x_n : n \in \mathbb{Z}^d\}$ in H is a frame for H if and only if there is $C > 0$ such that for all $y \in H$

$$\sum |\langle y, x_n \rangle|^2 < \infty,$$

$$\exists c_y = \{c_n\} \in \ell^2(\mathbb{Z}^d), \text{ such that } y = \sum c_n x_n \text{ in } H,$$

and

$$\|c_y\|_{\ell^2(\mathbb{Z}^d)} \leq C\|y\|.$$

Part a of *Theorem 2.5* is proved in [BF94, Theorem 7.15]; and part b, which we observed with David Walnut, is proved in [BF94, Remark 3.9]. The "sampling map" L of *Theorem 2.5a* is the *Bessel map* associated with $\{x_n\}$.

Example 2.6 (Gabor and Wavelet Systems)

a. Let $g \in L^2(\mathbb{R})$ and let $a, b > 0$. The *Gabor system* for g and (a, b) is the sequence $\{\varphi_{m,n} : (m, n) \in \mathbb{Z} \times \mathbb{Z}\}$, where

$$\varphi_{m,n}(t) = e^{2\pi i t m b} g(t - na) = e_{mb}(t) \tau_{na} g(t),$$

and where $e_\gamma(t) = e^{2\pi i t \gamma}$ and $\tau_x g(t) = g(t - x)$.

b. Let $\psi \in L^2(\mathbb{R})$. The *affine system* or *wavelet system* for ψ is the sequence $\{\psi_{m,n} : (m, n) \in \mathbb{Z} \times \mathbb{Z}\}$, where

$$\psi_{m,n}(t) = 2^{m/2} \psi(2^m t - n).$$

c. The Fourier transforms of $\varphi_{m,n}$ and $\psi_{m,n}$ are easily computed to be

$$\widehat{\varphi}_{m,n}(\gamma) = e^{2\pi i n a m b} e^{-2\pi i n a \gamma} \widehat{g}(\gamma - mb) = \tau_{mb}(e_{-na}\widehat{g})(\gamma)$$

and

$$\widehat{\psi}_{m,n}(\gamma) = 2^{-m/2} e^{-2\pi i n(\gamma/2^m)} \widehat{\psi}(\gamma/2^m) = 2^{-m/2}(e_{-n}\widehat{\psi})(\gamma/2^m).$$

d. Let $H = L^2(\mathbb{R})$, $g \in L^2(\mathbb{R})$, and $a, b > 0$. If the Gabor system $\{\varphi_{m,n}\}$ for g and (a, b) is a frame, then it is a *Gabor frame*.

Let $H = L^2(\mathbb{R})$ and $\psi \in L^2(\mathbb{R})$. If the wavelet system $\{\psi_{m,n}\}$ for ψ is a frame, then it is a *wavelet frame*.

The verification of the following result is elementary.

Proposition 2.7
Let $\{x_n : n \in \mathbb{Z}^d\} \subseteq H$ be a frame for H with frame bounds A and B, and let L be the Bessel map associated with $\{x_n\}$. Then
i. $S = L^*L$, where $L^* : \ell^2(\mathbb{Z}^d) \longrightarrow H$ is the adjoint of L,
and
ii. For all $y \in H$,
$$y = (S^{-1}L^*)Ly. \tag{2.6}$$

Equation (2.6) can be viewed as a reconstruction formula (for signals y) in which discrete "sampled data" Ly is given. Equations (2.4) and (2.6) can be combined to provide an iterative reconstruction algorithm for signal reconstruction. Unfortunately, this procedure is neither canonically digitally implementable nor effective theoretically as far as analyzing S^{-1} in an applicable way.

Definition 2.8 (Gram Operator)
Let $\{x_n : n \in \mathbb{Z}^d\} \subseteq H$ be a frame for H, with frame operator S, frame bounds A and B, and Bessel map $L : H \longrightarrow \ell^2(\mathbb{Z}^d)$.
  a. The *Gram operator* associated with $\{x_n\}$ is the map $R = LL^* : \ell^2(\mathbb{Z}^d) \longrightarrow \ell^2(\mathbb{Z}^d)$. Because of its discrete nature it can be approximated by finite Gram matrices, which, in turn, can be stored off-line.
  b. Let L' and R' denote the Bessel map and Gram operator, respectively, for the (dual) frame $\{S^{-1}x_n\}$.
  c. It is easy to check that R restricted to $L(H)$ is a bijection onto $L(H)$. If R^{-1} denotes the inverse defined on $L(H)$, then we can extend R^{-1} to $\ell^2(\mathbb{Z}^d)$ by defining the *pseudo-inverse* $R^\dagger$ of R as
$$R^\dagger = R^{-1}P_{L(H)} : \ell^2(\mathbb{Z}^d) \longrightarrow L(H) \subseteq \ell^2(\mathbb{Z}^d),$$
where $P_{L(H)}$ is the orthogonal projection operator onto the image *of* L.

Lemma 2.9
Let $\{x_n : n \in \mathbb{Z}^d\} \subseteq H$ be a frame for H, with Gram operator R, frame bounds A and B, and Bessel map $L : H \longrightarrow \ell^2(\mathbb{Z}^d)$. If $0 < \lambda < 2/B$, then $\|I - \lambda R\|_{L(H)} < 1$. We may take $\lambda = 2/(A+B)$.

Proof: Since $(L')^*$ is surjective, for any $y \in H$ there is a $c \in L'(H)$ so that $y = (L')^*c$. This together with the fact that $\{S^{-1}x_n\}$ is a frame for H yields
$$B^{-1}\langle c, R'c\rangle \leq \langle c, (R')^2 c\rangle \leq A^{-1}\langle c, R'c\rangle.$$
Letting $c = (R')^\dagger d$ for some $d \in \ell^2(\mathbb{Z}^d)$, we have $B^{-1}\langle Rd, d\rangle \leq \langle d, d\rangle \leq A^{-1}\langle Rd, d\rangle$. For all nonzero $d \in L'(H)$ this means
$$A \leq \frac{\langle Rd, d\rangle}{\langle d, d\rangle} \leq B.$$

Thus we have for $\lambda > 0$

$$1 - \lambda B \leq \frac{\langle (I - \lambda R)d, d \rangle}{\langle d, d \rangle} \leq 1 - \lambda A,$$

and, since $I - \lambda R$ is self-adjoint,

$$\|I - \lambda R\|_{L(H)} = \sup_{c \in L(H)} \frac{|\langle (I - \lambda R)c, c \rangle|}{\langle c, c \rangle} \leq \max\{|1 - \lambda A|, |1 - \lambda B|\}. \quad (2.7)$$

We would like to find λ such that $\|I - \lambda R\|_{L(H)} < 1$. This condition is satisfied for all $\lambda \in (0, 2/B)$. In particular, if $\lambda = 2/(A+B)$ then $|1 - \lambda A| = |1 - \lambda B| = (B - A)/(A + B) < 1$. For this choice of λ we have proved that $\|I - \lambda R\|_{L(H)} < 1$. □

Proposition 2.10
 Let $\{x_n : n \in \mathbb{Z}^d\} \subseteq H$ be a frame for H, with frame operator S, Gram operator R, frame bounds A and B, and Bessel map $L : H \longrightarrow \ell^2(\mathbb{Z}^d)$. If $\lambda \in (0, 2/B)$, e.g., if $\lambda = 2/(A+B)$, then

$$\forall y \in H, \quad y = \lambda \sum_{j=0}^{\infty} L^*(I - \lambda R)^j Ly, \quad (2.8)$$

where $L^*c = \sum c_n x_n$ for $c = \{c_n\} \in \ell^2(\mathbb{Z}^d)$.

Proof: Since $\langle Lx, c \rangle = \langle x, L^*c \rangle$ and
$\langle Lx, c \rangle = \sum \bar{c}_n \langle x, x_n \rangle = \langle x, \sum c_n x_n \rangle$, we obtain the formula for L^*c.
Because of (2.4) and the fact that $S = L^*L$, it is sufficient to prove

$$\lambda \sum_{j=0}^{\infty} L^*(I - \lambda R)^j Ly = \sum_{j=0}^{\infty} (I - \lambda L^*L)^j (\lambda L^*L)y, \quad (2.9)$$

where the sums are well-defined by *Lemma 2.9*. The $j = 0$ terms are clearly the same in (2.9). Assume

$$\lambda L^*(I - \lambda R)^j Ly = (I - \lambda L^*L)^j (\lambda L^*L)y. \quad (2.10)$$

Then, using (2.10), compute

$$\lambda L^*(I - \lambda R)^{j+1} Ly = \lambda L^*(I - \lambda R)^j Ly - \lambda L^*(I - \lambda R)^j \lambda R Ly$$
$$= \lambda (I - \lambda L^*L)^j L^* Ly - \lambda (I - \lambda L^*L)^j L^* L(\lambda L^* Ly)$$
$$= \lambda (I - \lambda L^*L)^j (I - \lambda L^*L) L^* Ly = \lambda (I - \lambda L^*L)^{j+1} L^* Ly,$$

and the result follows by induction. □

Proposition 2.10 leads directly to the following theorem (*Algorithm 2.11*), which provides an iterative reconstruction procedure for the recovery of a signal y from its "sampled values" Ly. This iterative procedure converges at an exponential rate.

Algorithm 2.11

Let $\{x_n : n \in \mathbb{Z}^d\} \subseteq H$ be a frame for H, with Gram operator R, frame bounds A and B, and Bessel map L. Let $y \in H$ and set $c_{(0)} = Ly \in \ell^2(\mathbb{Z}^d)$, $y_0 = 0$, $\lambda = 2/(A+B)$, and $\alpha = \|I - \lambda R\|_{L(H)} < 1$. Define $u_m, y_m \in H$ and $c_{(m)} \in L(H)$, $m = 0, 1, \ldots$, recursively as

$$u_m = \lambda L^* c_{(m)}, \quad c_{(m+1)} = c_{(m)} - Lu_m,$$

and

$$y_{m+1} = y_m + u_m.$$

Then

$$\forall m \in \mathbb{N}, \quad \|y - y_m\| < \alpha^m \frac{B}{A} \|y\|,$$

and, in particular, $\lim_{m \to \infty} y_m = y$ in H.

Proof: a. An elementary induction argument shows that

$$\forall m = 0, 1, \ldots, \quad y_{m+1} = \lambda L^* \left(\sum_{j=0}^{m} (I - \lambda R)^j \right) c_{(0)}.$$

Consequently, by *Proposition 2.10*, we have $\lim_{m \to \infty} y_m = y$ in H.

b. For any fixed $m \geq 0$ and for any $k \in \mathbb{N}$,

$$\|y - y_n\| = \|y + (y_{m+1} - y_m) + (y_{m+2} - y_{m+1})$$

$$+ \cdots + (y_{m+k+1} - y_{m+k}) - y_{m+k+1}\|$$

$$\leq \|y - y_{m+k+1}\| + \sum_{j=0}^{k} \|y_{m+j+1} - y_{m+j}\|.$$

Using part *a* and taking the lim sup as $k \to \infty$, e.g., [Ben96, page 278], we obtain

$$\|y - y_m\| \leq \sum_{k \geq m} \|y_{k+1} - y_k\|,$$

from which we compute, using part *a* for the first step and (2.7) for the last, that

$$\|y - y_m\| \leq \sum_{k \geq m} \|\lambda L^* (I - \lambda R)^k Ly\|$$

$$\leq \sum_{k \geq m} \lambda \|L^*\| \|(I - \lambda R)^k\|_{L(H)} \|L\| \|y\|$$

$$\leq \lambda B \left(\sum_{k \geq m} \alpha^k \right) \|y\| = \left(\frac{\alpha^m}{1 - \alpha} \right) \lambda B \|y\| \leq \alpha^m \frac{B}{A} \|y\|.$$

Algorithm 2.11 underscores the importance of the discrete nature of the Gram operator R in the reconstruction process.

Formally, we may rewrite (2.8) as

$$y = (L^* R^{-1}) L y, \qquad (2.11)$$

cf., (2.6).

A crucial element in the proof of *Algorithm 2.11* is the fact that the sampled data $c_{(0)}$ has the form $c_{(0)} = Ly$. If $c_{(0)}$ is not entirely in $L(H)$, then the algorithm will not converge. An analysis of this latter situation is found in [TB95, Section 6], and it is related to noise reduction.

In the Introduction we mentioned the relationship between frames and sampling. We shall explain that comment now in terms of *Example 2.6*.

The following sampling theorem for nonbandlimited functions was first proved in [BH90], cf., [Ben92], [BF94, Chapter 7]; and the proof depends on showing that the Gabor system $\{e_{na} \tau_{mb} \hat{g}\}$ is a frame.

Theorem 2.12 (Nonbandlimited Gabor Sampling Formula)
Let $T, \Omega > 0$ be constants for which $0 < 2T\Omega \leq 1$, and let $g \in PW_{1/(2T)}$ have the properties that $\hat{g} \in L^\infty(\widehat{\mathbb{R}})$, $\hat{g} = 1$ on $[-\Omega, \Omega]$, and, in case $2T\Omega < 1$, $\hat{g}$ is continuous and

$$|\hat{g}| > 0 \text{ on } (-\frac{1}{2T}, -\Omega] \cup [\Omega, \frac{1}{2T}).$$

Set

$$G(\gamma) = \sum_{m \in \mathbb{Z}} |\hat{g}(\gamma - mb)|^2 \text{ and } s(t) = \left(\frac{\hat{g}}{G} \right)^{\vee}(t),$$

where $\Omega + 1/(2T) \leq b < 1/T$ in case $2T\Omega < 1$ and $\Omega + 1/(2T) = b$ if $2T\Omega = 1$. Then

$$\exists A, B > 0 \text{ such that } A \leq G(\gamma) \leq B \text{ a.e.,}$$

$$s \in PW_{1/(2T)}, \hat{s} \in L^\infty(\widehat{\mathbb{R}}), \hat{s} = 1 \text{ on } [-\Omega, \Omega],$$

$$\forall f \in L^2(\mathbb{R}), \quad f = T \sum_{m,n \in \mathbb{Z}} \langle \hat{f}, e_{nT} \tau_{mb} \hat{g} \rangle \tau_{-nT}(e_{mb} s) \text{ in } L^2(\mathbb{R}), \qquad (2.12)$$

and

$$\forall f \in PW_\Omega, \quad f = T \sum f(nT) \tau_{nT} s \text{ in } L^2(\mathbb{R}). \qquad (2.13)$$

Equation (2.13) is the Classical Sampling Theorem for bandlimited functions f and for the sampling function s.

Equation (2.12), from which (2.13) follows for elements of PW_Ω, can be viewed as a sampling formula for nonbandlimited elements of $L^2(\mathbb{R})$. Also, the terms of the sum in (2.12) for $n \in \mathbb{Z}$ and $m \neq 0$ can be used to quantify aliasing error. Expanding on this comment, we know that in dealing with high frequency time series it is necessary to sample closely in order to capture all of the fluctuations. Thus, in the case of very high frequency information f, thought of as "infinite frequencies" and hence nonbandlimited, we cannot reconstruct f with a discrete set of samples. In this context, the terms of the sum in (2.12) for $n \in \mathbb{Z}$ and $m \neq 0$ can be thought of as dealing with the "infinite frequencies" associated with an arbitrary signal $f \in L^2(\mathbb{R})$.

We now present the wavelet analogue of *Theorem 2.12*, cf., [Dau92], [Mey90]. To this end, let $\Omega > 0$, let

$$\varphi = \varphi_{(\Omega)} = \frac{1}{\sqrt{2\Omega}} d_{2\pi\Omega}, \tag{2.14}$$

where $d_{2\pi\Omega}(t) = (\sin 2\pi\Omega t)/(\pi t)$, and define the *Shannon wavelet*

$$\psi = \psi_{(\Omega)} = \varphi_{(2\Omega)} - \varphi_{(\Omega)}. \tag{2.15}$$

The use of φ and ψ is compatible with the notation for multiresolution wavelet orthonormal bases, of which Shannon wavelets are a special example, for the case of orthonormal bases. Clearly

$$\sqrt{2\Omega}\hat{\psi} = \mathbf{1}_{[-2\Omega,-\Omega)} + \mathbf{1}_{[\Omega,2\Omega)},$$

and, in particular, $\int \psi(t)dt = 0$.

The proof of *Theorem 2.13* is elementary, and involves the interplay of Fourier series and Fourier transforms, just as in the proof of the Classical Sampling Theorem, e.g., [Ben96].

Theorem 2.13 (Nonbandlimited Shannon Wavelet Sampling Formula)
Let $\Omega > 0$, let φ be defined by (2.14), and let $\phi = \phi_{(\Omega)}$ be the corresponding Shannon wavelet (2.15).

a. The wavelet system $\{\phi_{m,n/(2\Omega)}\}$ is an orthonormal basis for $L^2(\mathbb{R})$.

b. The wavelet system $\{\phi_{m,n/(4\Omega)}\}$ is a nonorthogonal tight frame for $L^2(\mathbb{R})$ with frame constants $A = B = 2$. In particular, each $f \in L^2(\mathbb{R})$ can be written as

$$f = \sum_n f * \varphi\left(\frac{n}{2\Omega}\right) \tau_{n/(2\Omega)}\varphi + \sum_{m=0}^{\infty} \sum_n d_{m,n}\phi_{m,n/(4\Omega)} \quad \text{in } L^2(\mathbb{R}),$$

where

$$d_{m,n} = \frac{1}{\sqrt{2\Omega}2^{(m/2)+1}} \int_{-2^{m+1}\Omega}^{2^{m+1}\Omega} \hat{f}(\gamma)\left(\mathbf{1}_{[-2^{m+1}\Omega,-2^m\Omega)}(\gamma)\right. \\ \left. + \mathbf{1}_{[2^m\Omega,2^{m+1}\Omega)}(\gamma)\right) \times e^{2\pi i n\gamma/(2^{m+2}\Omega)} d\gamma.$$

c. In the case of $f \in PW_\Omega$, part b is the Classical Sampling Theorem formula,

$$f = \sum \frac{1}{2\Omega} f\left(\frac{n}{2\Omega}\right) \tau_{n/(2\Omega)} d_{2\pi\Omega} \quad \text{in } L^2(\mathbb{R}).$$

3 Uniform sampling

We begin with a statement of Cauchy's theorem (1841) [Cau1841].

Theorem 3.1 (Cauchy's Cardinal Series Theorem (1841))
Let
$$f(t) = \sum_{|n| \leq M} c_n e^{2\pi i t n},$$
and set $N = 2M + 1$. Then
$$f(t) = \sin(\pi t N) \sum_{m=0}^{N-1} \frac{1}{N} f\left(\frac{m}{N}\right) \frac{(-1)^m}{\sin \pi(t - \frac{m}{N})}. \tag{3.1}$$

Cauchy used Lagrange's interpolation formula in one of his proofs of (3.1). Although probably not universally accepted, it can be argued that equation (3.1) was the first Classical Sampling Theorem *since the polynomial f can be thought of as an M-bandlimited function*. Theorem 3.1 also has a formulation for the case of primitive Nth roots of unity.

Remark 3.2 (Interpolation and Number Theory)
a. Cauchy's Cardinal Series Theorem influenced some developments in interpolation theory and number theory at the end of the 19th century. For the former subject there is the work of Borel, Hadamard, and de la Vallée-Poussin during the period 1898–1908, e.g., [HM26, page 50], [Hig85, pages 49–50], [Whi35].

b. Let $\pi(x)$ denote the number of primes less than or equal to x. The prime number theorem (PNT) asserts that $\pi(x) \sim x/\log x$, $x \to \infty$; and the PNT was proved independently in 1896 by Hadamard and de la Vallée-Poussin. Earlier, in 1894 (in a note in the Comptes Rendus Acad. Sci., Paris), von Koch constructed entire functions f having the form of nonperiodic versions on $\mathbb{R}$ of Cauchy's formula (3.1). Using such functions f as building blocks for more complicated expressions F, he was able to write $\pi(n)$ in terms of values $F(m)$ for $m \leq n$.

von Koch quoted the work of Hadamard and Poincaré on entire functions, and, although his approach plays a role in Steffensen's point of view [Ste14], one suspects it was not pursued after 1896.

It should be pointed out that the PNT is equivalent to the fact that
$$\forall t \in \mathbb{R}, \quad \zeta(1 + it) \neq 0, \tag{3.2}$$
where ζ is the Riemann zeta function; and both Hadamard and de la Vallée-Poussin proved (3.2). A reasonable way to prove that (3.2) implies the PNT is to use Wiener's Tauberian Theorem [Ben75, pages 128–136] or the so-called Wiener-Ikehara Tauberian Theorem. We mention this because of an application of the Tauberian Theorem in interpolation theory that we shall mention in part d.

c. In 1915, E.T. Whittaker [Whi15, page 187] introduced the terminology *cardinal function* in the context of interpolation theory.

For a given sequence $\{c_n\} \subseteq \mathbb{C}$ and a fixed value of $T > 0$, Whittaker considered a class $X(\mathbb{R})$ of functions on $\mathbb{R}$, each of whose elements f has the property that
$$\forall n \in \mathbb{Z}, \quad Tf(nT) = c_n. \tag{3.3}$$
He referred to $X(\mathbb{R})$ as a *cotabular set of functions*. For a given $f \in X(\mathbb{R})$ he then posed the problem of finding $f_c \in X(\mathbb{R})$ which is Ω-bandlimited, where $2T\Omega = 1$. His goal was to replace a "given function $f \in X(\mathbb{R})$ by a cotabular function in such a way as to remove all the rapid oscillations from it" [Whi15, page 184]. He referred to his solution
$$f_c(t) = (\sin 2\pi t\Omega) \sum c_n \frac{(-1)^n}{\pi(t-nT)}, \tag{3.4}$$
as the *cardinal function* of $X(\mathbb{R})$. (Whittaker's use of the adjective "cardinal" is in the sense of "primary" or "to hinge upon" from *cardo, cardinis*, the Latin noun for hinge.)

Whittaker's paper was written in 1915 and contains no references. This is amazing in light of part *a* and Steffensen's 1914 paper [Ste14] in Acta Mathematica which deals with (3.4).

d. Let $T = 1$ and $\Omega = 1/2$. The *Newton-Gauss Interpolation Formula* for a function f on $\mathbb{R}$, whose values $\{f(n) : n \in \mathbb{Z}\}$ are known, is

$$f(0) + \left\{ t\Delta f(0) + \frac{t(t-1)}{2!}\Delta^2 f(-1) \right\}$$
$$+ \left\{ \frac{t(t^2-1^2)}{3!}\Delta^3 f(-1) + \frac{t(t^2-1^2)(t-2)}{4!}\Delta^4 f(-2) \right\}$$
$$+ \ldots,$$

where $\Delta f(0) = f(1) - f(0)$, $\Delta^2 f(0) = \Delta f(1) - \Delta f(0) = f(2) - 2f(1) + f(0), \ldots$, see [Whi35, Section 11]. Both Steffensen [Ste14] and E.T. Whittaker [Whi15, pages 190–193] showed that the Newton Gauss Interpolation Formula and the cardinal function f_c converge to the same sum for certain classes of functions. Further and interesting work on this relationship is due to Ferrar (1925), who required Littlewood's Tauberian Theorem in a fundamental way in his analysis, see [Whi35, page 64], cf., [Ben75, Chapter 2] for Tauberian theorems.

Since
$$\tau_{nT} d_{2\pi\Omega}(t) = \frac{(-1)^n \sin(2\pi t\Omega)}{\pi(t-nT)}$$
when $2T\Omega = 1$, we know that the cardinal function is related to the Classical Sampling Theorem referred to at the end of *Section 2*. With all of the background information in *Theorem 3.1* and *Remark 3.2*, let's now give a

clear statement of the Classical Sampling Theorem for PW_Ω, even though we have already addressed the nonbandlimited case in *Theorems 2.12* and *2.13*.

Theorem 3.3 (Classical Sampling Theorem)
Let $T, \Omega > 0$ satisfy the condition that $0 < 2T\Omega \leq 1$, and let $s \in PW_{1/(2T)}$ satisfy the condition that $\hat{s} = 1$ on $[-\Omega, \Omega]$ and $\hat{s} \in L^\infty(\widehat{\mathbb{R}})$. Then
$$\forall f \in PW_\Omega, \quad f = T\sum f(nT)\tau_{nT}s, \qquad (3.5)$$
where the convergence in (3.5) is in $L^2(\mathbb{R})$ norm and uniformly on $\mathbb{R}$, see [Ben96, pages 256–257] for a proof.

Example 3.4 (Poisson Summation Formula)
a. The Poisson Summation Formula (PSF) is
$$T\sum f(t+nT) = \sum \hat{f}(\frac{n}{T})e^{2\pi i t n/T}. \qquad (3.6)$$

Equation (3.6) is true for well-behaved functions such as continuous, integrable functions of bounded variation, e.g., [Ben96, pages 255–256]. More often than not, versions of (3.6) are true which require approximate identities or a.e. convergence. The idea behind the validity of a PSF for a function space on $\mathbb{R}$ is that the periodization operation
$$f \longrightarrow T\sum f(t+nT)$$
be a well-defined continuous surjection onto an appropriate function space on the torus $\mathbb{R}/T\mathbb{Z}$. Many PSFs based on this philosophy are proved in [BZ97].

b. It is well-known that versions of the PSF are equivalent to versions of the sampling formula. For example, suppose we want to verify the sampling formula (3.5). First, note that the right side of (3.5) is
$$\left[(\sum T\delta_{nT})f\right] * s,$$
which, by the dual "δ-train" version of the PSF (3.6), is
$$\left(\sum \tau_{n/T}\hat{f}\right)\hat{s}. \qquad (3.7)$$
By the hypotheses on s in *Theorem 3.3* and the hypotheses,
$$2T\Omega \leq 1 \text{ and } f \in PW_\Omega, \qquad (3.8)$$
we see that (3.7) is $\hat{f}$ and so (3.5) is true.

c. Less well-known and less-developed than the equivalence of sampling and PSF in part *b*, is the equivalence between sampling formulas (or PSFs) and (weighted) uncertainty principle inequalities in special cases. This point of view can be expected to evolve and depends on the idea from part *a*, quadrature formulas, and Fourier transform weighted norm inequalities, e.g., [BF94, Chapter 7], [Bou88], [dBr67], [HJ94].

In generalizing the uniform sampling theorem, the critical notion is that we deal with discrete subgroups of the original group, whether it be $\mathbb{R}$ or $\mathbb{R}^d$ or something more general. It is natural then to generalize *Theorem 3.3* to the setting of locally compact abelian groups (LCAGs). This was accomplished by Kluvánek in 1965 [Klu65]. Of course, the PSF for LCAGs has been known from early-on, see [Loo53, page 153], [Rei68]; and it can be used to prove Kluvánek's Theorem. Further, since $\mathbb{R}/\mathbb{Z}$ is a (compact) LCAG, *Theorem 3.1* is a consequence of Kluvánek's Theorem.

For the purpose of some comments we wish to make after stating Kluvánek's result (*Theorem 3.6*), we shall introduce *Theorem 3.6* in the following way.

Let G be a LCAG with dual group $\widehat{G}$, and let $H \subseteq G$ be a discrete subgroup. $H^\perp$ denotes the annihilator subgroup

$$H^\perp = \{\gamma \in \widehat{G} : \forall x \in H, \quad (x,\gamma) = 1\}.$$

Haar measure on a locally compact abelian group X is denoted by m_X. In our case, we adjust the Haar measures on $\widehat{G}$, $H^\perp$, and $\widehat{G}/H^\perp$ so that

$$\int_{\widehat{G}} = \int_{\widehat{G}/H^\perp} \int_{H^\perp}. \tag{3.9}$$

In this case, the aforementioned PSF is

$$\int_H f dm_H = \int_{H^\perp} \hat{f} dm_{H^\perp}. \tag{3.10}$$

The canonical mapping

$$h : \widehat{G} \longrightarrow \widehat{G}/H^\perp \tag{3.11}$$

is a surjection. We take the situation in which $H^\perp$ is discrete, from which we conclude that $\widehat{G}/H^\perp$ is compact since $\widehat{G}/H^\perp$ can be identified with the dual group of H, e.g., [Rud62, Chapter 2.1]. As such, we normalize Haar measure on $\widehat{G}/H^\perp$ so that

$$m_{\widehat{G}/H^\perp}(\widehat{G}/H^\perp) = 1.$$

Further, along with (3.9), we normalize H and $H^\perp$ so that $m_H(\{x\}) = 1 = m_{H^\perp}(\{\gamma\})$ for all $x \in H$ and $\gamma \in H^\perp$. At this point we have fixed Haar measure $\widehat{G}/H^\perp$, H, $H^\perp$, and $\widehat{G}$. Finally, we normalize Haar measure on G so that if the Fourier transform is formally defined as

$$\hat{f}(\gamma) = \int_G f(x)(-x,\gamma) dm_G(x)$$

then the formal inversion formula is

$$f(x) = \int_{\widehat{G}} \hat{f}(\gamma)(x,\gamma) dm_{\widehat{G}}(\gamma).$$

With this standard background, let $E \subseteq \widehat{G}$ be any subset of finite Haar measure for which the map h of (3.11) restricted to E is a bijection. Next, define the sampling function

$$\forall x \in G, \quad s_E(x) = \int_E (x,\gamma) dm_{\widehat{G}}(\gamma),$$

where $m_{\widehat{G}}$ is adjusted to have the property that $m_{\widehat{G}}(E) = 1$. Clearly, s_E is the analogue of the Dirichlet function $d_{2\pi\Omega}$ defined after (2.14). Of course, E can be more complicated than the interval $[-\Omega, \Omega)$ for the case $\widehat{G} = \widehat{\mathbb{R}}$.

Lemma 3.5

The function $s_E : G \to \mathbb{C}$ has the following properties:
i. The function s_E is a continuous, positive definite element of $L^2(G)$ for which $\|s_E\|_{L^2(G)} = 1$;
ii. $s_E(0) = 1$ and $s_E(x) = 0$ for all $x \in H \setminus \{0\}$; and
iii. For all $x \in H \setminus \{0\}$, $s_E * \tilde{s}_E(x) = 0$, where $\tilde{s}_E(z) = \overline{s_E(-z)}$.

The proof of *Lemma 3.5* is elementary and it is used to prove Kluvánek's result.

Theorem 3.6 (Kluvánek's Uniform Sampling Theorem for LCAGs)

Let $f \in L^2(G)$ and assume $\hat{f} = 0$ a.e. off of E.
a. f is equal a.e. to a continuous function.
b. If f is continuous on G then

$$f = \sum_{y \in H} f(y) \tau_y s_E \quad (3.12)$$

uniformly on G and in $L^2(G)$. Further, the "Gaussian quadrature" formula

$$\|f\|^2_{L^2(G)} = \sum_{y \in H} |f(y)|^2$$

is valid.

Example 3.7 (A Comparison of Theorems 3.3 and 3.5 on $\mathbb{R}$)

a. The Nyquist hypothesis $2T\Omega \leq 1$ and bandlimited hypothesis $f \in PW_\Omega$ of *Theorem 3.3* are replaced in *Theorem 3.6* by the pairing $H, H^\perp$ (where $H^\perp$ is essential for choosing some E) and the hypothesis that $\hat{f} = 0$ off of E, respectively.

b. Suppose $f \in PW_{5\Omega}$ and $2T(5\Omega) = 1$. Then the Classical Sampling Theorem asserts that

$$f = \frac{1}{10\Omega} \sum f\left(\frac{n}{10\Omega}\right) \tau_{n/10\Omega} s; \qquad (3.13)$$

and, in particular, sampling takes place on the set $\{n/10\Omega : n \in \mathbb{Z}\}$. If f has the further property that $\hat{f}$ vanishes off of $E = [-5\Omega, -4\Omega) \cup [4\Omega, 5\Omega) \subseteq \widehat{\mathbb{R}}$, then the Classical Sampling Theorem (3.13) still requires sampling on the set $\{n/10\Omega\}$. On the other hand, we can consider $E \subseteq \widehat{\mathbb{R}}$ as the domain of the function h (of (3.11)) with the property that $h(E) = \widehat{\mathbb{R}}/H^\perp$ so that $H^\perp = \{2\Omega n : n \in \mathbb{Z}\}$ and $H = \{n/2\Omega : n \in \mathbb{Z}\}$. Thus, in this case, *Theorem 3.6* allows us to write

$$f = \frac{1}{2\Omega} \sum f\left(\frac{n}{2\Omega}\right) \tau_{n/2\Omega} s_E;$$

and in particular we need only sample from the set $\{n/2\Omega\}$.

Remark 3.8 (Wavelets and Tiling)

The notion of congruence inherent in the canonical function h and our choice of E plays a role in many areas of mathematics including several closely related to the present topic.

For example, in 1990 Albert Cohen used such congruence criteria to establish orthonormality of scaling functions (associated with wavelet multiresolution analysis (MRA)), for a given quadrature mirror filter (QMF), e.g., [Dau92, pages 182-186].

Until recently, it had been thought that dilates and translates of more than one function were required to generate a wavelet orthonormal basis for $\mathbb{R}^d$, $d \geq 2$. Such is the case when such bases are generated from a MRA. However, in 1992, using the notion of congruence and other methods, David Larson and Xingde Dai constructed (non-MRA) wavelet orthonormal bases for $\mathbb{R}^d$, $d \geq 2$, from a single function [DL97], [DLS97], cf., [HWW97].

As a final example we mention the fundamental role of congruence in the recent deep work on self-similar tilings of $\mathbb{R}^d$ by K. Gröchenig, A. Haas, J.C. Lagarias, W. Madych, Yang Wang, et al., e.g., [GM92], [LW97].

4 Seizure prediction

In *Section 2* we discussed the theory of frames. This theory can be viewed as an *oversampling* technique, and, as such, as a method which can be used, at least theoretically, in noise reduction problems (since the non-surjectivity of the Bessel map provides a place, viz., $\ell^2(\mathbb{Z}^d) \backslash L(H)$, to put noise separated from a sampled intelligent message of the form Ly). In *Section 3* we discussed *uniform sampling*. In this section we shall discuss epileptic seizure prediction which we feel is best addressed by *wavelet uniform oversampling* methods. The epileptic prediction problem is to predict seizure activity as far in advance as possible of the seizure occurrence.

We shall devote most of this section to the spectral analysis of epileptic seizure prediction. Although interesting in its own right as well as being a natural introduction to seizure prediction problems, our spectral methods generally yield unsatisfactory solutions. This quandary, of having a natural method leading to inadequate prediction times, led us to the approach of wavelet uniform oversampling. At the end of this section we shall expand briefly on the basic oversampling idea which we developed with Pfander in [BP97].

The data we analyze are signals derived from brain potentials. The most common signals derived from brain potentials are electroencephalograms (EEGs). Theoretically, EEG time series should provide quantitative data to fathom and describe normal brain rhythms, as well as aberrations such as some epileptic seizures.

Since EEG signals are measured on the scalp and potentials are on the order of microvolts, electroencephalograms are subject to many complicated influences such as head geometry, propagation of brain waves through the skull, and muscle movement. These effects are often regarded as "noise", and, consequently, the signal-to-noise ratio of EEG time series can be quite low. Time series which have a much higher signal-to-noise ratio are obtained by measuring potentials directly on the surface of the brain. These are called electrocorticograms (ECoGs), and can only be obtained by invasive procedures.

The bioelectric trace f at the top of *Figure 1* is electrocorticogram (ECoG) data. The analysis of this data was made at The MITRE Corporation, e.g., [BC95], [B-KCJ94].

The length of the trace is 240 seconds, and it includes (epileptic) seizure activity as well as significant nonseizure activity. Typical of the latter type is activity f_{ns} in the time interval from 25 to 30 seconds, extracted from the original trace f; "ns" designates "nonseizure". f_{ns} is reproduced in the left graph on the second line of *Figure 1*. The ordinate is a microvolt measurement of the potential of the electric field at the surface of the brain relative to a referential electrode. The right graph on the second line of *Figure 1* is activity f_s extracted from seizure data in the original trace f; "s" designates "seizure". (We admit to begging the question of what precisely is seizure data.) The third line of *Figure 1* contains absolute values of the fast Fourier transforms (FFTs) of f_{ns} and f_s, respectively. The *dc* components "$\int f_{ns}(t)dt$" and "$\int f_s(t)dt$" have been removed from the FFTs since, as can be seen from line 2 of *Figure 1*, these values are so large as to affect the readability of nonzero-frequency information. The left graph on the third line of *Figure 1* is indicative of $1/f$ noise, whereas the right graph exhibits some periodic behavior. We shall now look at this latter issue a little more closely.

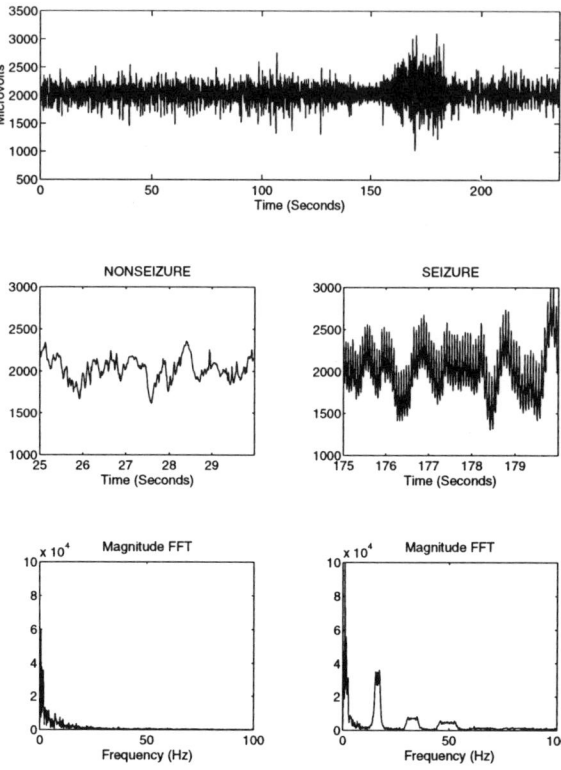

FIGURE 1. Top: Time series trace of electrocorticogram data. Middle: Time series for nonseizure epoch (left) and seizure epoch (right). Bottom: Magnitude FFT for the time series shown in the middle.

The right graph on the second line of *Figure 1* exhibits a high frequency almost periodic signal f "riding-on" a low frequency wave. This latter wave is usually in the θ-band $(4-8Hz)$ or the α-band $(8-13Hz)$ of brain activity. The θ-band is the spectral range of drowsiness or light sleep, and the α-band is the spectral range of rhythmic activity in an awake person. The amplitude of the bioelectric trace for the α-band is typically between 5 and 100 microvolts, e.g., [Nun81].

A close look at f justifies our labeling of f as an *almost periodic* signal. To see this, first let f_1, resp., f_2, be the restriction of f to a time interval at the beginning, resp., the end, of the seizure. Then, by counting peaks of the trace over the duration of the seizure activity, we find that the period p_1 of f_1 is less than the period p_2 of f_2.

Assuming each $f_j, j = 1, 2$, is periodic on $\mathbb{R}$, even though it is only periodic on a portion of the seizure interval, we have the Fourier series representation,

$$Sf_j(t) = \sum_n a_{j,n} e^{2\pi i n t/p_j}; \tag{4.1}$$

and we further assume that $f_j = Sf_j$. Because of (4.1), the Fourier transform of each f_j, as a distribution on $\mathbb{R}$, is

$$\widehat{f_j} = \sum_n a_{j,n} \delta_{n/p_j}. \tag{4.2}$$

As such, if we consider the (t, γ)-phase plane as the domain of the spectrogram, and if we take a value t_1 near the beginning of the seizure, then because of (4.2) we expect nontrivial spectrogram amplitudes at the points $\{(t_1, n/p_1) : n \in \mathbb{Z}\}$. A similar remark holds for other parts of the seizure. In particular, if t_2 is a time value at the end of the seizure, we expect nontrivial spectrogram amplitudes at the points $\{t_2, n/p_2 : n \in \mathbb{Z}\}$.

Finally, for a fixed $n > 0$, we expect the graph of this spectral data in the seizure interval to be a decreasing function of t since $1/p_1 > 1/p_2$, cf., (4.2). This time varying spectral activity can be described in terms of elementary chirps, e.g., [Ben96, Section 2.10].

If the previous discussion is too discursive, then experimental data reflected by *Figure 2* tell the same story pictorially. In fact, the region of the (t, γ)-phase plane determined by the seizure time interval [170,180] bears out the previous analysis (and handwaving).

Recall that the *epileptic seizure prediction problem* is to predict the onset of seizure sufficiently far in advance in order to take remedial action. Such action can only be effective if there is also a solution to the *localization problem*. The localization problem is to find the region of the brain responsible for the onset of seizure activity. In principle, solutions to the prediction and localization problems would allow chemical response to specific local parts of the brain in time to temper seizure intensity.

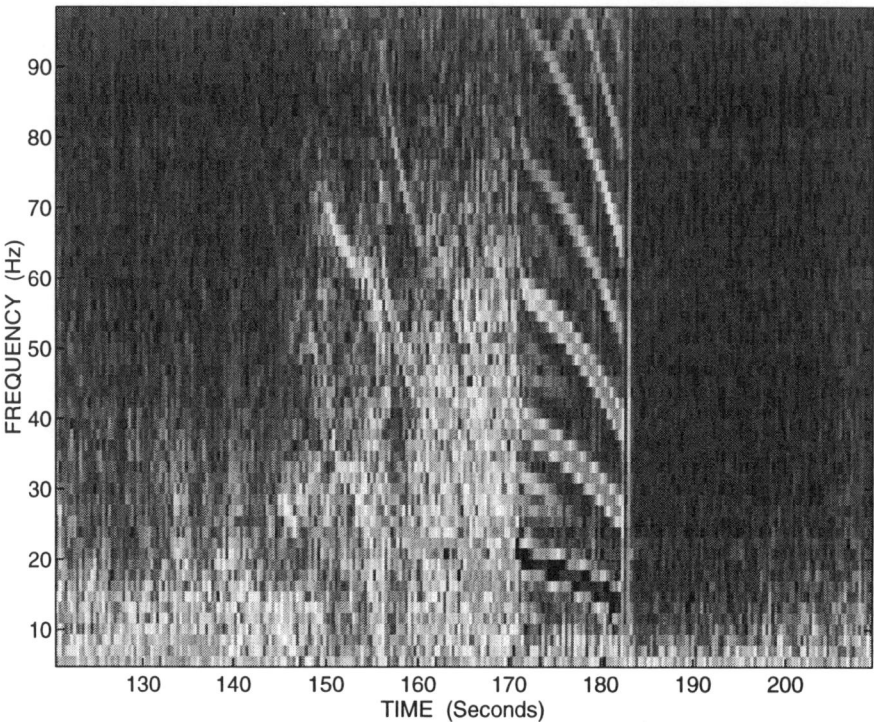

FIGURE 2. Spectrogram for the ECoG time series shown in Figure 1.

The spectrogram in *Figure 2* provides information about the prediction problem. For example, the definitive chirps in the seizure time interval [170,180] of *Figure 2* have as precursors the periodic chirp-like activity in the time interval [155,162]. One can even detect some such activity in the time interval [150,155]. A close look at the trace on the top line of *Figure 1* shows that these precursors are embedded in low amplitude data. Realistically, even a 20 second prediction time is not sufficient in order to take effective remedial action.

Our wavelet uniform oversampling approach to lengthen prediction times begins by sampling the given data as much as possible and to take advantage of any cancellation that might occur. This leads directly to the introduction of wavelet frames. At this point our implementation of this approach has been almost exclusively with Haar wavelet frames, see [BC95]. Then, using such oversampling, we have quantified the following 3-step procedure.

1. Invasive ECoG data of an individual patient are analyzed by spectral methods, e.g., *Figure 2*, and oversampling wavelet frame methods to extract the periodic patterns we have associated with epileptic seizures of a specific

patient.

2. Using this knowledge of seizure periodicity, we construct an *optimal piecewise constant wavelet* designed to detect the epileptic periodic patterns of the patient.

3. We introduce a fast wavelet frame transform, as well as waveletgram averaging techniques, to detect occurrence and period of the seizure periodicities in preseizure noninvasive EEG data of the patient.

Thus, step 1 obtains data by invasive means. By the oversampling method, optimal data is obtained, and the periodic nature of the patient's seizure behavior is quantified. Using this periodic function, our mathematical theory of piecewise constant wavelets gives rise to a wavelet which will detect epileptic periodic patterns which may exist in noisy environments such as preseizure EEG data. This is the thrust of step 2. Step 3 is the implementation of step 2 on noninvasive EEG data. Because of the use of wavelet frames, the optimal data collection of step 1 and the mathematical results of step 2, we can expect step 3 to pick out periodicities (when it can be done at all) in the presence of preseizure $1/f$ noise as well as cranial noise, etc. that arise in dealing with EEGs. The theory and algorithms for our 3-step procedure are contained in [BP97].

Periodicity detection has other applications including ocean wave prediction and cockpit motion sickness prediction.

5 Notation

Besides the standard notation found in the books by Hörmander [Hör83], Schwartz [Sch66], and Stein and Weiss [SW71], we shall also use the following notation and conventions.

The *Fourier transform* $\hat{f}$ of $f \in L^1(\mathbb{R}^d)$ is defined by

$$\hat{f}(\gamma) = \int f(t) e^{-2\pi i t \cdot \gamma} dt,$$

where $\mathbb{R}$ is the real line, $\mathbb{R}^d$ is d-dimensional Euclidean space, "$\int$" designates integration over $\mathbb{R}^d$, and $\gamma \in \widehat{\mathbb{R}}^d (= \mathbb{R}^d)$. Similarly, "$\sum$" designates summation over $\mathbb{Z}^d$, where $\mathbb{Z}$ is the ring of integers. $F^\vee$ designates the *inverse Fourier transform* of F. Formally, if $\hat{f} = F$, then

$$f(t) = F^\vee(t) = \int F(\gamma) e^{2\pi i t \cdot \gamma} d\gamma, \tag{5.1}$$

where integration is over $\widehat{\mathbb{R}}^d$, see [Ben96, Chapter 1], [SW71] for criteria for the validity of (5.1).

Finally, we use the following notation:

$$(\tau_x f)(t) = f(t-x);$$

$$e_\gamma(t) = e^{2\pi i t \gamma};$$

$$\mathbf{1}_X(t) = \begin{cases} 1, & \text{if } t \in X \\ 0, & \text{if } t \notin X; \end{cases}$$

$$\delta(m,n) = \begin{cases} 1, & \text{if } m = n \\ 0, & \text{if } m \neq n; \end{cases}$$

supp f is the support of f;

$|X|$ is the Lebesgue measure of $X \subseteq \mathbb{R}^d$

$$PW_\Omega = \{f \in L^2(\mathbb{R}) : \hat{f} = 0 \text{ off of } [-\Omega, \Omega]\}.$$

6 References

[BC95] J. J. Benedetto and D. Colella, *Wavelet analysis of spectrogram seizure chirps*, SPIE, San Diego, 1995.

[BF94] J. J. Benedetto and M. W. Frazier, editors, *Wavelets: Mathematics and Applications*, CRC Press, Boca Raton, FL, 1994.

[BH90] J. J. Benedetto and W. Heller. Irregular sampling and the theory of frames. *Mat. Note,* **10**(Suppl. 1)(1990), 103–125.

[BL97] J. J. Benedetto and S. Li, *The theory of multiresolution analysis frames and applications to filter banks*, Applied and Computational Harmonic Analysis, to appear.

[BM62] A. Beurling and P. Malliavin, *On Fourier transforms of measures with compact support*, Acta Math. **107** (1962), 291–309.

[BM67] A. Beurling and P. Malliavin, *On the closure of characters and the zeros of entire functions*, Acta Math. **118** (1967), 79–93.

[BP97] J. J. Benedetto and G. E. Pfander, *Wavelet detection of periodic behavior in EEG and ECoG data*, IMACS, Berlin, 1997.

[BT93] J. J. Benedetto and A. Teolis, *A wavelet auditory model and data compression*, Applied and Computational Harmonic Analysis **1** (1993), 3–28.

[BZ97] J. J. Benedetto and G. Zimmermann, *Sampling multipliers and the Poisson summation formula*, J. Fourier Analysis and Applications **3** (1997).

[B-KCJ94] M. Bozek-Kuzmicki, D. Colella, and G. Jacyna, *Feature-based epileptic seizure detection and prediction from ECoG recordings*, IEEE-SP Int'l Symposium on Time-Frequency and Time-Scale Analysis, Philadelphia, 1994.

[Ben75] J. J. Benedetto, *Spectral Synthesis*, Academic Press, New York, 1975.

[Ben76] J. J. Benedetto, *Real Variable and Integration*, B. G. Teubner, Stuttgart, 1976.

[Ben92] J. J. Benedetto, *Irregular sampling and frames*, in Wavelets-A Tutorial in Theory and Applications, C. K. Chui, editor, Academic Press, Boston, 1992, pp.445–507.

[Ben96] J. J. Benedetto, *Harmonic Analysis and Applications*, CRC Press, Boca Raton, FL, 1996.

[Bou88] J. Bourgain, *A remark on the uncertainty principle for Hilbertian basis*, J. Funct. Analysis **79** (1988), 136–143.

[CT92] J. C. Candy and G. C. Temes, Oversampling Delta-Sigma Data Converters, IEEE Press, Piscataway, NJ, 1992.

[Cau1841] A. Cauchy, *Mémoire sur diverses formules d'analyse*, Comptes Rendus Acad. Sci., Paris, **12** (1841), 283–298.

[DGM86] I. Daubechies, A. Grossmann, and Y. Meyer, *Painless nonorthogonal expansions*, J. Math. Physics **27** (1986), 1271–1283.

[DL97] X. Dai and D.R. Larson, *Wandering vectors for unitary systems and orthogonal wavelets*, Memoirs Amer. Math. Soc. Providence, RI, 1997.

[DLS97] X. Dai, D.R. Larson and D.M. Speegle, *Wavelet sets in $\mathbb{R}^d$*, J. Fourier Analysis and Applications, **3** (1997).

[DS52] R. J. Duffin and A. C. Schaeffer, *A class of nonharmonic Fourier series*, Trans. Amer. Math. Soc. **72** (1952), 341–366.

[Dau92] I. Daubechies, *Ten Lectures on Wavelets*, CBMS-NSF Conf. Lect. Notes Ser. Appl. Math., vol. 61, Soc. Ind. Appl. Math., Philadelphia, PA, 1992.

[dBr67] N. G. de Bruijn, *Uncertainty principles in Fourier analysis*, in Inequalities, O. Shiska, editor, Academic Press, New York, 1967, pp.57–71.

[FG86]   H. G. Feichtinger and K. Gröchenig, *A unified approach to atomic decompositions through integrable group representations*, Proc. Lund Conference 1986, Lecture Notes in Mathematics, Springer Verlag, New York.

[FJ85]   M. W. Frazier and B. Jawerth, *Decomposition of Besov spaces*, Indiana Univ. Math. J., **34** (1985), 777–799.

[GG81]   I. Gohberg and S. Goldberg, *Basic Operator Theory*, Birkhäuser, Boston, 1981.

[GM92]   K. Gröchenig and W. Madych, *Multiresolution analysis, Haar bases, and self-similar tilings*, IEEE Trans. Information Theory, **38** (1992), 558–568.

[HJ94]   V. Havin and B. Jöricke, The Uncertainty Principle in Harmonic Analysis, Springer-Verlag, New York, 1994

[HM26]   J. Hadamard and S. Mandelbrojt, La Série de Taylor, second edition, Gauthier-Villars, Paris, 1926.

[HWW97]   E. Hernandez, X. Wang, and G. Weiss, *Smoothing minimally supported frequency wavelets, Part II*, J. Fourier Analysis and Applications, **3** (1997), 23–41.

[Hig85]   J. R. Higgins, Five short stories about the cardinal series, *Bull. Amer. Math. Soc.*, **12** (1985), 45–89.

[Hör83]   L. Hörmander, *The Analysis of Linear Partial Differential Operators, Volumes I and II*, Springer-Verlag, New York, 1983.

[Kla96]   J. R. Klauder, *Optical coherence before and after Wiener*, Proc. Symposia Applied Math., **52** (1997), 195–211. (Proceedings of the Norbert Wiener Centenary Congress, 1994).

[Klu65]   I. Kluvánek, *Sampling theorem in abstract harmonic analysis*, Mat.-Fyz. Časopis Sloven. Akad. Vied, **15** (1965), 43–48.

[LW97]   J. C. Lagarias and Y. Wang, *Integral self-affine tiles in $\mathbb{R}^n$, Part II: Lattice tilings*, J. of Fourier Analysis and Applications, **3** (1997), 83–102.

[Loo53]   L. H. Loomis, *An Introduction to Abstract Harmonic Analysis*, D. Van Nostrand Co., New York, 1953.

[Mey90]   Y. Meyer, *Ondelettes et Opérateurs*, Hermann, Paris, 1990.

[Nun81]   P. Nunez, *Electric Fields of the Brain: The Neurophysics of EEG*, Oxford University Press, 1981.

[PW34]   R.E.A.C. Paley and N. Wiener, *Fourier Transforms in the Complex Domain*, vol. 19, Amer. Math. Soc. Colloq. Publ., Providence, R.I., 1934.

[Rei68] M. Reiter, *Classical Harmonic Analysis and Locally Compact Groups*, Oxford University Press, 1968.

[Rud62] W. Rudin, *Fourier Analysis on Groups*, John Wiley and Sons, New York, 1962.

[SW71]   E. Stein and G. Weiss, *An Introduction to Fourier Analysis on Euclidean Spaces*, Princeton University Press, 1971.

[Sch66] L. Schwartz, *Théorie des Distributions*, Hermann, Paris, 1966. (1st ed., 1950).

[Ste14] J. F. Steffensen, *Uber eine Klasse von ganzen Funktionen und ihre Anwendung auf die Zahlentheorie*, Acta. Math., **37** (1914), 75–112.

[TB95]   A. Teolis and J. J. Benedetto, *Local frames and noise reduction*, Signal Processing, **45** (1995), 369–387.

[Whi15] E. T. Whittaker, *On the functions which are represented by the expansions of the interpolation theory*, Proc. Roy. Soc. Edinburgh, **35** (1915), 181–194.

[Whi35] J. M. Whittaker, *Interpolatory Function Theory*, Cambridge University Press, 1935.

[You80] R. M. Young, *An Introduction to Nonharmonic Fourier Series*, Academic Press, New York, 1980.

2
Construction of Compactly Supported Affine Frames in $L_2(\mathbf{R}^d)$

Amos Ron and Zuowei Shen[1]

ABSTRACT This paper reviews the recent articles [18-20] and [22] on the theory of wavelet frames, with an emphasis on the derived construction algorithms: the unitary extension principle, and the mixed extension principle. The practical usefulness of that theory is demonstrated with the aid concrete univariate and multivariate constructions of tight frames and bi-frames, some appear here for the first time.

1 Wavelet Frames: What and Why?

Since the publication, less than ten years ago, of Mallat's paper on Multiresolution Analysis [14], and Daubechies' paper on the construction of smooth compactly supported wavelets [8], wavelets had gained enormous popularity in mathematics and in the application domains. It is sufficient to note that there are currently more than 10,000 subscribers to the monthly *Wavelet Digest*. At the same time, tailoring concrete wavelet systems to specific applications is still a challenge, especially in more than one dimension (although a few constructions are available, such as tensor products, or the methods developed in [15] and [12]). The main search is for simple and feasible constructions of orthonormal and bi-orthogonal systems of wavelets with small (and of desirable shape) support, high smoothness and many symmetries.

In a series of recent articles ([16]–[22] and [10]), a theory that changes the previous state-of-the-art had been developed. That theory makes wavelet

---

[1]Computer Science Department, University of Wisconsin-Madison, 1210 West Dayton Street, Madison, Wisconsin 53706, USA, (e-mail: amoscs.wisc.edu) Department of Mathematics, National University of Singapore, 10 Kent Ridge Crescent, Singapore 119260, Singapore, (e-mail: matzuowsleonis.nus.edu.sg) This work was partially sponsored by the National Science Foundation under Grants DMS-9102857, DMS-9224748, and DMS-9626319, by the United States Army Research Office under Contracts DAAL03-G-90-0090, DAAH04-95-1-0089, and by the Strategic Wavelet Program Grant from the National University of Singapore.

constructions simple and feasible, and it is the intent of the present article to provide a brief glance into it, with an emphasis on particular examples of univariate and multivariate constructs.

We want to start with somewhat philosophical discussion: anyone who is familiar with wavelets knows that the simplest wavelet system is the Haar family. The Haar function is piecewise-constant, has a very small support, and the algorithms based on it are fast and simple. Had the Haar wavelet been found satisfactory, other wavelet constructions, together with the MRA framework, would have been superfluous. However, the frequency localization (read: the smoothness) of this wavelet is so bad, that improvements had been sought for at the outset. It is reasonable to argue that if piecewise-constants are rejected, then continuous piecewise-linears are next in line: this is exactly the line of development in *spline theory*. Indeed, even before MRA was introduced, Battle [1], and Lemarié [13], constructed (independently) a piecewise-linear continuous spline with orthonormal dilated shifts (and knots at the half-integers only). Alas, that spline is of *global* support, and even its exponential decay at ∞ did not attract the masses, who deserted it in favor of Daubechies' orthogonal wavelets and their bi-orthogonal off-springs (cf. [5]). The simplest wavelets constructed from Daubechies' family [8] of refinable functions (i.e., that with support $[0,3]$) is not piecewise-linear, but is related to piecewise-linears in some weak sense (the shifts of the refinable function reproduce all linear polynomials, just as the the shifts of the piecewise-linear hat function do); in any event, the question whether the corresponding wavelet is a 'natural' or 'unnatural' replacement for the Haar wavelet was not on the agenda anymore; rather, this wavelet is considered next in line to Haar because it is the continuous orthonormal wavelet with *shortest support*.

Before we get to the main point of the present discussion, we need to introduce the notion of a *tight frame*. For that, we recall that, given any orthonormal system X for $L_2(\mathbb{R}^d)$, we have

$$f = \sum_{x \in X} \langle f, x \rangle x, \quad \text{all } f \in L_2(\mathbb{R}^d).$$

More concretely, the above identity states that we may use the same system X during the *decomposition process* $f \mapsto \{\langle f, x \rangle\}$, and during the *reconstruction process* $c \mapsto \sum_{x \in X} c(x) x$ (here, c is an arbitrary sequence defined on, and labeled by the elements of X). However, the property just expressed does not *characterize* orthonormality:

Definition: tight frames. A system $X \subset L_2(\mathbb{R}^d)$ is called a **tight frame** if the equality

$$f = \sum_{x \in X} \langle f, x \rangle x, \quad \text{all } f \in L_2(\mathbb{R}^d)$$

holds.

FIGURE 1. The generators of the piecewise-linear tight frame

While piecewise-linear compactly supported orthonormal wavelet systems (generated by a single wavelet) do not exist, the elements depicted in Figure 1 were shown in [18] to generate a tight frame (using dyadic dilations and integer translations) and may be viewed as a natural extension of the Haar wavelet. More importantly, it is followed by a wealth of constructions of affine (tight) frames. Examples of this class are given in §2 (univariate), and in §3 (multivariate). A glimpse into the theory that leads to such and other constructs is the goal of §4.

We have explained so far what tight frames are. We 'almost' explained why they are needed: the main reason is that it is significantly simpler to construct tight wavelet frames (or, more, generally, bi-frames, a notion that is defined in §2) as compared to orthonormal wavelets systems or biorthogonal ones. This is largely due to the fact that the latter constructions require refinable functions with properties similar to the desired properties of the sought-for wavelets: e.g., a refinable function with orthonormal shifts is required for the construction of an orthonormal wavelet system. In contrast, compactly supported tight wavelet frames can be derived from *any* refinable function, including splines in one dimension and box splines in higher dimensions. We do not even need to assume that the shifts of the refinable function form a Riesz basis, or a frame. Of course, one should still keep in mind that tight frames do not form an orthonormal system (they can be essentially regarded as 'redundant orthonormal systems'), and for certain applications (primarily data compression) the oversampling that is inherent in frames may be undesired. At the same time, other applications, such as noise reduction and/or feature detection may find the redundancy of frames a plus, and some other applications may find that a neutral feature.

2 Examples of Univariate Tight Frames

As we mentioned in the previous section, it is possible, at least in theory, to derive wavelet frames from *any* refinable function. We have defined in the previous section the notion of a *tight frame*, and explained that they should be considered as 'redundant orthonormal bases'. In a similar way, we define now the notion of *bi-frames*, which are the redundant analog of bi-orthogonal Riesz bases.

Definition 2.1 *Let X be a countable collection of functions in L_2. Let $R : X \to L_2$ be some map. We call the pair (X, RX) **bi-frames** if the following two conditions are satisfied:*

(i) The identity $\sum_{x \in X} \langle f, Rx \rangle x = f$, holds for every $f \in L_2$, and

(ii) There exists a constant $C < \infty$, such that for every $f \in L_2$, the inequality $\sum_{x \in X} |\langle f, x \rangle|^2 + \sum_{x \in X} |\langle f, Rx \rangle|^2 \leq C\|f\|_{L_2}$ is valid.

In the above definition, the second property (which implies that X and RX are *Bessel systems*) is technical and mild. (Recall that a collection of functions X in L_2 is a **Bessel system** if there exists a constant $C < \infty$ such that, for every $f \in L_2$, the inequality $\sum_{x \in X} |\langle f, x \rangle|^2 \leq C\|f\|_{L_2}$ holds.) The major property in the definition of bi-frames is the first one. That property tells us that we may use the system RX for *decomposition* and then use the dual system X during the reconstruction.

We now provide various examples of univariate tight frames and bi-frames. All the constructs in the examples are derived from a *Multiresolution Analysis*. We recall in that context that a function $\phi \in L_2$ is called a (dyadic) **refinable function** or a **scaling function** if there exists a mask $a_\phi : \mathbb{Z} \to \mathbb{C}$ such that

$$\phi = 2 \sum_{\alpha \in \mathbb{Z}} a_\phi(\alpha) \phi(2 \cdot - \alpha). \tag{2.1}$$

Sometimes, it is easier to express a_ϕ in terms of its symbol

$$\tau_\phi(\omega) := \sum_{\alpha \in \mathbb{Z}} a_\phi(\alpha) e^{-i\alpha\omega}.$$

2. Construction of affine frames

In the examples we discuss, the mask a_ϕ is finite (which implies that ϕ is compactly supported), hence τ_ϕ is a trigonometric polynomial. The refinement equation (2.1) can be written in the Fourier domain as

$$\widehat{\phi}(2\cdot) = \tau_\phi \widehat{\phi}.$$

For notational convenience, when sequentially listing the entries of a sequence $a : \mathbb{Z} \to \mathbb{C}$, we put in **boldface** the entry $a(0)$, thus

$$a = (\ldots, 0, 1, 2, \mathbf{3}, 4, 0, \ldots)$$

means that $a(0) = 3$, $a(1) = 4$, $a(-1) = 2$, $a(-2) = 1$, and all other entries are 0.

In fact, in all our examples, the refinable function is chosen to be the B-spline of order k, with k varying from one example to another. Recall that the B-spline is a C^{k-2} piecewise-polynomial of local degree $k-1$, which is supported in an interval of length k and has its knots at the integers only. Suppose that k is even. Then, the Fourier transform of that B-spline if given by

$$\widehat{\phi}(\omega) = \left(\frac{\sin(\omega/2)}{\omega/2}\right)^k.$$

The support of ϕ is $[-k/2, k/2]$. The B-spline ϕ is dyadically refinable with mask

$$\tau_\phi(\omega) = \cos^k(\omega/2).$$

When k is odd, one needs to insert a factor $\omega \to e^{i\omega/2}$ into the definitions of $\widehat{\phi}$ and τ_ϕ.

Example 2.1 *(Piecewise-linear tight frame)* We choose ϕ to be the B-spline of order 2, i.e., the hat function. The generators of the tight frame are drawn in Figure 1. The refinement mask is

$$a_\phi = (\ldots, 0, \frac{1}{4}, \frac{\mathbf{1}}{\mathbf{2}}, \frac{1}{4}, 0, \ldots).$$

The two wavelet masks are

$$a_{\psi_1} = (\ldots, -\frac{1}{4}, \frac{\mathbf{1}}{\mathbf{2}}, -\frac{1}{4}, 0, \ldots),$$

and

$$a_{\psi_2} = (\ldots, -\frac{\sqrt{2}}{4}, \mathbf{0}, \frac{\sqrt{2}}{4}, 0, \ldots),$$

This example is the simplest in a general construction of tight spline wavelet frames that was described in [18]. In that construction, the number of wavelets is k (with k the order of the B-spline which is used as a refinable function). The details of the piecewise-cubic case are as follows.

Example 2.2 *(Piecewise-cubic tight frame)* We choose ϕ to be the B-spline of order 4. The generators of the tight frame are shown in Figure 2. The refinement mask is

$$a_\phi = (\ldots, 0, \frac{1}{16}, \frac{1}{4}, \frac{3}{8}, \frac{1}{4}, \frac{1}{16}, 0, \ldots).$$

The four wavelets have masks as follows:

$$a_{\psi_1} = (\ldots, \ 0, \ -\tfrac{1}{8}, \ -\tfrac{1}{4}, \ \mathbf{0}, \ \tfrac{1}{4}, \ \tfrac{1}{8}, \ 0, \ \ldots)$$
$$a_{\psi_2} = (\ldots, \ 0, \ \tfrac{1}{16}, \ 0, \ -\tfrac{1}{8}, \ 0, \ \tfrac{1}{16}, \ 0, \ \ldots) * \sqrt{6}$$
$$a_{\psi_3} = (\ldots, \ 0, \ -\tfrac{1}{8}, \ \tfrac{1}{4}, \ \mathbf{0}, \ -\tfrac{1}{4}, \ \tfrac{1}{8}, \ 0, \ \ldots)$$
$$a_{\psi_4} = (\ldots, \ 0, \ \tfrac{1}{16}, \ -\tfrac{1}{4}, \ \tfrac{3}{8}, \ -\tfrac{1}{4}, \ \tfrac{1}{16}, \ 0, \ \ldots)$$

It is also possible to construct bi-frames where the two frames involved are derived from B-splines of different orders. In the next example, we derive the frame X from cubic splines, while its dual is derived from piecewise-linear splines.

Example 2.3 *(Bi-frames: cubics and linears mixed.)* We choose one refinable function to be the B-spline of order 4 (whose mask is already listed in Example 2.2), and the other B-spline to be of order 2, (i.e., it is the hat function of Example 2.1). There are two sets of wavelets now: those that generate the wavelet system X, and those that generate the dual wavelet system RX. The piecewise-linear wavelets (that can be used, say, during the decomposition step) are depicted in Figure 3. They are supported on the intervals $[.5, 3.5], [.5, 3], [1, 3.5]$ respectively. Note that, essentially, there are only two wavelets: the left-most one (together with its integer shifts) and the middle one (together with its half integer shifts). The masks of these three elements (ordered from left to right) are:

$$(\ldots, \ \mathbf{0}, \ 1, \ -4, \ 6, \ -4, \ 1, \ 0, \ \ldots) * \tfrac{1}{16\sqrt{2}}$$
$$(\ldots, \ \mathbf{0}, \ 0, \ -1, \ -1, \ 1, \ 1, \ 0, \ \ldots) * \tfrac{\sqrt{3}}{8}$$
$$(\ldots, \ \mathbf{0}, \ -1, \ -1, \ 1, \ 1, \ 0, \ 0, \ \ldots) * \tfrac{\sqrt{3}}{8}$$

The masks of the cubic dual frame are (in the same order)

$$(\ldots, \ \mathbf{0}, \ \tfrac{1}{\sqrt{8}}, \ -\tfrac{1}{\sqrt{2}}, \ \tfrac{1}{\sqrt{8}}, \ 0, \ \ldots)$$
$$(\ldots, \ \mathbf{0}, \ -\tfrac{1}{2}, \ \tfrac{1}{2}, \ 0, \ 0, \ \ldots) * \tfrac{\sqrt{3}}{4}$$
$$(\ldots, \ \mathbf{0}, \ 0, \ -\tfrac{1}{2}, \ \tfrac{1}{2}, \ 0, \ \ldots) * \tfrac{\sqrt{3}}{4}$$

Note that in the last example two wavelets are used for creating the system (one is shifted along integer translations, while the other ones along the denser half-integer translations). Examples of that sort are the rule rather than the exception. For example, it is possible to derive from the B-spline of order k a tight compactly supported spline frame with similarly two generators (however, the wavelets, in general, of those constructions are not symmetric.)

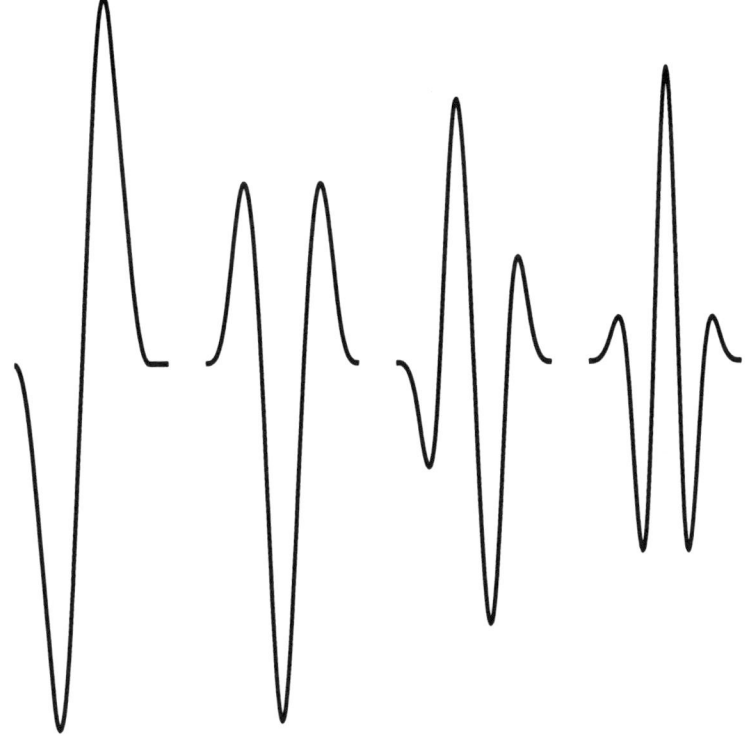

FIGURE 2. The four piecewise-cubic wavelets

3 Examples of Multivariate Wavelet Frames

Our examples of univariate wavelet frames in the previous section were derived from the multiresolution analysis generated by the B-spline. This ensured us, e.g., that the wavelets are smooth piecewise-polynomials. An attempt to extend this approach to the multivariate setup requires multivariate analogs of B-splines, i.e., smooth compactly supported refinable piecewise-polynomials. Fortunately, such functions exist and are known as 'box splines'. However, in contrast with the univariate cardinal B-splines that have only one 'degree of freedom', i.e., their order, a d-variate box spline is determined by a set of *directions*. Here, a **direction** is a non-zero vector in $\mathbb{Z}^d$. We stress that the 'sets' of directions below are not actually sets but multi-sets, i.e., a direction may appear several times in it. We do assume (without further notice) that each direction set to be considered spans the entire $\mathbb{R}^d$ space.

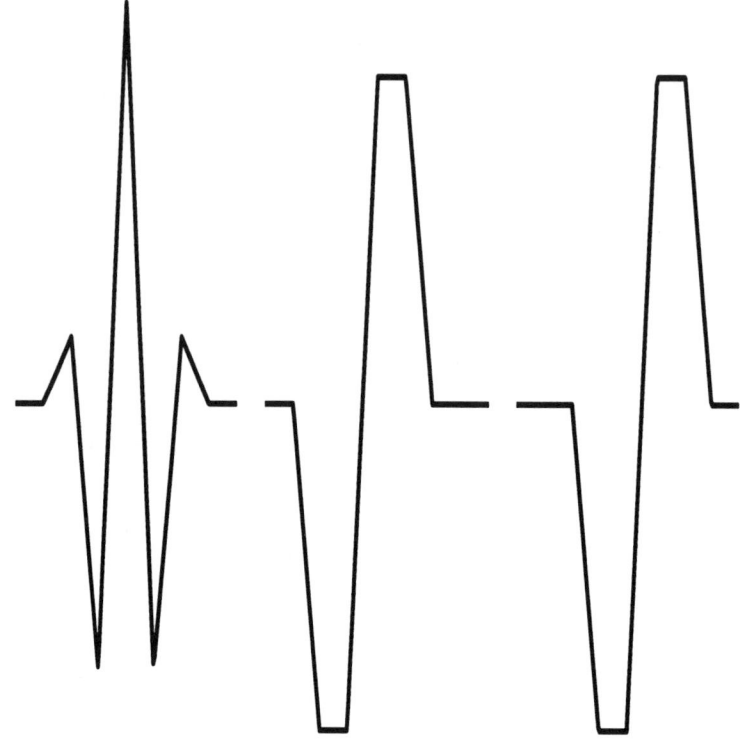

FIGURE 3. The generators of the piecewise-linear frame of Example 2.3

Definition 3.1 *Let $\Xi \subset \mathbb{Z}^d$ be a direction set. The **box spline** $\phi := \phi_\Xi$ is the function whose Fourier transform is*

$$\widehat{\phi}(\omega) = \prod_{\xi \in \Xi} \frac{1 - e^{-i\xi \cdot \omega}}{i\xi \cdot \omega}.$$

The box spline ϕ is a piecewise-polynomial of local degree $n := \#\Xi - d$ (i.e., each of the polynomial pieces is of degree $\leq n$). It lies in $C^{k-1} \backslash C^k$, with

$$k := \max\{\#Y : Y \subset \Xi,\ \mathrm{span}(\Xi \backslash Y) = \mathbb{R}^d\}.$$

Its support is the convex polyhedron

$$[0,1]^\Xi \Xi := \{\sum_{\xi \in \Xi} t_\xi \xi :\ t \in [0,1]^\Xi\}.$$

Much of the basic theory of box splines can be found in the book [4].

2. Construction of affine frames   35

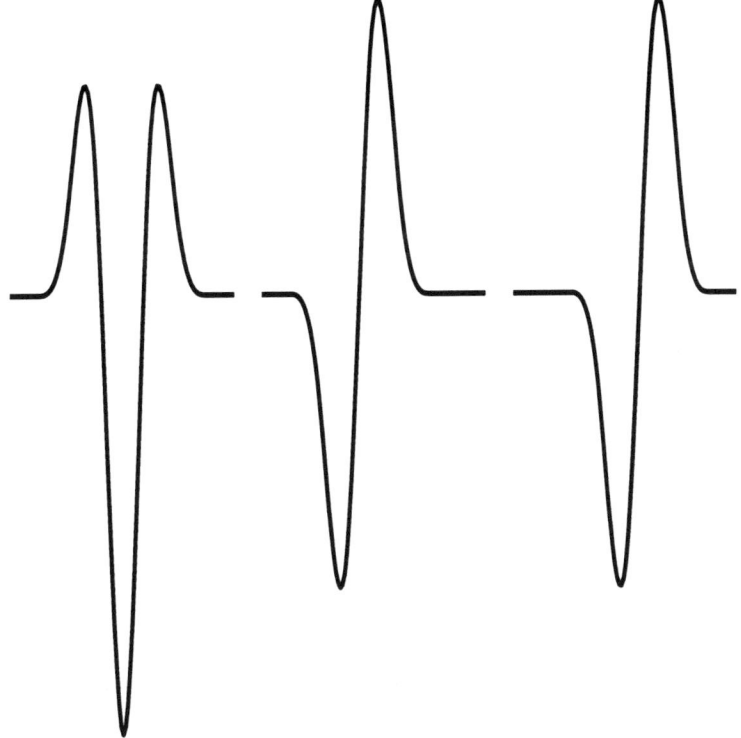

FIGURE 4. The generators of the piecewise-cubic frame of Example 2.3

We will be interested primarily in the 4-*direction bivariate box splines*. These box splines correspond to a direction set Ξ which consists of the four vectors

$$(\xi_1, \xi_2, \xi_3, \xi_4) := \begin{pmatrix} 1 & 0 & 1 & -1 \\ 0 & 1 & 1 & 1 \end{pmatrix},$$

each appearing with a certain multiplicity. We set $m = (m_1, m_2, m_3, m_4) \in \mathbb{Z}_+^4$ for the vector of multiplicities (i.e., $\xi_1 = (1\ 0)'$ appears in Ξ m_1 times, etc.) The support of the 4-direction box spline is an octagon, four of its vertices are $(0,0), (m_1, 0), (m_1 + m_3, m_3), (m_1 + m_3, m_2 + m_3), (m_1 + m_3 - m_4, m_2 + m_3 + m_4)$. Four direction box splines possess a wealth of symmetries; nonetheless, prior to [18], [20] there were hardly any wavelet constructions based on such splines. The reason for that is that the shifts (i.e., integer translates) of the 4-direction box spline are always *linearly dependent* (unless $m_3 m_4 = 0$, but then the box spline is not truly 4-directional);

indeed, we always have that

$$\sum_{\alpha \in \mathbb{Z}^2} (-1)^{\alpha_1 + \alpha_2} \phi(\cdot - \alpha) = 0, \qquad (3.1)$$

for a 4-direction box spline; the major previous algorithms for deriving wavelets from multiresolution all required, at a minimum, that the shifts of the underlying refinable function form a Riesz basis or a frame for V_0 (the latter being the closed *shift invariant* space generated by the shifts of ϕ). However, the dependence relation (3.1) implies that the shifts of ϕ form neither a Riesz basis nor a frame for V_0. (The reader is warned that the last statement is more subtle than it may look like: first, the shifts of $\phi \in L_2$ can form a Riesz basis while being linearly dependent. However, in such a case the coefficient sequence of each dependence relation is *unbounded*. Second, the elements of a frame can certainly be, and usually are, linearly dependent. However, a frame which consists of the shifts of a *single compactly supported* function is necessarily a Riesz basis, cf. [16]).

The box spline ϕ is dyadically refinable with mask whose symbol is

$$\prod_{j=1}^{4} e^{-im_j \xi_j \cdot \omega / 2} \cos^{m_j}(\xi_j \cdot \omega / 2).$$

Moreover, if we restrict our attention to 4-direction box splines whose multiplicities satisfy $m_1 = m_3$, $m_2 = m_4$, then those box splines are also refinable with respect to the dilation matrix

$$s = \begin{pmatrix} 1 & 1 \\ 1 & -1 \end{pmatrix}, \qquad (3.2)$$

and the symbol τ in this case is simpler:

$$\tau(\omega) = e^{-i(m_1, m_2) \cdot \omega / 2} \cos^{m_1}(\omega_1 / 2) \cos^{m_2}(\omega_2 / 2).$$

Warning: the above τ is also the symbol of the *tensor product* B-spline. This of course is possible: it is the symbol of the 4-direction box spline, when we use the above dilation matrix, and it is the symbol of the tensor B-spline when we use the more standard dyadic dilation (another way to view that: the 4-direction box spline is the convolution product of the tensor B-spline with its s-dilate). This coincidence enables us to convert standard construction techniques of tensor-product wavelets to the 4-direction box spline setup.

In what follows we discuss masks of bivariate refinable functions and masks of the corresponding wavelets. Until further notice, the dilation matrix is always assumed to be

$$\begin{pmatrix} 1 & 1 \\ 1 & -1 \end{pmatrix}.$$

2. Construction of affine frames   37

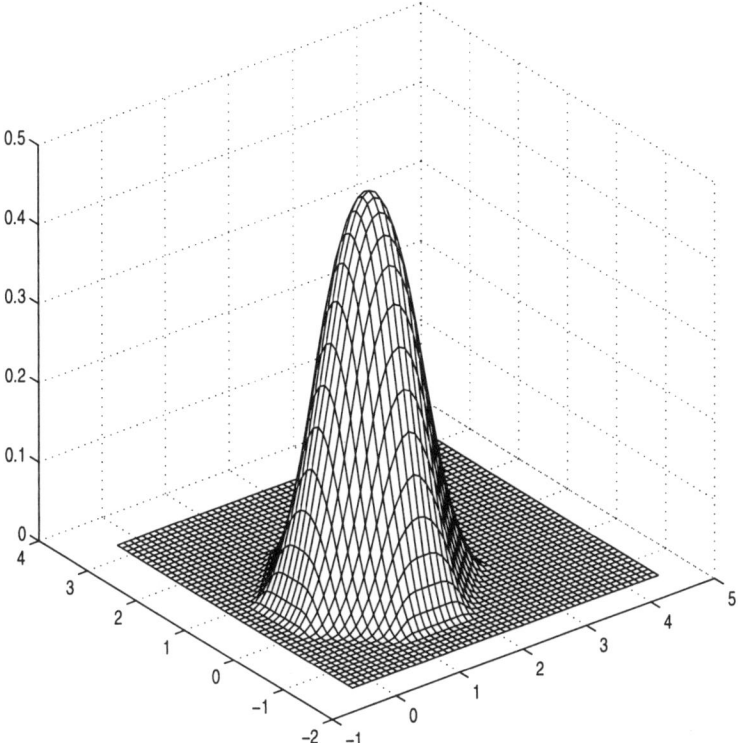

FIGURE 5. *The Powell-Zwart element*

We adopt the following convention concerning the mask discussed: given a finitely supported sequence on $\mathbb{Z}^2$, we simply display its non-zero values against the background of an (invisible) integer mesh. We mark with **boldface** the location of the origin, which is always displayed (even when its value is 0). For example, the notation

$$\begin{matrix} 4 & \\ \mathbf{0} & -1 \end{matrix}$$

stands for a sequence that takes the value 4 at $(0,1)$, the value -1 at $(1,0)$, and the value 0 anywhere else (on $\mathbb{Z}^2$).

Example 3.1 *Let ϕ be the 4-direction box spline whose multiplicity vector is $(1,1,1,1)$. This box spline is known in the finite element literature as the Powell-Zwart element, and its graph is drawn in Figure 5.*

The Powell-Zwart element is refinable with mask

$$a = \begin{matrix} .25 & .25 \\ .25 & .25 \end{matrix}.$$

It is a C^1 piecewise-quadratic spline, and its support is the smallest octagon with integer vertices (those vertices are $(.5, 1.5) + (\pm 1.5, \pm .5)$ and $(.5, 1.5) + (\pm .5, \pm 1.5)$). A tight frame that is generated by three wavelets can be derived from the multiresolution of the Powell-Zwart element. The three wavelet masks are

$$\begin{matrix} -.25 & -.25 \\ .25 & .25 \end{matrix} , \quad \begin{matrix} .25 & -.25 \\ .25 & -.25 \end{matrix} , \quad \begin{matrix} -.25 & .25 \\ .25 & -.25 \end{matrix}$$

Note that these masks are identical to those used in the construction of the bivariate dyadic orthonormal Haar system. That latter system is derived from the multiresolution analysis of the support function χ of the unit square, and our refinable function here is indeed related to χ: the Powell-Zwart element is the convolution product of χ and $\chi(t_1 + t_2, t_1 - t_2)$. The graphs of the three wavelets are drawn in Figures 6-8. All the wavelets have the same octagonal support as that of the Powell-Zwart element.

Since the dilation matrix s has determinant -2, one expects to use a single wavelet in the construction of irredundant wavelet systems (that are based on s). Since we used in Example 3.1 three wavelets, it seems reasonable to assert that the system there has 'a 3-fold rate of oversampling'. It is possible to modify the construction and to obtain a tight frame generated by *two* compactly supported wavelets. We refer to [20] for the details of that modified construction, but, for the reader convenience, list in the next example the corresponding masks.

Example 3.2 *(C^1 piecewise-quadratic compactly supported tight frame generated by two wavelets)* In this case the refinable function is slightly changed, and the refinement mask becomes:

$$\begin{matrix} & .25 & \\ .25 & .25 \\ .25 & & \end{matrix}.$$

The masks of the two wavelets are

$$\begin{matrix} & -.25 & \\ .25 & -.25 & \\ .25 & & \end{matrix} \qquad \begin{matrix} .5 \\ -.5 \end{matrix}$$

Note that the second wavelet has a smaller support than the first. Indeed, while in the previous example each of the three wavelets is supported in a domain of area 7, the two wavelets here are supported in domains of areas 10 and 7 respectively.

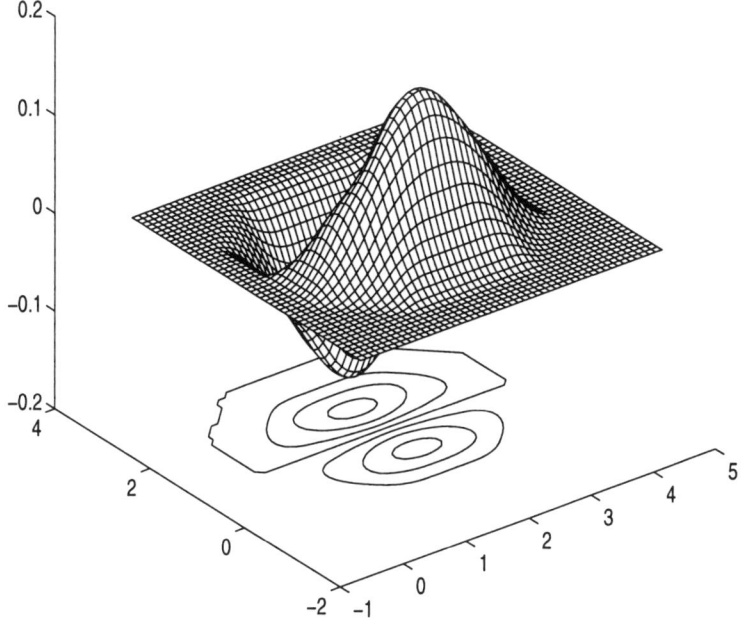

FIGURE 6. The first wavelet in Example 3.1

Algorithms for constructing compactly supported tight spline frames from box splines of higher smoothness are detailed in [20]. These algorithms work, essentially, with *any* box spline (though they may require to modify somewhat the magnitude of the directions that define the box spline as was actually done in the last example). However, in all these algorithms the number of wavelets that are used increases with the increase of the smoothness of the box spline (the determining factor is the degree of the mask, viewed as a trigonometric polynomial, and that degree must increase together with the smoothness). In what follows, we describe a general algorithm that applies to 4-direction box splines whose multiplicity vector is of the form (m_1, m_2, m_1, m_2). Recall that box spline is refinable with respect to the dilation matrix s of (3.2), and its mask, on the Fourier domain, is

$$\tau(\omega) = e^{-i(m_1, m_1)\cdot\omega/2} \cos^{m_1}(\omega_1/2) \cos^{m_2}(\omega_2/2). \qquad (3.3)$$

The algorithm can be extended to more general box splines (provided that those box splines are also refinable with respect to a dilation matrix whose

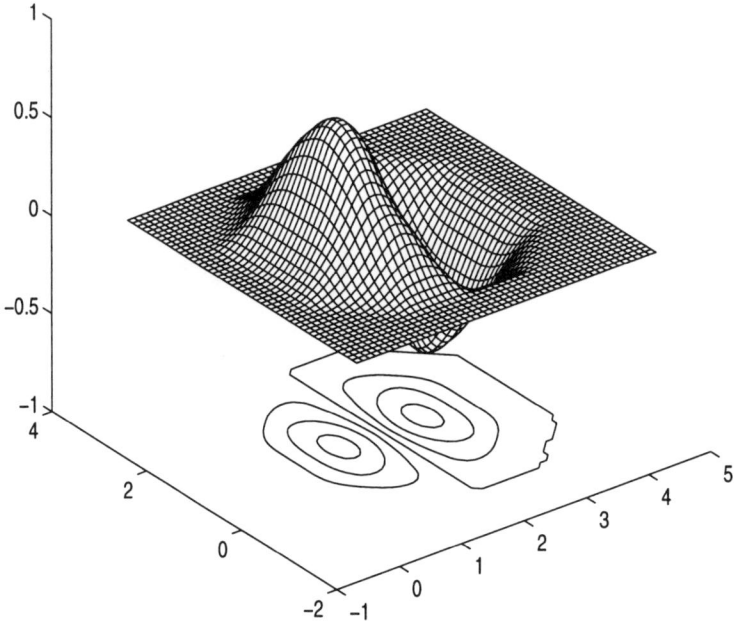

FIGURE 7. The second wavelet in Example 3.1

determinant is ±2) and is new, i.e., appears here for the first time. In contrast with the previous constructions, it yields bi-frames rather than tight frames. On the other hand, the number of wavelets is 3 regardless of the values of m_1, m_2 (i.e, regardless of the smoothness of the resulting wavelet system). We describe below the algorithm in general terms, and then provide the details of one of its special cases.

Algorithm: 4-directional compactly supported bi-frames of arbitrary smoothness generated by three wavelets. We need here *two* refinable functions, and assume both of them to be 4-direction box splines which are refinable with respect to the dilation matrix s, hence with masks of the form (3.3). We set ϕ for one of these functions, and ϕ^d for the other, set also τ and τ^d for their masks, and denote their multiplicity vectors by $m = (m_1, m_2, m_1, m_2)$ and $n = (n_1, n_2, n_1, n_2)$, respectively. We assume that all the entries of $r := (m+n)/2$ are (positive) *integers*. Under these mere assumptions, we derive two wavelet systems that form a bi-frame in

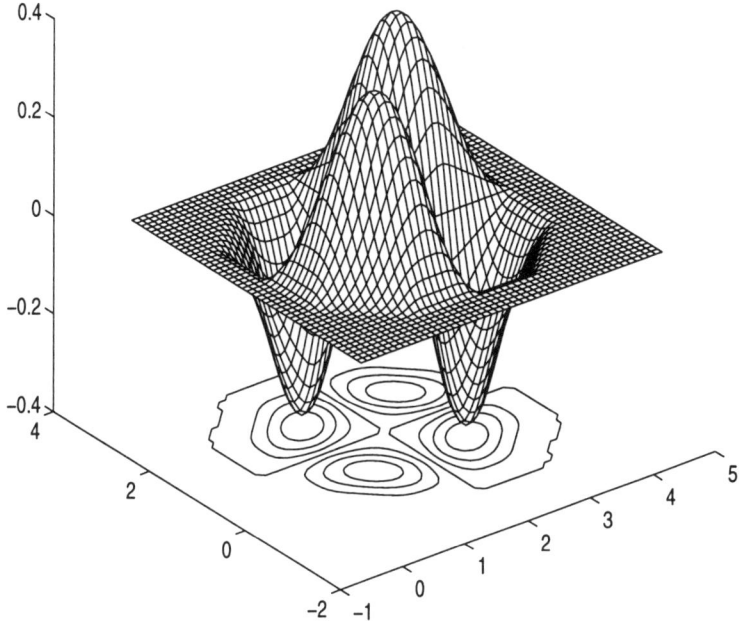

FIGURE 8. The third wavelet in Example 3.1

the following way. We first expand the expression

$$1 = (\cos^2(\omega_1/2) + \sin^2(\omega_1/2))^{r_1}(\cos^2(\omega_2/2) + \sin^2(\omega_2/2))^{r_2}, \tag{3.4}$$

and group the various summands into four groups. The first two groups are the singletons:

$$R_1(\omega) := \cos^{2r_1}(\omega_1/2)\cos^{2r_2}(\omega_2/2),$$
$$R_2(\omega) := \sin^{2r_1}(\omega_1/2)\sin^{2r_2}(\omega_2/2).$$

Since $R_2 = R_1(\cdot + (\pi, \pi))$, it is possible then to divide the other terms into two groups, R_3 and R_4, such that $R_4 = R_3(\cdot + (\pi, \pi))$. This can be done in many different ways, and the only condition we need is that R_3 is divisible by $\cos^2(\omega_1/2)\sin^2(\omega_2/2)$ (something that can be achieved by, e.g., putting all terms that are divisible by $\cos^{2r_1}(\omega_1/2)$ into R_3 and all terms that are divisible by $\cos^{2r_2}(\omega_2/2)$ into R_4). Observing that $R_1 =$

$\tau\tau^d$, we factor R_3 into $\tau_1\tau_1^d$ in a way that both τ_1 and τ_1^d are divisible by $\cos(\omega_1/2)\sin(\omega_2/2)$ (and both are 2π-periodic). We then define two wavelet systems, each consists of three wavelets. In the first system, the three wavelets masks are

$$t_1(\omega) := e^{i\omega_1}\overline{\tau^d(\omega+(\pi,\pi))}, \quad t_2(\omega) := \tau_1(\omega),$$

and

$$t_3(\omega) := e^{i\omega_1}\overline{\tau_1^d(\omega+(\pi,\pi))},$$

and in the second system the wavelet masks are

$$t_1^d(\omega) := e^{i\omega_1}\overline{\tau(\omega+(\pi,\pi))}, \quad t_2^d(\omega) := \tau_1^d(\omega),$$
$$t_3^d(\omega) := e^{i\omega_1}\overline{\tau_1(\omega+(\pi,\pi))}.$$

Since $t_j\overline{t_j^d} = R_{j+1}$, $j=1,2,3$, we conclude that $\tau\overline{\tau^d} + \sum_{j=1}^{3} t_j\overline{t_j^d} = 1$. At the same time, we have that $t_2\overline{t_2(\cdot+(\pi,\pi))} + t_3\overline{t_3(\cdot+(\pi,\pi))} = 0$, and also $\tau\overline{\tau(\cdot+(\pi,\pi))} + t_1\overline{t_1(\cdot+(\pi,\pi))} = 0$, and we thus conclude that the wavelets are constructed according to the mixed extension principle (see Theorem 4.4). Moreover, each of the wavelets in either system has a sin-factor in its mask, hence has a zero mean-value, which, together with its compact support assumption, implies that the wavelet system is Bessel (cf. [22]). Altogether, the two wavelet systems generated as above are bi-frames.

Example 3.3 We let ϕ and ϕ^d be, both, the 4-direction box splines with multiplicity $(2,2,2,2)$; the refinement masks (up to an exponential factor) are then $\tau(\omega) = \tau^d(\omega) = \cos^2(\omega_1/2)\cos^2(\omega_2/2)$. Also, $r = (2,2,2,2)$, and the expression in (3.4) is

$$(\cos^2(\omega_1/2) + \sin^2(\omega_1/2))^2(\cos^2(\omega_2/2) + \sin^2(\omega_2/2))^2.$$

After defining

$$R_1(\omega) = \cos^4(\omega_1/2)\cos^4(\omega_2/2),$$

and

$$R_2(\omega) = \sin^4(\omega_1/2)\sin^4(\omega_2/2),$$

we are left with seven additional terms that should be split between R_3 and R_4. One possibility is to define, with $b_j := \cos^2(\omega_j/2)$, $j=1,2$, (and after performing some straightforward simplifications)

$$R_3(\omega) := b_1(1-b_2)(2-b_1(1-b_2)),$$

and hence

$$R_4(\omega) := b_2(1-b_1)(2-b_2(1-b_1)).$$

There are then many ways to construct the wavelets. For example, we can define the generators of the first system to be

$$t_1(\omega) = e^{i\omega_1} \sin^2(\omega_1/2) \sin^2(\omega_2/2),$$
$$t_2(\omega) = e^{i\omega_1/2} \cos(\omega_1/2) \sin(\omega_2/2),$$
$$t_3(\omega) = e^{-i\omega_1/2} \sin(\omega_1/2) \cos(\omega_2/2)(2 - \sin^2(\omega_1/2) \cos^2(\omega_2/2)),$$

and, correspondingly,

$$t_1^d(\omega) = e^{i\omega_1} \sin^2(\omega_1/2) \sin^2(\omega_2/2),$$
$$t_2^d(\omega) = e^{i\omega_1/2} \cos(\omega_1/2) \sin(\omega_2/2)(2 - \cos^2(\omega_1/2) \sin^2(\omega_2/2)),$$
$$t_3^d(\omega) = e^{-i\omega_1/2} \sin(\omega_1/2) \cos(\omega_2/2).$$

4 The Theory of Affine Frames

In this section, we review the theory that led to the constructions detailed in the previous sections, and explain the basic principles behind the actual constructions.

The analysis of wavelet frames in [18] and [19] is based on the theory of *shift-invariant systems* that was developed in Approximation Theory (box splines, [4], form a special case of shift-invariant systems). A system $X \subset L_2$ is **shift-invariant** if there exists $F \subset X$ such that

$$X = (f(\cdot + \alpha) : f \in F, \ \alpha \in \mathbb{Z}^d).$$

A systematic study of the "frame properties" of a shift-invariant X can be found in [16], and the results that were subsequently applied in [17] to *Gabor systems* (which, indeed, are shift-invariant). Wavelet systems, on the other hand, are *not* shift-invariant (the negative dilation levels are invariant under translations that become sparser as the dilation level decreases). The main effort of [18] was devoted, indeed, to circumventing that obstacle, i.e., finding a way to apply the "shift-invariant methods" of [16] to the 'almost shift-invariant' wavelet systems.

This was achieved in [18] and [19] with the aid of the new notion of **quasi-affine system**, that we describe here (for the dyadic dilation case only; the development in [18] and [19] is valid with respect to general dilation matrices with integer entries). Let the affine system X be a wavelet system generated by a finite number of wavelets $\Psi \subset L_2(\mathbb{R}^d)$. The affine system X is the disjoint union of $D^k E(\Psi)$ where $E(\Psi) = \cup_{\psi \in \Psi} E(\psi)$ with $E(\psi) := \{\psi(\cdot - \alpha) : \alpha \in \mathbb{Z}^d\}$, the shift invariant set generated by ψ, and D is the dyadic dilation operator $D : f \mapsto 2^{d/2} f(2\cdot)$. That is

$$X = \bigcup_{k \in \mathbb{Z}} D^k E(\Psi).$$

The **quasi-affine** system associated with X (denoted by X^q) is, roughly speaking, the smallest shift-invariant set containing X. It is obtained from X by replacing, for each $k < 0$, the set of the functions $2^{kd/2}\psi(2^k \cdot + j)$, $\psi \in \Psi$, $j \in \mathbb{Z}^d$ that appears in X, by the larger shift-invariant set of functions
$$2^{kd}\psi(2^k \cdot + j), \quad \psi \in \Psi, \quad k < 0, \quad j \in 2^k \mathbb{Z}^d.$$

Note that, while the affine system is dilation-invariant, the quasi-affine X^q is shift-invariant, but is not dilation invariant.

While the "basis properties" of X (such as the Riesz basis property) are not preserved when passing to X^q, the "frame properties" of X are preserved. The following result is a special case of Theorem 5.5 of [18].

Theorem 4.1 *An affine system X is a frame for $L_2(\mathbb{R}^d)$ if and only if its quasi-affine counterpart X^q is one. Furthermore, the two systems have the same frame bounds. In particular, the affine frame X is tight if and only if the corresponding quasi-affine system X^q is tight.*

The theorem allows one to analyze the 'frame properties' of the affine X via a study of its quasi-affine counterpart. The latter is more mathematically accessible, by virtue of its shift-invariance. Specifically, [18] employs the so-called "dual Gramian" analysis of [16] (which is a 'shift-invariance method') to this end. The result is a complete characterization of all wavelet frames that we now describe.

The characterization is in terms of certain bi-infinite matrices, dubbed 'fibers'. The matrices and their entries are best described in terms of the following *affine product*:

$$\Psi[\omega, \omega'] := \sum_{\psi \in \Psi} \sum_{k=\kappa(\omega-\omega')}^{\infty} \widehat{\psi}(2^k \omega) \overline{\widehat{\psi}(2^k \omega')}, \quad \omega, \omega' \in \mathbb{R}^d,$$

where κ is the dyadic valuation:

$$\kappa : \mathbb{R} \to \mathbb{Z} : \omega \mapsto \inf\{k \in \mathbb{Z} : 2^k \omega \in 2\pi \mathbb{Z}^d\}.$$

(Thus, $\kappa(0) = -\infty$, and $\kappa(\omega) = \infty$ unless ω is 2π-dyadic.) Our convention is that $\Psi[\omega, \omega'] := \infty$ unless we have absolute convergence in the corresponding sum. We assume here that

$$|\widehat{\psi}(\omega)| = O(|\omega|^{-1/2-\delta}), \quad \text{near } \infty, \quad \text{for some } \delta > 0, \qquad (4.1)$$

for every wavelet $\psi \in \Psi$. This smoothness assumption on Ψ is mild, still the actual assumption in [18], [19] is even milder (multivariate Haar wavelets do not satisfy the smoothness assumption here, but do satisfy the milder assumption of [18], [19]). Theorem 4.1 is originally proved in [18] under this latter smoothness assumption; the subsequent proof in [7] avoids that assumption.

2. Construction of affine frames

The fibers (i.e., matrices) in the 'dual Gramian fiberization' are indexed by $\omega \in \mathbb{R}^d$. Each fiber is a non-negative definite self-adjoint matrix $\tilde{G}(\omega)$ whose rows and columns are indexed by $2\pi\mathbb{Z}^d$, and whose (α, β)-entry is

$$\tilde{G}(\omega)(\alpha, \beta) = \Psi[\omega + \alpha, \omega + \beta].$$

The matrix $\tilde{G}(\omega)$ is interpreted then as an endomorphism of $\ell_2(2\pi\mathbb{Z}^d)$ with norm denoted by $\mathcal{G}^*(\omega)$ and inverse norm $\mathcal{G}^{*-}(\omega)$. It is understood that $\mathcal{G}^*(\omega) := \infty$ whenever $\tilde{G}(\omega)$ does not represent a bounded operator, and a similar remark applies to $\mathcal{G}^{*-}(\omega)$. Theorem 4.1 together with the general 'shift-invariance tools' of [16] lead to the following characterization of wavelet frames.

Theorem 4.2 *Let X be an affine system generated by the Ψ. Let $\mathcal{G}^*$ and $\mathcal{G}^{*-}$ be the dual Gramian norm functions defined as above. Then X is a frame for $L_2(\mathbb{R}^d)$ if and only if $\mathcal{G}^*, \mathcal{G}^{*-} \in L_\infty$. Furthermore, the frame bounds of X are $\|\mathcal{G}^*\|_{L_\infty}$ and $1/\|\mathcal{G}^{*-}\|_{L_\infty}$.*

The theorem sheds new light on various previous studies of wavelet frames. For example, the estimates for the frames bounds of a wavelet frame (cf., e.g., [9]) can be reviewed as an attempt to estimate the norm and/or inverse norm of a matrix (viz., $\tilde{G}(\omega)$) in terms of its entries. 'Oversampling principles' (originated in the work of Chui and Shi, cf. e.g., [6]) are derived from the fact that the fibers of the oversampled systems are submatrices of the fibers of the original system.

The above theorem leads to the following characterization of *tight wavelet frames* (cf. Corollary 5.7 of [18]. Part (a) of that result was independently established in [11]):

Corollary 4.1 *(a) An affine system X generated by Ψ is a tight frame for $L_2(\mathbb{R}^d)$ with frame bound C if and only if*

$$\Psi[\omega, \omega] = C, \tag{4.2}$$

$$\Psi[\omega, \omega + 2\pi + 4\pi j] = 0, \tag{4.3}$$

for a.e. $\omega \in \mathbb{R}$ and $j \in \mathbb{Z}^d$.

(b) An affine system X is an orthonormal basis of $L_2(\mathbb{R}^d)$ if only if (4.3) holds, (4.2) holds with $C = 1$, and Ψ lies on the unit sphere of L_2.

We now show how the above theory leads to concrete algorithms for constructing wavelet frames. Assume that ϕ is a compactly supported refinable function with $\hat{\phi}(0) = 1$ (and satisfies (4.1)). Note that, in contrast with most of the wavelet literature, we are not making *a-priori* any assumption on the shifts of the refinable function: these shifts may not be orthonormal, nor they need to form a Riesz basis, nor even a frame. (Furthermore, we actually need only the condition $\widehat{\phi}(0) = 1$; the other assumptions are made here for convenience.)

We denote by V_0 the closed linear span of the shifts of ϕ and by V_j the 2^j-dilate of V_0. The assumption that ϕ is **refinable** is defined here to merely mean that $V_0 \subset V_1$. We remark in passing that (cf. §4 of [3]) $\cap_{j \in \mathbb{Z}} V_j = 0$ and that $\cup_{j \in \mathbb{Z}} V_j$ is dense in $L_2(\mathbb{R}^d)$ (the latter follows from the compact support assumption on ϕ, while the former holds for *any* refinable L_2-function, compactly supported or not); however, we will not need these two properties for the subsequent development.

In classical MRA constructions of orthogonal wavelets, prewavelets, bi-orthogonal wavelets, and frames, one starts with one or two refinable function(s) ϕ (and ϕ^d) that has certain properties (e.g., the shifts of ϕ are orthonormal, or form a Riesz basis; the shifts of ϕ^d are bi-orthogonal to those of ϕ, etc.) Then, one carefully selects a set of wavelets Ψ from the space V_1 in a way that makes the span W_0 of $E(\Psi)$ complementary (in some suitable sense) to V_0 in V_1; for example, W_0 may be the orthogonal complement of V_0 in V_1. The cardinality of the wavelet set Ψ is always $2^d - 1$.

In these classical constructions, we encounter difficulties in one (or both) of the following two major steps: (i) finding refinable functions with desired properties (the main difficulty being the deduction of the properties of ϕ from its refinement mask), and (ii) constructing the corresponding wavelet masks when the masks of the refinable functions are given.

Our MRA constructions in [18]–[20] deviate from this classical approach in the following way: while still selecting the wavelets Ψ from V_1, we allow the cardinality of the wavelet set Ψ to exceed the traditional number $2^d - 1$. We use these acquired degrees of freedom to construct affine frames with desired properties without requiring the underlying scaling function(s) to satisfy any substantial property. The examples in the previous sections demonstrate this point.

All the constructions of wavelet systems in this paper are based on two closely related algorithms for the derivation of wavelet frames from MRA. The first is the *(rectangular) unitary extension principle*, [18], which is used in the construction of tight wavelet frames, and the other is the *mixed extension principle*, [19], which is used in the construction of wavelet bi-frames. The unitary extension principle (Theorem 4.3 below) is derived in [18] as follows: assuming that ϕ is refinable and that Ψ is any finite subset of V_1, one rewrites first the conditions in Corollary 4.1 in terms of the various masks and the scaling function ϕ only. This leads, [18], to a complete characterization of all tight wavelet frames which can be constructed from any MRA, in terms of the underlying masks only. The following algorithm then follows easily from that general characterization. In its statement, we

relation
$$\widehat{\psi}(2\cdot) = \tau_\psi \widehat{\phi}.$$
We then construct a rectangular matrix Δ whose rows are indexed by Ψ', and whose columns are indexed by $\mathcal{Z} := \{0, \pi\}^d$:

$$\Delta := (\tau_\psi(\cdot + \nu))_{\psi \in \Psi', \nu \in \mathcal{Z}}. \tag{4.4}$$

Theorem 4.3 *(the unitary extension principle) Let ϕ a refinable function corresponding to MRA $(V_j)_j$ and Ψ be a finite subset of V_1. Let Δ be the matrix (4.4) that corresponds to $\Psi' := \Psi \cup \phi$, and X the affine systems generated by Ψ. If $\Delta^* \Delta = I$, a.e., then X is a tight frame for L_2.*

In [19], the above algorithm was extended to include bi-frames.

Theorem 4.4 *(the mixed extension principle) Let ϕ and ϕ^d be two refinable functions corresponding to MRAs $(V_j)_j$ and $(V_j^d)_j$, respectively. Let Ψ be a finite subset of V_1, and let $\mathrm{R} : \Psi \to V_1^d$ be some map. Let Δ be the matrix (4.4) that corresponds to $\Psi' := \Psi \cup \phi$, and let Δ^d be the matrix of (4.4) that corresponds to $\Psi' := \mathrm{R}\Psi \cup \phi^d$. Finally, let X and $\mathrm{R}X$ be the affine systems generated by Ψ and $\mathrm{R}\Psi$, respectively. If*

(a) X and $\mathrm{R}X$ are Bessel, and

(b) $\Delta^* \Delta^d = I$, a.e.,

then X and $\mathrm{R}X$ are frames for L_2 that are dual one to the other.

5 References

[1] G. Battle, A block spin construction of ondelettes. Part I: Lemarie Functions, Communications Math. Phys. **110** (1987), 601–615.

[2] C. de Boor, R. A. DeVore and A. Ron, Approximation from shift-invariant subspaces of $L_2(\mathbb{R}^d)$, Transactions of Amer. Math. Soc. **341** (1994), 787–806.

[3] C. de Boor, R. DeVore and A. Ron, On the construction of multivariate (pre) wavelets, Constr. Approx. **9** (1993), 123–166.

[4] C. de Boor, K. Höllig and S. D. Riemenschneider, Box splines, Springer Verlag, New York, (1993).

[5] A. Cohen, I. Daubechies and J.C. Feauveau, Biorthogonal bases of compactly supported wavelets, Comm. Pure. Appl. Math. **45** (1992), 485-560.

[6] C. K. Chui and X. Shi, Inequalities on matrix-dilated Littlewood-Paley functions and oversampled affine operators, CAT Report #337, Texas A&M University, 1994, SIAM J. Math. Anal., to appear.

[7] C. K. Chui, X. L. Shi and J. Stöckler, Affine frames, quasi-affine frames, and their duals, preprint, 1996.

[8] I. Daubechies, Orthonormal bases of compactly supported wavelets, Comm. Pure and Appl. Math., **41** (1988), 909–996.

[9] I. Daubechies, The wavelet transform, time-frequency localization and signal analysis, IEEE Trans. Inform. Theory **36** (1990), 961–1005.

[10] K. Gröchenig and A. Ron, Tight compactly supported wavelet frames of arbitrarily high smoothness, Proc. Amer. Math. Soc., to appear,
Ftp site: ftp://ftp.cs.wisc.edu/Approx file cg.ps.

[11] B. Han, On dual wavelet tight frames, ms., (1995).

[12] H. Ji, S. D. Riemenschneider and Z. Shen, Multivariate Compactly supported fundamental refinable functions, duals and biorthogonal wavelets, (1997) Studies in Appl. Math., to appear.

[13] P. G. Lemarié, Ondelettes à localisation exponentielle, J. de Math. Pures et Appl. **67** (1988), 227–236.

[14] S. G. Mallat, Multiresolution approximations and wavelet orthonormal bases of $L^2(\mathbb{R})$, Trans. Amer. Math. Soc. **315**(1989), 69–87.

[15] S. D. Riemenschneider and Z. Shen, Construction of biorthogonal wavelets in $L_2(\mathbb{R}^s)$, preprint (1997).

[16] A. Ron and Z. Shen, Frames and stable bases for shift-invariant subspaces of $L_2(\mathbb{R}^d)$, Canad. J. Math., **47** (1995), 1051–1094.
Ftp site: ftp://ftp.cs.wisc.edu/Approxfile frame1.ps.

[17] A. Ron and Z. Shen, Weyl-Heisenberg frames and Riesz bases in $L_2(\mathbb{R}^d)$, Duke Math. J., **89** (1997), 237–282.
Ftp site: ftp://ftp.cs.wisc.edu/Approx file wh.ps.

[18] A. Ron and Z. Shen, Affine systems in $L_2(\mathbb{R}^d)$: the analysis of the analysis operator, J. Functional Anal., **148** (1997), 408–447.
Ftp site: ftp://ftp.cs.wisc.edu/Approx file affine.ps.

[19] A. Ron and Z. Shen, Affine systems in $L_2(\mathbb{R}^d)$ II: dual system, J. Fourier Anal. App., **3** (1997), 617–637.
Ftp site: ftp://ftp.cs.wisc.edu/Approx file dframe.ps.

[20] A. Ron and Z. Shen, Compactly supported tight affine spline frames in $L_2(\mathbb{R}^d)$, Math. Comp., **67** (1998), 191–207.
Ftp site: ftp://ftp.cs.wisc.edu/Approx file tight.ps.

[21] A. Ron and Z. Shen, Frames and stable bases for subspaces of $L_2(\mathbb{R}^d)$: the duality principle of Weyl-Heisenberg sets, *Proceedings of the Lanczos International Centenary Conference*, Raleigh NC, 1993, D. Brown, M. Chu, D. Ellison, and R. Plemmons eds., SIAM Pub. (1994), 422–425

[22] A. Ron and Z. Shen, Gramian analysis of affine bases and affine frames, *Approximation Theory VIII, Vol. 2: Wavelets and Multilevel Approximation*, C.K. Chui and L.L. Schumaker eds, World Scientific Publishing, New Jersey, 1995, 375–382.

3
Understanding Wavelet Image Coding

Stéphane Mallat [1] and Frédéric Falzon

ABSTRACT It is shown that the low bit rate compression performance of image transform coders depends mostly on the ability to efficiently approximate images with few basis vectors. In wavelet bases, for less than 1 bit per pixel the distortion rate $D(\bar{R})$ varies like $\bar{R}^{1-2\gamma}$ instead of $2^{-2\bar{R}}$ as predicted by classical transform code theory, with γ of the order of 1 for most natural images. The improved performance of embedded coding in wavelet bases is also studied.

1 Introduction

Research on image compression led to a culture shock between mathematicians who became interested in this topic through the discovery of wavelet orthonormal bases, and signal processing engineers who had been developing efficient algorithms for many years. Many mathematicians thought that if one could approximate an image with a small number of wavelet coefficients, then these bases should be remarkably well adapted to performing bit compression. Most signal processing engineers replied that the problem is much more difficult than this since it also involves a careful analysis of the performance of scalar quantizations and bit allocations. If the signals to be encoded are realizations of a Gaussian process, under the high resolution quantization hypothesis, we know nearly everything about the performance of a transform coder. For an average of $\bar{R}$ bits per pixel, the mean-square error $D(\bar{R})$ varies proportionally to $2^{-2\bar{R}}$ with a constant that depends upon the bit allocation and the basis.

Current image transform coders operate below 1 bit per pixel. For such low bit rates, classical transform coding theory yields an incorrect estimate of the distortion rate $D(\bar{R})$. In this range, we show that $D(\bar{R})$ depends

---

[1]This work was supported by the French Centre National d'Etudes Spatiales and the AFOSR grant F49620-96-1-0455.
Stéphane Mallat is with Ecole Polytechnique, 91128 Palaiseau, France and with the Courant Institute, New York University, NY 10012.
Frédéric Falzon was at INRIA, Sophia Antipolis, France, and is now at Alcatel Alsthom Recherche, 91460 Marcoussis, France.

mostly on the error D_0 when approximating signals with a limited number of vectors selected from the orthogonal basis. From that point of view, the intuition of mathematicians was right. We verify that in wavelet and block cosine bases $D(\bar{R})$ varies like $\bar{R}^{1-2\gamma}$, where γ is of the order of 1 for most "natural" images.

Transform coding algorithms can be improved by an embedding strategy which sends the larger amplitude coefficients first and then progressively refines their quantization. This improvement is analyzed mathematically and evaluated numerically for the wavelet zero-tree algorithm of Shapiro [Sha93]. Embedded coders outperform classical transform coders when there is some prior information on which basis vectors produce large, average or small decomposition coefficients for typical signals.

This chapter begins with a brief review of high resolution quantization and high bit rate transform coding. Section 3 analyzes the distortion rate at low bit rates and gives numerical examples with a wavelet transform coder. The performance of embedded transform coders and their applications to wavelets is studied in Section 4.

2 High Bit Rate Compression

The class of signals to be encoded is represented by a random vector Y of size N. Although these signals may be multidimensional like images, they are indexed by an integer n to simplify notations: $Y[n]$. A transform coder decomposes these signals in an orthonormal basis $\mathcal{B} = \{g_m\}_{0 \leq m < N}$

$$Y = \sum_{m=0}^{N-1} A[m]\, g_m.$$

Each coefficient $A[m]$ is a random variable defined by

$$A[m] = \langle Y, g_m \rangle = \sum_{n=0}^{N-1} Y[n]\, g_m^*[n].$$

To construct a finite code, each coefficient $A[m]$ is approximated by a quantized variable $\hat{A}[m]$. We concentrate on scalar quantizations which are most often used for transform codes. The next section reviews important results concerned with minimizing the quantization error.

2.1 Optimized Scalar Quantization

A scalar quantizer Q approximates a real random variable X by a quantized variable $\hat{X} = Q(X)$, which takes its values in a finite set. Suppose that X takes its values in $[a, b]$, which may correspond to the whole real axis. We decompose $[a, b]$ into K intervals $(y_{k-1}, y_k]_{1 \leq k \leq K}$ of variable lengths, with $y_0 = a$ and $y_K = b$. If $x \in (y_{k-1}, y_k]$ then $Q(x) = x_k$.

The goal is to minimize the number of bits required to encode the quantized values $\hat{X} = Q(X)$ for a fixed mean-square distortion $D = \mathsf{E}\{(X - \hat{X})^2\}$. We denote

$$p_k = \Pr\{X \in (y_{k-1}, y_k]\} = \Pr\{\hat{X} = x_k\}.$$

The Shannon theorem [GG92] proves that the entropy

$$\mathcal{H}(\hat{X}) = -\sum_{k=1}^{K} p_k \log_2 p_k$$

is a lower bound of the average number of bits per symbol used to encode the values of $\hat{X}$. Arithmetic entropy codes [WNC87] achieve an average bit rate that can be arbitrarily close to the entropy lower bound, so we shall consider that this lower bound is reached.

Let $p(x)$ be the probability density of the random source X. The differential entropy of X is defined by

$$\mathcal{H}_d(X) = -\int_{-\infty}^{+\infty} p(x) \log_2 p(x) \, dx.$$

A quantizer is said to have a *high resolution* if $p(x)$ is approximately constant on each quantization bin $(y_{k-1}, y_k]$ of size $\Delta_k = y_k - y_{k-1}$. This is the case if the sizes Δ_k are sufficiently small relative to the rate of variation of $p(x)$. The following well known theorem [GG92] proves that uniform quantizers are optimal among high resolution quantizers.

Theorem 1 *If Q is a high resolution quantizer with respect to $p(x)$ then*

$$\mathcal{H}(\hat{X}) \geq \mathcal{H}_d(X) - \frac{1}{2}\log_2(12D). \tag{1}$$

This inequality is an equality if and only if Q is a uniform quantizer, in which case $D = \frac{\Delta^2}{12}$.

This theorem proves that for a fixed distortion D, under the high resolution quantization hypothesis the minimum average bit rate $R_X = \mathcal{H}(\hat{X})$ is achieved by a uniform quantizer and

$$R_X = \mathcal{H}_d(X) - \log_2 \Delta. \tag{2}$$

The distortion rate is obtained by taking the inverse of (1):

$$D(R_X) = \frac{1}{12} 2^{2\mathcal{H}_d(X)} 2^{-2R_X}. \tag{3}$$

2.2 Distortion Rate

Let us optimize the transform coding of $Y = \sum_{m=0}^{N-1} A[m] \, g_m$. The average bit budget to encode $\hat{A}[m] = Q(A[m])$ is $R_m = \mathcal{H}(\hat{A}[m])$. For a high resolution quantization, Theorem 1 proves that the error $D_m = \mathsf{E}\{|A[m] - \hat{A}[m]|^2\}$ is minimized when using a uniform scalar quantization. An optimal bit allocation minimizes the total number of bits $R = \sum_{m=0}^{N-1} R_m$ for a specified total error $D = \sum_{m=0}^{N-1} D_m$. Let $\bar{R} = \frac{R}{N}$ be the average number of bit per sample. With Lagrange multipliers we verify that $\bar{R}$ is minimum if all D_m are equal, in which case

$$D(\bar{R}) = \frac{N}{12} 2^{2\overline{\mathcal{H}}_d} 2^{-2\bar{R}}, \qquad (4)$$

where $\overline{\mathcal{H}}_d$ is the averaged differential entropy

$$\overline{\mathcal{H}}_d = \frac{1}{N} \sum_{m=0}^{N-1} \mathcal{H}_d(A[m]).$$

The performance of this transform coding depends on the choice of the orthonormal basis $\mathcal{B}$ through $\overline{\mathcal{H}}_d$. In general, it is difficult to find $\mathcal{B}$ which minimizes $\overline{\mathcal{H}}_d$ because the probability density of $A[m] = \langle Y, g_m \rangle$ may depend on g_m in a complicated way. If Y is a Gaussian random vector then the coefficients $A[m]$ are Gaussian random variables in any basis. In this case, the probability density of $A[m]$ depends only on the variance σ_m^2 and one can verify that

$$\mathcal{H}_d\Big(A[m]\Big) = \log_2 \sigma_m + \log_2 \sqrt{2\pi e}.$$

Inserting this expression in (4) yields

$$D(\bar{R}) = N \frac{\pi e}{6} \left(\prod_{m=0}^{N-1} \sigma_m^2 \right)^{1/N} 2^{-2\bar{R}}. \qquad (5)$$

One can prove that $\prod_{m=0}^{N-1} \sigma_m^2$ is minimum if and only if $\mathcal{B}$ is a Karhunen-Loeve basis of Y [GG92], which means that $\mathcal{B}$ diagonalizes the covariance matrix of Y. The transform coding of a Gaussian process is thus optimized in a Karhunen-Loeve basis. When Y is not Gaussian, the Karhunen-Loeve basis is *a priori* not optimal anymore. This is the case for images which cannot be considered as realizations of a Gaussian process.

Let us describe a simple wavelet transform coder for images. Separable wavelet bases of images include three wavelets with horizontal, vertical or diagonal orientations [Mal89], indexed by $1 \leq k \leq 3$. At an orientation k and scale 2^j, the wavelet vector $g_m = \psi_{j,p,q}^k$ is approximately centered at $(2^j p, 2^j q)$, with a square support whose size is proportional to 2^j. At high bit rates, we saw that the distortion rate is optimized by quantizing uniformly all decomposition coefficients. The domains where the image has smooth grey level variations yield small amplitude wavelet coefficients that are quantized to zero. To improve the efficiency of this transform coding, the wavelet coefficients are scanned in a predefined order and the position of zero versus non-zero quantized coefficients is recorded with a run-length coding that is entropy encoded. In the same scanning order, the amplitude of the non-zero quantized coefficients are also entropy encoded with a Huffman or an Arithmetic coding.

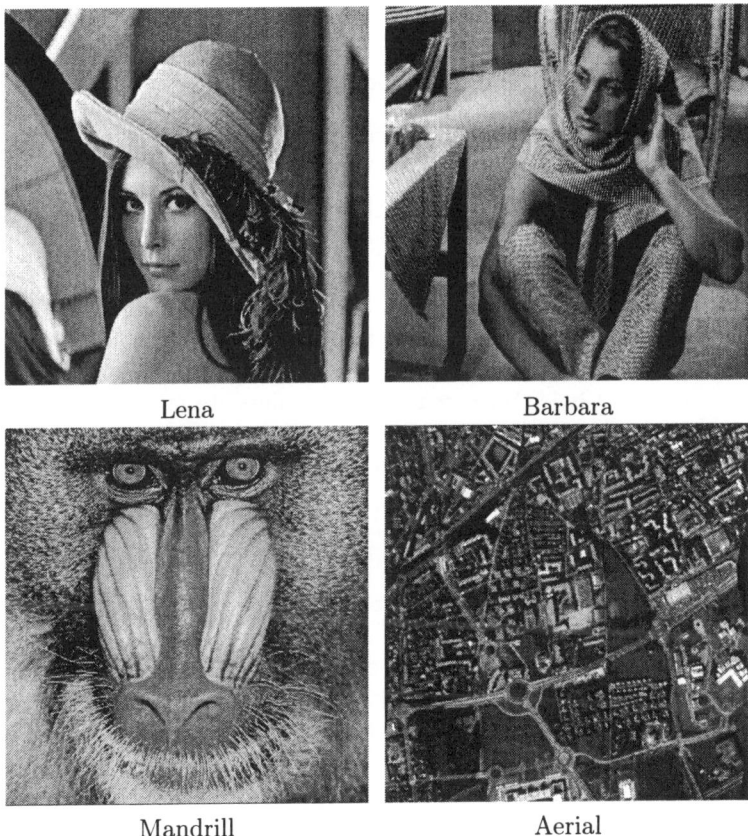

FIGURE 1. Four test images for numerical experiments.

Figure 2 gives the distortion $\log_2 D(\bar{R})$ of this wavelet transform coding for the test images shown in Figure 1. These numerical experiments are performed with an orthogonal cubic spline Battle-Lemarie wavelet [Mal89].

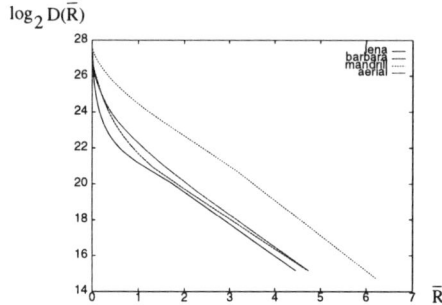

FIGURE 2. Log distortion rate curve for the wavelet transform coding of each test image.

The distortion rate formula (4) predicts that

$$\log_2 D(\bar{R}) = 2\overline{\mathcal{H}_d} + \log_2(\frac{N}{12}) - 2\bar{R}$$

and $\log_2 D(\bar{R})$ should therefore decay with a slope of -2 as a function of $\bar{R}$. This is indeed verified in Figure 2 for $\bar{R} \geq 1$, but not for $\bar{R} < 1$ where $\log_2 D(\bar{R})$ has a much faster decay. At low-bit rates $\bar{R} < 1$ the distortion rate formula (4) is not valid because the high resolution quantization assumption does not hold. Wavelet transform coders are most often used for $\bar{R} < 1$ because they recover images of nearly perfect visual quality up to $\bar{R} = 0.5$ bits per pixel. The next section studies the distortion rate at these low bit rates.

3 Low Bit Rate Compression

At low bit rates, the decomposition coefficients of an image in an orthonormal basis are coarsely quantized. Since many coefficients are set to zero, the positions of zero versus non-zero quantized coefficients are stored in a binary significance map, which is recorded with a run-length coding or a more sophisticated zero-tree algorithm. The distortion rate theory previously described does not apply for two reasons. First, the high resolution quantization hypothesis does not hold because the quantization bins are large. Second, one cannot treat the total bit budget R as a sum of bits R_m allocated independently to each decomposition coefficient. Indeed, the encoding of the zero quantized coefficients through a significance map is a form of vector quantization, which relates the encoding of different coefficients.

To evaluate the distortion rate we cannot rely on a precise stochastic model for images. There is as yet no model that incorporates the full diversity of image structures, such as non-stationary textures and edges. To avoid this difficulty, as in the min-max non-linear estimation theory of Donoho and Johnstone [DJ94], we shall consider the signals as deterministic vectors whose decomposition coefficients in the basis $\mathcal{B}$ have a parameterized decay. The distortion rate is therefore not calculated with an ensemble average but for each signal f. The key result shows that this distortion rate depends mostly on the ability to precisely approximate f with a small number of vectors selected from $\mathcal{B}$. Low-bit rate image compressions in wavelet bases and block cosine bases illustrate the distortion rate results.

3.1 Distortion Rate

Let f be a signal decomposed in an orthonormal basis $\mathcal{B} = \{g_m\}_{0 \leq m < N}$:

$$f = \sum_{m=0}^{N-1} a[m] \, g_m \quad \text{with } a[m] = \langle f, g_m \rangle.$$

The transform coder quantizes all coefficients and reconstructs

$$\hat{f} = \sum_{m=0}^{N-1} Q(a[m]) \, g_m.$$

The coding error is

$$D = \|f - \hat{f}\|^2 = \sum_{m=0}^{N-1} |a[m] - Q(a[m])|^2. \tag{6}$$

We denote by $h[x]$ the discrete histogram of the N coefficients $a[m]$, normalized so that $\sum_x h[x] = 1$. The values of this histogram are interpolated to define a function $p(x) \geq 0$ for all $x \in \mathbf{R}$ such that $\int_{-\infty}^{+\infty} p(x) \, dx = 1$. This $p(x)$ is the probability density of a random variable X. We suppose N to be sufficiently large and the histogram sufficiently regular so that for all functions $\phi(x)$ that appear in our calculations one has :

$$\frac{1}{N} \sum_{m=0}^{N-1} \phi(a[m]) = \sum_x \phi(x) \, h[x] \approx \int_{-\infty}^{+\infty} \phi(x) \, p(x) \, dx = \mathsf{E}\{\phi(X)\} \, . \tag{7}$$

This hypothesis holds for the histograms of the test images shown in Figure 3, as well as for most "natural" images. It is equivalent to the coefficients $a[m]$ being successive values of the random variable X. Applied to $\phi(x) = |x - Q(x)|^2$, (7) yields

$$\frac{D}{N} = \frac{1}{N} \sum_{m=0}^{N-1} |a[m] - Q(a[m])|^2 = \mathsf{E}\{|X - Q(X)|^2\}.$$

Let $\bar{R}$ be the average number of bits per coefficient to encode the $Q(a[m])$. If Q is a high resolution uniform quantizer with step size Δ then (3) implies a distortion rate formula similar to (4):

$$\frac{D(\bar{R})}{N} = \frac{1}{12} 2^{2\mathcal{H}_d(X)} 2^{-2\bar{R}} . \qquad (8)$$

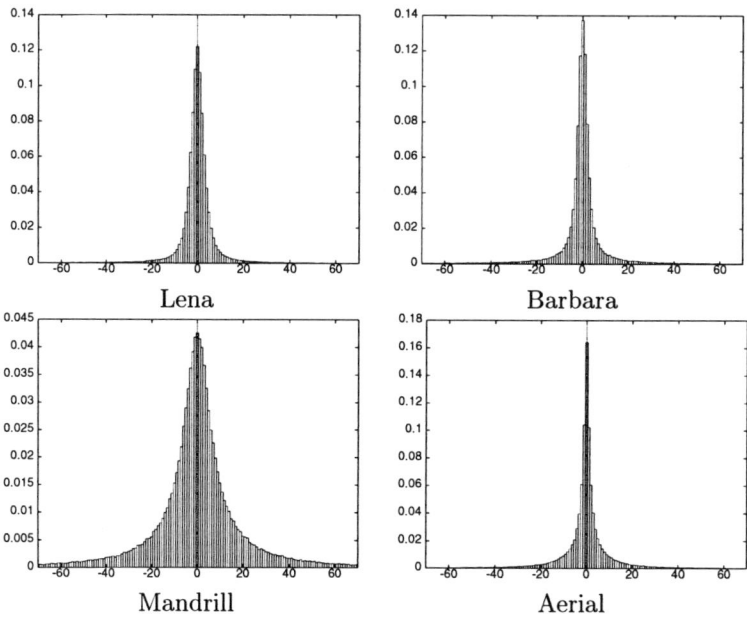

FIGURE 3. Normalized histograms of the cubic-spline wavelet coefficients of the test images.

If the basis $\mathcal{B}$ is chosen so that many coefficients $a[m] = \langle f, g_m \rangle$ are close to zero and few have a large amplitude then $p(x)$ has a sharp high peak at $x = 0$. This is the case for the histograms of wavelet coefficients shown in Figure 3. If Δ is large then $p(x)$ has important variations in the bin $[-\frac{\Delta}{2}, \frac{\Delta}{2}]$ where coefficients are quantized to zero. Hence the high resolution quantization hypothesis does not apply in this zero bin. This explains why $\log_2 D(\bar{R})$ shown in Figure 2 decays like $-2\bar{R}$ only for $\bar{R} \geq 1$.

Low bit rate image transform coders are often implemented with a scalar quantizer that modifies the size of the zero quantization bin. Let $[-T, T]$ be the zero bin where Q sets all coefficients to zero. In this bin, we cannot apply the high-resolution quantization hypothesis because $p(x)$ varies too much. Outside $[-T, T]$, we shall consider that the high resolution assumption holds. Although this assumption is only an approximation, it captures sufficiently precisely the quantizer properties in this domain. As a consequence, Theorem 1 proves that outside $[-T, T]$ the optimal quantizer Q has bins of constant size Δ. The ratio $\theta = \frac{T}{\Delta}$ is a parameter that must be adjusted to minimize the overall distortion D.

Any coefficient $|a[m]| > T$ that is not quantized to zero is called a *significant coefficient*. Coding the position of non-zero quantized coefficients is equivalent to storing a binary *significance map* defined by

$$b[m] = \begin{cases} 0 & \text{if } |a[m]| \leq T \\ 1 & \text{if } |a[m]| > T \end{cases}. \tag{9}$$

The wavelet image coder of Section 2.2 uses a run-length coding to store this significance map. More efficient zero-tree encoding techniques may also be used for wavelet significance maps [LK92].

Let R_0 be the total number of bits required to encode the significance map. Let M be the number of significant coefficients. There is a proportion $p = \frac{M}{N}$ of indices m such that $b[m] = 1$. An upper bound for R_0 is computed by supposing that there is no redundancy in the position of the 0 and the 1 in the significance map. The average number of bits to encode the position of one coefficient is then the entropy of a binary source with a probability $p = \frac{M}{N}$ to be equal to 1 and $1 - p$ to be equal to 0:

$$\frac{R_0}{N} \leq -p \log_2 p - (1-p) \log_2(1-p).$$

For $x \in (0, 1]$ then $-x \log_2 x \leq (1-x) \log_2 e$ so the average number of bits per significant coefficient to encode the significance map is

$$r_0 = \frac{R_0}{M} \leq \log_2 \frac{N}{M} + \log_2 e. \tag{10}$$

For wavelet coefficients, when the proportion of significant coefficients $\frac{M}{N}$ is small, a run-length coding yields an average bit rate $r_0 = \frac{R_0}{M}$ which is much smaller than the upper bound (10), because of the redundancy in the positions of the zero coefficients. For large classes of images, numerical calculations show that $r_0 = \frac{R_0}{M}$ varies slowly relative to $\frac{M}{N}$.

The amplitude of the M significant coefficients is uniformly quantized with a step Δ and these quantized values are entropy encoded. Let us compute the total number of bits R_1 of the resulting entropy coding. For $M \gg 1$, the M significant coefficients $a[m]$ above T have a normalized histogram that is interpolated by

$$p_T(x) = \frac{N}{M} p(x) \mathbf{1}_{\{|x|>T\}}.$$

Let X_T be the random variable whose probability density is $p_T(x)$. Since the high resolution quantization hypothesis applies to significant coefficients, the average number of bits to encode the amplitude of each quantized significant coefficient, denoted r_1, is calculated from (2):

$$r_1 = \frac{R_1}{M} = \mathcal{H}_d(X_T) - \log_2 \Delta. \tag{11}$$

Overall, the transform coding requires $R = R_0 + R_1$ bits.

To estimate the quantization error, $D = \|f - \hat{f}\|^2$ in (6), insignificant coefficients quantized to zero are separated from significant coefficients: $D = D_0 + D_1$, where

$$D_0 = \sum_{|a[m]| \leq T} |a[m]|^2 \tag{12}$$

is the error due to quantizing insignificant coefficients, and

$$D_1 = \sum_{|a[m]| > T} |a[m] - Q(a[m])|^2 \tag{13}$$

is the error due to quantizing significant coefficients. The average quantization error $\frac{D_1}{M}$ per significant coefficient is calculated with the high resolution quantization assumption

$$\frac{D_1}{M} = \frac{1}{M} \sum_{|a[m]| > T} |a[m] - Q(a[m])|^2 = \mathsf{E}\{|X_T - Q(X_T)|^2\} = \frac{\Delta^2}{12}. \tag{14}$$

To compute the error due to quantizing insignificant coefficients we denote by f_M the approximation of f using the M vectors g_m of $\mathcal{B}$ such that $|a[m]| = |\langle f, g_m \rangle| > T$

$$f_M = \sum_{|a[m]| > T} a[m]\, g_m.$$

The signal f_M can also be interpreted as an approximation of f from the M vectors of $\mathcal{B}$ whose inner products with f have the largest amplitude. The distortion D_0 can be rewritten

$$D_0\left(\frac{M}{N}\right) = \|f - f_M\|^2 = \sum_{|a[m]| \leq T} |a[m]|^2 \tag{15}$$

In approximation theory, D_0 is called a *non-linear approximation error* because the M vectors are selected depending upon f as opposed to linear algorithms that approximate all signals with the same M vectors. Clearly D_0 decays when M increases, but how fast? This issue is a central question that is studied in approximation theory in relation to particular functional spaces [DJL92, Mal98].

Let us sort the inner products $a[m]$ by their amplitude. The amplitude of the kth coefficient is written

$$x\left(\frac{k}{N}\right) = |a[m_k]| \leq x\left(\frac{k+1}{N}\right) = |a[m_{k+1}]| \quad \text{for } 1 \leq k < N. \tag{16}$$

3. Understanding Wavelet Image Coding 61

The approximation error D_0 is the sum of the $N - M$ squared coefficients of smaller amplitude

$$D_0\left(\frac{M}{N}\right) = \sum_{k=M+1}^{N} \left|x\left(\frac{k}{N}\right)\right|^2. \tag{17}$$

The error $D_0(\frac{M}{N})$ has a fast decay when $\frac{M}{N}$ increases if $x(z)$ decreases quickly when z increases.

Observe that $p(x)$ is related to the inverse $z(x)$ of $x(z)$ by

$$z(x) = 1 - \int_{-x}^{x} p(u)\,du. \tag{18}$$

The probability density $p(x)$ is defined as a function of a continuous variable by interpolating the normalized histogram of the decomposition coefficients $a[m]$ of f. This also defines a function $x(z)$ for any $z \in [0,1]$, which interpolates the values $x(\frac{k}{N})$.

To estimate the decay of $D_0(z)$ when z increases, a standard approximation theory approach computes the rational decay of the sorted coefficients, and hence compares $x(z)$ with $z^{-\gamma}$ for some $\gamma > 0$. For functions $f \in \mathbf{L^2(R)}$ decomposed in a wavelet orthonormal basis, the exponent γ characterizes particular functional spaces called Besov spaces [DJL92]. To suppose that $x(z) = C\,z^{-\gamma}$ would clearly be too restrictive to model interesting classes of signals. We shall rather suppose that this exponent is slowly varying and define

$$\gamma(z) = -\frac{d\log_2 x(z)}{d\log_2 z}. \tag{19}$$

Figure 4 plots $\log_2 x(z)$ as a function of $\log_2 z$ for the wavelet coefficients and the block cosine coefficients of the test images. Observe that in both cases the slope $\gamma(z)$ varies slowly for $z \leq 2^{-1}$. This behavior is further discussed in Section 3.2.

To compute the distortion rate we shall assume that the second order derivative is bounded by a small $\epsilon > 0$

$$\left|\frac{d^2 \log_2 x(z)}{(d\log_2 z)^2}\right| \leq \epsilon \quad \text{for } z \in (0, \tfrac{1}{2}]. \tag{20}$$

We also suppose that

$$\inf_{z \in [0,1]} \gamma(z) > 0, \tag{21}$$

$$\frac{d^2 \log_2 x(z)}{(d\log_2 z)^2} \leq 0 \quad \text{for } z \in (0,1), \tag{22}$$

and that $p(x)$ is symmetric

$$p(x) = p(-x). \tag{23}$$

The four test images as well as most natural images have wavelet coefficients that satisfy (20-23). The concavity (22) is a technical condition that is used to control corrective terms in distortion rate calculations, but which is generally satisfied. The symmetry (23) of the probability density is verified in Figure 3. Assuming (20-23), the following theorem relates the distortion rate $D(\bar{R})$ to the approximation error D_0, through parameters that are evaluated as a function of $\gamma(z)$ and the number M of significant coefficients.

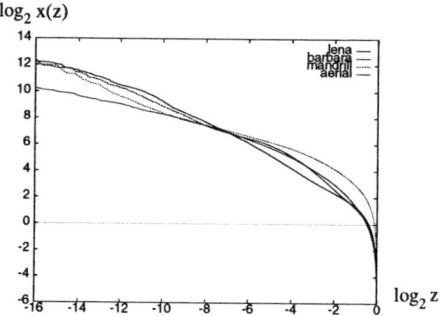

FIGURE 4. Amplitude of the sorted decomposition coefficients of the test images in a cubic spline wavelet basis.

Theorem 2 *Suppose that $x(z)$ satisfies (20-23). Let $\gamma_M = \gamma(\frac{M}{N}) > \frac{1}{2}$. If $\frac{M}{N} \leq \epsilon$ and $M \geq \frac{1}{\epsilon}$ then*

$$D(\bar{R}) = (1+K) D_0 \left(\frac{\bar{R}}{r_1 + r_0} \right) \quad (24)$$

with

$$K = \frac{D_1}{D_0} = \frac{2\gamma_M - 1}{12\,\theta^2} \left[1 + O\!\left(\epsilon | \log_2 \epsilon |^2 + \epsilon^{2\gamma_M - 1} \right) \right], \quad (25)$$

and

$$r_1 = \frac{R_1}{M} = 1 + (1 + \gamma_M) \log_2 e + \log_2 \gamma_M + \log_2 \theta + O(\epsilon). \quad (26)$$

Moreover, the derivative of $D_0(z)$ satisfies

$$\frac{d \log_2 D_0(\frac{M}{N})}{d \log_2 z} = (1 - 2\gamma_M) \left[1 + O\!\left(\epsilon | \log_2 \epsilon |^2 + \epsilon^{2\gamma_M - 1} \right) \right]. \quad (27)$$

The proof is in Appendix I. To understand the implications of this theorem, these formulae are simplified with an approximation, and we neglect the corrective terms in ϵ. Since the second order derivative (20) remains small, the slope

$$\gamma_M = \frac{-d \log_2 x(\frac{M}{N})}{d \log_2 z} \quad (28)$$

varies slowly as a function of $\log_2 \frac{M}{N}$. In the compression range of interest, it will be considered constant: $\gamma_M \approx \gamma$. It follows from (25) and (26) that $K = \frac{D_1}{D_0}$ and $r_1 = \frac{R_1}{M}$ are also constant. We have already mentioned that this is also the case for $r_0 = \frac{R_0}{M}$. Hence $D(\bar{R})$ is calculated in (24) by scaling and multiplying the non-linear approximation error $D_0(z)$ by constant factors

$$D(\bar{R}) = (1+K) D_0 \left(\frac{\bar{R}}{r_1 + r_0} \right). \tag{29}$$

Since $\gamma_M \approx \gamma$, (27) implies that $D_0(z) \sim z^{2\gamma-1}$ so $D(\bar{R}) \sim \bar{R}^{1-2\gamma}$. This distortion rate decay is very different from the high resolution formula (8) where $D(\bar{R}) \sim 2^{-2\bar{R}}$.

The distortion D in (29) depends essentially upon the approximation error D_0 of f from $M = \frac{\bar{R}}{r_1+r_0}$ vectors selected in the basis $\mathcal{B}$. To optimize the transform coding, the basis $\mathcal{B}$ must be able to approximate precisely each signal f with a small number of basis vectors. If we consider f as a realization of a random vector Y then we may wonder which is the basis which minimizes $\mathsf{E}\{D_0(\frac{M}{N})\}$ over all realizations. This problem is difficult because the M basis vectors are selected depending upon each realization f of Y. They are the ones which have the largest inner products with f. The energy compaction theorem [GG92] proves that the Karhunen-Loeve basis is optimal for approximating Y from M vectors chosen once and for all, but it has no optimality property in this non-linear setting where the vectors depend upon each realization. In some cases, we know how to find bases that minimize the maximum error $D_0(\frac{M}{N})$ over a whole signal class. For example, wavelet bases are optimal in this min-max sense for piece-wise regular signals that belong to Besov spaces [DJL92].

To optimize the quantization, the size of the zero bin $[-T,T]$ must be adjusted with respect to the other quantization bins of size Δ. To minimize the distortion D, we want to find $\theta = \frac{T}{\Delta}$ such that for a fixed $\bar{R}$

$$\frac{\partial D(\bar{R}, \theta)}{\partial \theta} = 0. \tag{30}$$

With few calculations, one can derive [MF98] the following theorem that gives an analytic formula for θ.

Theorem 3 *Suppose that $x(z)$ satisfies (20-23) and that $r_0 = \frac{R_0}{M}$ is a constant independent of M. Let $\gamma_M = \gamma(\frac{M}{N}) > \frac{1}{2}$. If $\frac{M}{N} \leq \epsilon$ and $M \geq \frac{1}{\epsilon}$ then the optimal zero bin ratio is*

$$\theta = \sqrt{\frac{r_1 + r_0}{6 \log_2 e} - \frac{2\gamma_M - 1}{12}} \left[1 + \mathcal{O}(\epsilon) \right]. \tag{31}$$

3.2 Wavelet Transform Coding

Wavelet bases are known to efficiently approximate piece-wise regular functions with a small number of non-zero wavelet coefficients [Mal98]. Since images often include piecewise regular structures, wavelet bases are good candidates for building efficient image transform coders. The central assumption of Theorem 2 is that the sorted decomposition coefficients $x(z)$ of f in the basis $\mathcal{B}$ have a rational decay. In wavelet bases, this is in accordance with asymptotic image models based on Besov spaces introduced for compression by DeVore, Jawerth and Lucier [DJL92], and further studied in [CDGO97]. Let $f \in \mathbf{L}^2[0,N]^2$. If there exists $C > 0$ and $\gamma > \frac{1}{2}$ so that for all $k \geq 0$ the wavelet coefficient of f of rank k is bounded by $C k^{-\gamma}$ then f belongs to a family of Besov spaces whose indexes depend upon γ. Let us consider a piecewise regular image f, which is uniformly regular (Lipschitz $\alpha \geq 1$) inside the regions $\{\Omega_i\}_{1 \leq k \leq K}$ which partition $[0,N]^2$. This image has discontinuities along the borders of the Ω_i, which have a finite length. One can prove then [Mal98] that the sorted wavelet coefficients decay like $C k^{-\gamma}$, with $\gamma = 1$. The discontinuities create large amplitude wavelet coefficient which are responsible for this decay exponent. This piecewise regular model applies to an image such as Lena because even the fur texture does not create enough large wavelet coefficients to modify the exponent $\gamma = 1$. On the other hand, the Mandrill image is composed of regions with highly irregular textures which create enough high amplitude wavelet coefficients to reduce the exponent γ. For finite images $k \leq N$, which is why we renormalize $z = \frac{k}{N}$ and compare the decay of $x(z)$ with $z^{-\gamma}$ when z increases in $[0,1]$. Theorem 2 does not require that γ is a constant but (20) assumes that it varies slowly as a function of $\log_2 z$.

If $x(z) \sim z^{-\gamma}$ for x large enough then one can derive from (43) that $p(x) \sim x^{-1-\frac{1}{\gamma}}$. However, Figure 3 shows that $p(x)$ has an exponential decay when x is small. This can be explained by looking at the normalized histograms $p_j(x)$ of the wavelet coefficients $\langle f, \psi_{j,p,q}^k \rangle$ at a fixed scale 2^j, for all positions $0 \leq p, q \leq 2^{-j} N$ and orientations $1 \leq k \leq 3$. Such histograms are well modeled by generalized Gaussian distributions [Mal89], which have an exponential decay and a variance which increases with the scale 2^j. It is the aggregation of these histograms that yield a global histogram $p(x)$ which has a rational decay for x sufficiently large [LeP97]. However, the finite image resolution implies that wavelet coefficients are zero for $j < 0$. One can verify that the "border effect" created by the absence of finer scale wavelet coefficients implies that $p(x) \approx p_1(x)$ for x small. This explains the exponential decay of $p(x)$ when x is small, and the rapid variation of the slope γ for $\log_2 z \geq -1$ in Figure 4.

To simplify the expression of the distortion rate, the slope γ_M is approximated by a constant $\gamma_M \approx 1$, which corresponds to piecewise regular image models. Although γ_M can differ from 1 in many images such as Mandrill, this approximation is justified by the small sensitivity of K and $r_1 + r_0$

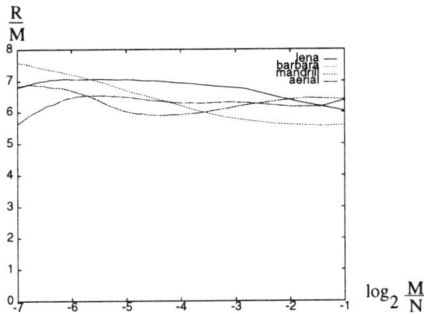

FIGURE 5. Variations of $\frac{R}{M}$ for the wavelet transform coding of the test images.

with respect to fluctuations of γ_M around 1. Since $\theta = 1$, we get $K \approx \frac{1}{12}$ and $r_1 \approx 3.9$. Figure 5 displays

$$\frac{R}{M} = \frac{R_0}{M} + \frac{R_1}{M} = r_0 + r_1$$

which was computed numerically for the four test images. For $\frac{M}{N} \in [2^{-7}, 2^{-1}]$, the ratio $\frac{R}{M}$ can be approximated by a constant $r_0 + r_1 \approx 6.5$. The distortion D calculated in (24) is thus approximated by

$$\hat{D}(\bar{R}) = \left(1 + \frac{1}{12}\right) D_0\left(\frac{\bar{R}}{6.5}\right). \tag{32}$$

Figure 6 compares the Peak Signal to Noise Ratio

$$P_{SNR}(D) = 10 \log_{10} \frac{N\,255^2}{D},$$

which was calculated numerically for the four test images, with the approximated $P_{SNR}(\hat{D})$ derived from (32). Observe that $\hat{D}(\bar{R})$ gives a remarkably precise evaluation of the true distortion rate $D(\bar{R})$, despite the fact that γ_M is not exactly equal to 1.

The increment of $P_{SNR}(D)$ for each additional bit is calculated by inserting (27) in (31) while neglecting the variations of K and $r_1 + r_0$

$$\frac{dP_{SNR}(D)}{d\log_2 \bar{R}} \approx (2\gamma_M - 1)\,10\,\log_{10} 2.$$

For Lena, $\gamma_M = 1$ so $P_{SNR}(D)$ increases by 3db for each additional bit, which is indeed verified in Figure 6. For the three other images, the variations of γ_M can not be neglected over the whole compression range $\bar{R} \in$

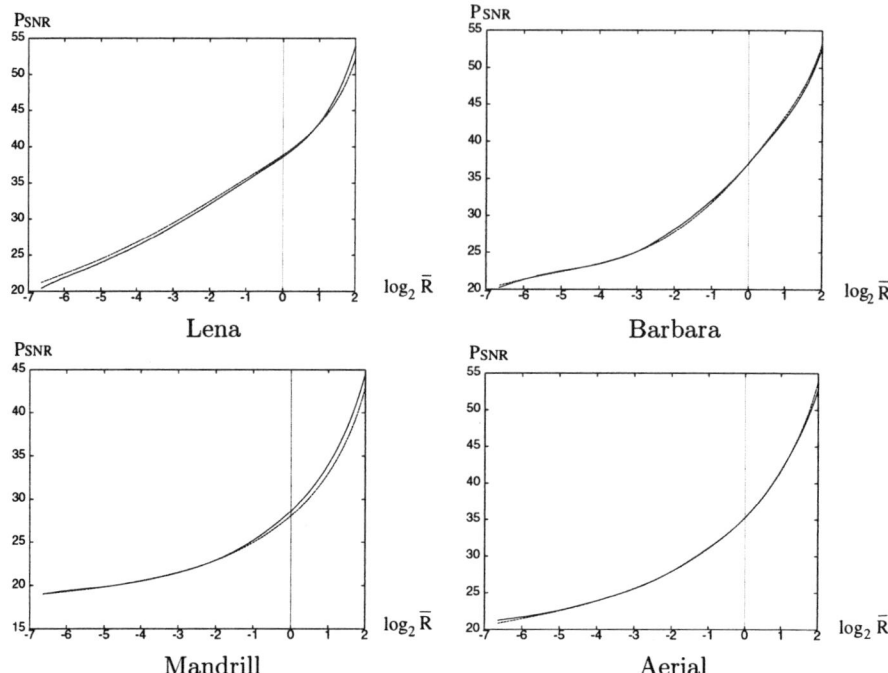

FIGURE 6. $P_{SNR}(D)$ (solid line) and $P_{SNR}(\hat{D})$ (dashed line) for a wavelet coder, with $\hat{D} = (1 + \frac{1}{12})D_0(\frac{\bar{R}}{6.5})$.

$[2^{-6}, 1]$ and Figure 6 shows that $P_{SNR}(D)$ has a slope which varies slowly with $\log_2 \bar{R}$.

Theorem 3 gives an analytical expression which computes the $\theta = \frac{T}{\Delta}$ that minimizes the distortion D. For $r_1 + r_0 = 6.5$ and $\gamma_M \approx 1$ we get $\theta = 0.81 \approx 1$. This theoretical estimate is precisely the choice that is most often used in wavelet compression softwares, after ad-hoc numerical trials.

To optimize the wavelet transform coder, the distortion rate (32) shows that one must choose a wavelet basis that gives precise approximations of images with few wavelet coefficients, in order to maintain a small approximation error $D_0(z)$. This essentially depends upon the support size and the number of vanishing moments of the wavelets [Mal98]. The optimization of the wavelet basis may depend upon the particular class of images to be encoded.

4 Embedded Transform Coding

For rapid transmission or fast image browsing from a data base, one should provide a coarse signal approximation quickly, and then progressively enhance it as more bits are transmitted. Embedded coders offer this flexibility by grouping the bits in order of significance. The decomposition coefficients are sorted and the first bits of the largest coefficients are sent first. An image approximation can be reconstructed at any time from the bits already transmitted. Embedded coders can take advantage of any prior information on the location of large versus small coefficients. Such prior information is available for natural images decomposed in wavelet bases. As a result, an implementation with zero-trees designed by Shapiro [Sha93] yields better compression rates than classical wavelet transform coders.

The decomposition coefficients $a[m] = \langle f, g_m \rangle$ are partially ordered by grouping them in sets $\mathcal{S}_k$ of indices defined for any $k \in \mathbf{Z}$ by

$$\mathcal{S}_k = \{m \; : \; 2^k \leq |a[m]| < 2^{k+1}\}.$$

The set $\mathcal{S}_k$ is encoded with a binary significance map $b_k[m]$:

$$b_k[m] = \begin{cases} 0 & \text{if } m \notin \mathcal{S}_k \\ 1 & \text{if } m \in \mathcal{S}_k \end{cases}. \tag{33}$$

An embedded algorithm quantizes $a[m]$ uniformly with a quantization step (bin size) $\Delta = 2^n$ that is progressively reduced. Let $m \in \mathcal{S}_k$ with $k \geq n$. The amplitude $|Q(a[m])|$ of the quantized number is represented in base 2 by a binary string with non-zero digits between the bit k and the bit n. The bit k is necessarily 1 because $2^k \leq |Q(a[m])| < 2^{k+1}$. Hence, $k - n$ bits are sufficient to specify this amplitude, to which one bit is added for the sign.

The embedded coding is initiated with the largest quantization step to produce at least one non-zero quantized coefficient. To refine the quantization step from 2^{n+1} to 2^n, the algorithm records the significance map $b_n[m]$ and the sign of $a[m]$ for $m \in \mathcal{S}_n$. This can be done by directly recording the sign of significant coefficients with a variable incorporated into the significance map $b_n[m]$. Afterwards, the code stores the bit n of all amplitudes $|Q(a[m])|$ for $m \in \mathcal{S}_k$ with $k > n$. If necessary, the coding precision is improved by decreasing n and continuing the encoding. The different steps of the algorithm can be summarized as follows [SP96]:

1. Store the index n of the first non-empty set $\mathcal{S}_n$, where

$$n = \left\lfloor \sup_m \log_2 |a[m]| \right\rfloor.$$

2. Store the significance map $b_n[m]$ and the sign of $a[m]$ for $m \in \mathcal{S}_n$.
3. Store the nth bit of all coefficients $|a[m]| > 2^{n+1}$. These are coefficients that belong to some set $\mathcal{S}_k$ for $k > n$, whose coordinates were already

stored. Their n^{th} bit is stored in the order in which their position was recorded in the previous passes.

4. Decrease n by 1 and go to step 2.

This algorithm may be stopped at any time in the loop, providing a code for any specified number of bits. The distortion rate is analyzed when the algorithm is stopped at step 4. All coefficients above $T = 2^n$ are uniformly quantized with a bin size $\Delta = 2^n$. The zero quantization bin $[-T, T]$ is therefore twice as big as the other quantization bins. This quantizer is the same as in the direct transform coding studied in Section 3.1, for a zero-bin ratio $\theta = \frac{T}{\Delta} = 1$. This value was shown to be nearly optimal for wavelet image coders. The total distortion $D = D_0 + D_1$ is therefore not modified by the embedding strategy.

Once the algorithm stops, we denote by M the number of significant coefficients above $T = 2^n$. The total number of bits of the embedded code is $R = R_0^e + R_1^e$, where R_0^e is the number of bits needed to encode all significance maps $b_k[m]$ for $k \geq n$, and R_1^e is the number of bits used to encode the amplitudes of the quantized significant coefficients $Q(a[m])$, knowing that $m \in \mathcal{S}_k$ for $k > n$.

To appreciate the efficiency of this embedding strategy, the bit budget $R_0^e + R_1^e$ is compared to the number of bits $R_0 + R_1$ used by the direct transform coder of Section 3.1. The value R_0 is the number of bits used to encode the overall significance map

$$b[m] = \begin{cases} 0 & \text{if } |a[m]| \leq T \\ 1 & \text{if } |a[m]| > T \end{cases} \quad (34)$$

and R_1 is the number of bits used to encode the quantized significant coefficients.

An embedded strategy encodes $Q(a[m])$ knowing that $m \in \mathcal{S}_k$ and hence that $2^k \leq |Q(a[m])| < 2^{k+1}$, whereas a direct transform coding knows only that $|Q(a[m])| > T = 2^n$. Thus fewer bits are needed for embedded codes: $R_1^e \leq R_1$. This improvement may be offset however by the supplement of bits needed to encode the significance maps $\{b_k[m]\}_{k>n}$ of the sets $\{\mathcal{S}_k\}_{k>n}$. A direct transform coder records a single significance map $b[m]$, which specifies $\cup_{k \geq n} \mathcal{S}_k$. It provides less information and is therefore encoded with fewer bits: $R_0^e \geq R_0$. An embedded coder brings an improvement over a direct transform coder if

$$R_0^e + R_1^e \leq R_0 + R_1.$$

This can happen if we have some prior information about the position of large decomposition coefficients $Q(a[m])$ versus smaller ones. It allows us to reduce the number of bits needed to encode the partial sorting of all coefficients provided by the significance maps $\{b_k[m]\}_{k>n}$. The use of such prior information produces an overhead of R_0^e relative to R_0 that is smaller than the gain of R_1^e relative to R_1. Let $x(z)$ be the sorted amplitudes of the coefficients $a[m]$. The following theorem computes R_1^e with the same hypotheses as Theorem 2, allowing us to compare it with R_1.

Theorem 4 *Suppose that $x(z)$ satisfies (20-23). Let $\gamma_M = \gamma(\frac{M}{N}) > \frac{1}{2}$. If $\frac{M}{N} \leq \epsilon$ and $M \geq \frac{1}{\epsilon}$ then*

$$r_1^e = \frac{R_1^e}{M} = 1 + \frac{1}{2^{\frac{1}{\gamma_M}} - 1}\left[1 + O(\epsilon|\log_2 \epsilon|)\right], \qquad (35)$$

and

$$D(\bar{R}) = (1+K)D_0\left(\frac{\bar{R}}{r_1^e + r_0^e}\right), \qquad (36)$$

where $r_0^e = \frac{R_0^e}{M}$ and $K = \frac{D_1}{D_0}$ is given by (25).

The proof is in Appendix II. In the following, we omit the corrective terms to simplify the notation. Let us compare $r_1^e = \frac{R_1^e}{M}$ calculated in (35) with the value $r_1 = \frac{R_1}{M}$ estimated in (26) for a direct transform coding, with $\theta = 1$. The bit budget of an embedded coding is smaller than that of a direct transform coding if $r_0^e + r_1^e \leq r_0 + r_1$ and hence

$$r_0^e - r_0 \leq r_1 - r_1^e = (1 + \gamma_M)\log_2 e + \log_2 \gamma_M - \frac{1}{2^{\frac{1}{\gamma_M}} - 1}. \qquad (37)$$

If $\gamma_M \approx 1$ then $r_1 - r_1^e \approx 1.9$. The inequality (37) is satisfied for embedded transform codings implemented in wavelet bases [Sha93] and in a block cosine basis [XGO97], by taking advantage of prior knowledge of the location of large versus small coefficients using zero-trees.

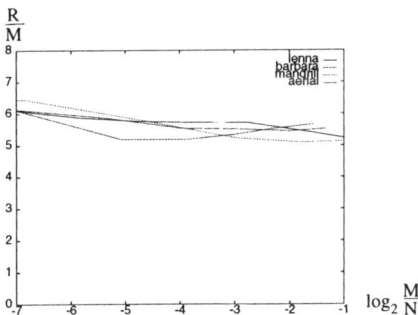

FIGURE 7. Variations of $\frac{R}{M}$ for an embedded wavelet coding of the test images.

A wavelet coefficient $\langle f, \psi_{j,p,q}^k \rangle$ has a large amplitude where the signal has sharp transitions. If an image f is Lipschitz α in the neighborhood of

(x_0, y_0), then for wavelets $\psi_{j,p,q}^k$ located in this neighborhood one can prove [Mal98] that there exists $A \geq 0$ such that

$$|\langle f, \psi_{j,p,q}^k \rangle| \leq A\, 2^{j(\alpha+1)}.$$

The worst singularities are often discontinuities which means that $\alpha \geq 0$. In the neighborhood of singularities without oscillations, the amplitudes of wavelet coefficients thus decrease when the scale 2^j decreases. This property is not valid for oscillatory patterns. High frequency oscillations create coefficients at large scales 2^j that are typically smaller than those at the fine scale that matches the period of oscillation. Such oscillatory patterns are not often encountered in images although they do appear as thin lines in the Barbara image.

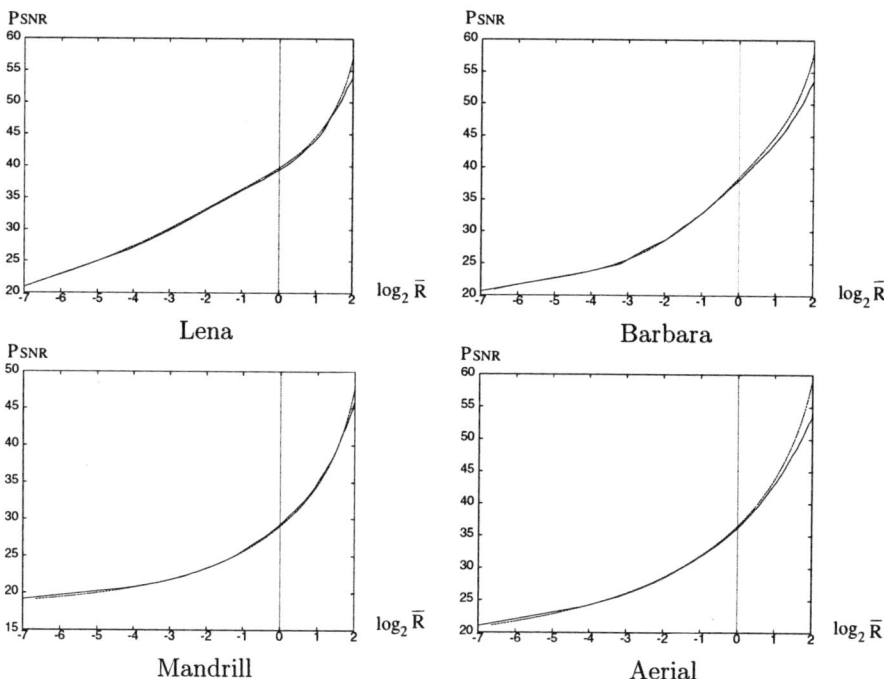

FIGURE 8. $P_{SNR}(D)$ (solid line) and $P_{SNR}(\hat{D})$ (dashed line) for an embedded wavelet coding, with $\hat{D} = (1 + \frac{1}{12})\, D_0(\frac{\bar{R}}{5.5})$.

Wavelet zero-trees, introduced by Lewis and Knowles [LK92], take advantage of the decay of wavelet coefficients by relating these coefficients across scales with quad-trees. These zero-trees take advantage of a partial self-similarity of the image [Dav97]. Shapiro [Sha93] used this zero-tree structure to encode the embedded significance maps of wavelet coefficients.

Numerical experiments were performed with Said and Pearlman's software [SP96], which improves Shapiro's zero-tree coder with a set partitioning technique. A cubic spline orthogonal wavelet basis was used. Figure 7 displays the value of $\frac{R}{M} = r_0^e + r_1^e$ as a function of $\log_2 \frac{M}{N}$ for the four test images. These curves have variations centered at 5.5. This graph should be compared with Figure 5 that shows $r_0 + r_1$ calculated with a direct wavelet image coder. An improvement of approximately 1 bit per significant coefficient is obtained.

The embedded distortion rate function is calculated with (36). Inserting $\gamma_M \approx 1$ in (25) yields $K \approx \frac{1}{12}$. Since $r_0^e + r_1^e \approx 5.5$, we get an approximate distortion rate formula

$$\hat{D}(\bar{R}) = (1 + \frac{1}{12}) D_0 \left(\frac{\bar{R}}{5.5}\right).$$

Figure 8 compares the $P_{SNR}(D)$ calculated numerically for the our test images and its theoretical approximation $P_{SNR}(\hat{D})$. Once more we verify that the distortion rate essentially depends only upon the approximation error function $D_0(z)$. The variations of the constants K and $r_0^e + r_1^e$ can be neglected. The embedding reduces $r_0 + r_1$ to $r_0^e + r_1^e$, but the variations of the distortion rate still depends on the non-linear approximation error $D_0(z)$.

Appendix I Proof of Theorem 2

Proof of (24). The distortion rate formula is derived by observing that $D = D_0 + D_1$ and $\bar{R} = \frac{R_0}{N} + \frac{R_1}{N}$. By definition $K = \frac{D_1}{D_0}$, $r_0 = \frac{R_0}{M}$ and $r_1 = \frac{R_1}{M}$. Inserting these variables in the equation $D = D_0 + D_1$ yields (24).

Proof of (25). To compute $\frac{D_1}{D_0}$ we evaluate the expression (12) using an integral approximation

$$\frac{D_0}{N} = \frac{1}{N} \sum_{|a[m]| \leq T} |a[m]|^2 = \frac{1}{N} \sum_{k=M}^{N-1} x^2(\frac{k}{N}) \approx \int_{\frac{M}{N}}^{1} x^2(z) \, dz. \quad (38)$$

The distribution $x(z)$ is approximated by the tangential exponential curve at $r = \frac{M}{N}$:

$$x_M(z) = T \left(\frac{Nr}{M}\right)^{-\gamma_M}. \quad (39)$$

It satisfies $x_M(\frac{M}{N}) = x(\frac{M}{N}) = T$ and

$$\frac{d \log_2 x(z)}{d \log_2 z} = \frac{d \log_2 x_M(z)}{d \log_2 z} = -\gamma_M.$$

Approximating $x(z)$ by $x_M(z)$ in (38) yields

$$D_0 \approx T^2 N \int_{\frac{M}{N}}^{1} \left(\frac{Nz}{M}\right)^{-2\gamma_M} dz$$

$$\approx \frac{T^2 M}{2\gamma_M - 1}\left(1 + \left(\frac{M}{N}\right)^{2\gamma_M - 1}\right). \qquad (40)$$

Since $T = \Delta\theta$ it follows from (14) that

$$D_1 = \frac{M\Delta^2}{12} = \frac{MT^2}{12\,\theta^2}.$$

For $\frac{M}{N} \leq \epsilon$, inserting this equation in (40) shows that

$$D_0 \approx \frac{12\,\theta^2 D_1}{2\gamma_M - 1}\left[1 + O(\epsilon^{2\gamma_M - 1})\right].$$

With the second order derivative properties (22,20), one can prove [MF98] that the approximation of $x(z)$ by $x_M(z)$ introduces an corrective term in $O(\epsilon|\log_2 \epsilon|^2)$, which yields (25).

Proof of (26). To compute

$$\frac{R_1}{M} = \mathcal{H}_d(X_T) - \log_2 \Delta \qquad (41)$$

we need to evaluate the differential entropy

$$\mathcal{H}_d(X_T) = -\int_{-\infty}^{+\infty} p_T(x)\,\log_2 p_T(x)\,dx,$$

where $p_T(x) = \frac{N}{M} p(x)\,\mathbf{1}_{\{|x|>T\}}(x)$. Since $p(x) = p(-x)$

$$\mathcal{H}_d(X_T) = -2\,\frac{N}{M}\int_T^{+\infty} p(x)\,\log_2 p(x)\,dx. \qquad (42)$$

This integral is calculated by using the relation (18) between $p(x)$ and $x(z)$. Since $p(x) = p(-x)$ we derive that

$$x'(z) = \frac{1}{z'(x)} = \frac{-1}{2p(x)}. \qquad (43)$$

By using $p(x)\,dx = -\frac{1}{2}dz$, the change of variable $z = z(x)$ in (42) yields

$$\mathcal{H}_d(X_T) = 2\,\frac{N}{M}\int_0^{\frac{M}{N}} \log_2 |2x'(z)|\,\frac{dz}{2},$$

and with a change of variable,

$$\mathcal{H}_d(X_T) = 1 + \int_0^1 \log_2\left(\frac{M}{N}\left|x'\left(\frac{M}{N}z\right)\right|\right) dz. \tag{44}$$

This integral is estimated by approximating $x'(z)$ with $x'_M(z)$

$$\begin{aligned}
\mathcal{H}_d(X_T) &\approx 1 + \int_0^1 \log_2\left(\frac{M}{N}\left|x'_M\left(\frac{M}{N}z\right)\right|\right) dz \\
&\approx 1 + \int_0^1 \log_2(T\gamma_M z^{-\gamma_M-1}) dz \\
&\approx 1 + \log_2 T + \log_2 \gamma_M + (\gamma_M + 1)\log_e 2. \tag{45}
\end{aligned}$$

Inserting this result in (41) with $T = \theta \Delta$ gives

$$\frac{R_1}{M} \approx \log_2 \theta + 1 + (1 + \gamma_M)\log_2 e + \log_2 \gamma_M.$$

Using (22,20), one can verify [MF98] that the approximation of $x'(z)$ by $x'_M(z)$ yields a corrective term that is $O(\epsilon)$, which finishes the proof of (26).

Proof of (27). Let us calculate

$$\frac{d\log_2 D_0(\frac{M}{N})}{d\log_2 z} = \frac{dD_0(\frac{M}{N})}{dz} \frac{M}{ND_0(\frac{M}{N})}. \tag{46}$$

We derive from (38) that

$$\frac{dD_0\left(\frac{M}{N}\right)}{dz} \approx -N\, x^2(\frac{M}{N}) = -N\, T^2.$$

Since

$$\frac{1}{D_0} \approx \frac{2\gamma_M - 1}{MT^2},$$

inserting these two equations in (46) yields (27) plus corrective terms.

Appendix II Proof of Theorem 4

The proof of the distortion rate formula (36) is identical to the proof of (24). We thus concentrate on (35).

Let $\Delta = 2^n$ be the quantization interval of the embedded code. We saw that if $2^k \leq |a[m]| < 2^{k+1}$, then the number of bits required to code $|Q(a[m])|$ is $k - n$, to which is added one bit for the sign. The inverse $z(x)$ of $x(z)$ gives the proportion of coefficients whose amplitude is above x.

Encoding the amplitudes of the M significant coefficients above $T = 2^n$ thus requires a number of bits

$$R_1^e = M + \sum_{k=n}^{+\infty} (k-n) N\left[z(2^k) - z(2^{k+1})\right].$$

As in Theorem 2, $x(z)$ is approximated by the tangential curve $x_M(z)$ defined in (39):

$$x(z) \approx x_M(z) = T\left(\frac{Nz}{M}\right)^{-\gamma_M} \quad \text{so} \quad z(x) \approx \frac{M}{N}\left(\frac{x}{T}\right)^{-\frac{1}{\gamma_M}}.$$

It follows that

$$R_1^e - M \approx M \sum_{k=n}^{+\infty} (k-n)\left(2^{\frac{n-k}{\gamma_M}} - 2^{\frac{n-k-1}{\gamma_M}}\right) = M\left(1 - 2^{-\frac{1}{\gamma_M}}\right) \sum_{i=0}^{+\infty} i\, 2^{-\frac{i}{\gamma_M}}.$$
(47)

One can verify that

$$\sum_{i=0}^{+\infty} i\, 2^{-\frac{i}{\gamma_M}} = \frac{2^{-\frac{1}{\gamma_M}}}{(1 - 2^{-\frac{1}{\gamma_M}})^2},$$
(48)

so

$$\frac{R_1^e}{M} \approx 1 + \frac{1}{2^{\frac{1}{\gamma_M}} - 1}.$$

As a consequence of (22,20), one can verify [MF98] that the approximation of $x(z)$ by $x_M(z)$ yields a corrective term that is $O(\epsilon|\log_2 \epsilon|)$, which proves (35).

5 References

[CDGO97] A. Cohen, I. Daubechies, O. Guleryuz, and M. Orchard. On the importance of combining wavelet-based non-linear approximation with coding strategies, 1997. Preprint.

[Dav97] G. M. Davis. A wavelet-based analysis of fractal image compression. *IEEE Trans. Image Proc.*, 1997.

[DJ94] D. Donoho and I. Johnstone. Ideal spatial adaptation via wavelet shrinkage. *Biometrika*, 81:425–455, December 1994.

[DJL92] R. A. DeVore, B. Jawerth, and B. J. Lucier. Image compression through wavelet transform coding. *IEEE Trans. Info. Theory*, 38(2):719–746, March 1992.

[GG92]   A. Gersho and R. M. Gray. *Vector Quantization and Signal Compression*. Kluwer Academic Publishers, Boston, 1992. Chapter 2.

[LeP97]   E. LePennec. Modélisation d'images par ondelettes, December 1997. Mémoire de DEA, CMAP, Ecole Polytechnique, Paris, France.

[LK92]   A. S. Lewis and G. Knowles. Image compression using the 2-D wavelet transform. *IEEE Trans. Image Proc.*, 1(2):244–250, April 1992.

[Mal89]   S. Mallat. A theory for multiresolution signal decomposition: the wavelet representation. *IEEE Trans. Patt. Recog. and Mach. Intell.*, 11(7):674–693, July 1989.

[Mal98]   S. Mallat. *A Wavelet Tour of Signal Processing*. Academic Press, New York, 1998.

[MF98]   S. Mallat and F. Falzon. Analysis of low bit rate image transform coding. *IEEE Trans. Signal Proc.*, 46(4), April 1998.

[Sha93]   J. M. Shapiro. Embedded image coding using zerotrees of wavelet coefficients. *IEEE Trans. Signal Proc.*, 41(12):3445–3462, December 1993.

[SP96]   A. Said and W. A. Pearlman. A new, fast, and efficient image codec based on set partitioning in hierarchical trees. *IEEE Trans. Circ. and Syst. for Video Tech.*, 6(3):243–250, June 1996.

[WNC87]   I. Witten, R. Neal, and J. Cleary. Arithmetic coding for data compression. *Comm. of the ACM*, 30(6):519–540, 1987.

[XGO97]   Z. X. Xiong, O. Guleryuz, and M. T. Orchard. Embedded image coding based on DCT. In *VCIP in Europeen Image Proc. Conf.*, 1997.

4
Multiplication of Short Wavelet Series Using Connection Coefficients

Valerie Perrier [1] and Mladen Victor Wickerhauser [2]

ABSTRACT Given two functions approximable with short wavelet series, we wish to find the short wavelet series representing their product. This can be done by pre-calculating the *connection coefficients* which express the product of two wavelets or scaling functions as a wavelet series. We follow a method suggested by Daubechies and also used by Dahmen, *et al.*, to rapidly compute these coefficients as elements of a matrix which solves a fixed-point problem, and derive some of the formulas and identities satisfied by the coefficients. We estimate the complexity of the connection coefficient multiplication algorithm by counting the number of terms, and then illustrate through a series of graphs how few of these terms are non-negligible.

1 Introduction

We are motivated in our present work by recent speedups in numerical simulations obtained by representing solutions to complicated problems as superpositions of relatively few basic functions. These good basic functions are called *wavelet packets*; they have three useful properties:

- They are almost as well localized in both position and wavenumber as the Heisenberg uncertainty inequality allows;

- They can be assembled into orthogonal bases;

- They come equipped with fast well-conditioned transformations: to compute N expansion coefficients of a function costs only $O(N \log N)$ operations.

---

[1]LMD–CNRS, Ecole Normale Supérieure, 24 rue Lhomond, 75231 Paris, France

[2]Dept. of Mathematics, Washington University, St. Louis, Missouri 63130 USA Research supported in part by NATO CRG-930456 and by the Southwestern Bell Telephone Company

The wavelet packet approximation scheme is nonlinear, since the function choice depends upon the solution at each time instant. It works by keeping only those component functions with significant amplitudes; the others are discarded. Making this choice to minimize a description length or information cost criterion is called a *best basis* algorithm [3].

Computed simulations of fully-developed turbulence in the bidimensional Navier–Stokes equation (2D-NSE) provide an example [7, 19]. 2D-NSE simulations on 10^4 to 10^6 grid points indicate that the number of components needed to obtain deterministic predictability for short times is about one-tenth the number of grid points. The number needed for statistical predictability, such as estimates of the slope of a line fitted to the vorticity power spectrum, is about one-hundredth the number of grid points. Turbulence simulations are thus an example "compressible" high-dimensional problem; the relevant features contain far fewer degrees of freedom than the original mathematical model. Transformation into wavelet packet coordinates holds the promise of reducing the number of parameters in large but compressible problems by many orders of magnitude.

To take advantage of this reduction, it is necessary to perform all of the numerical computations in wavelet packet coordinates. In a series of papers [1, 2, 8, 12, 15, 17, 18], we and others have developed some of the needed tools. The operations which are well understood and efficiently implemented include matrix-vector and matrix-matrix multiplication, numerical differentiation, and certain integral operators. Conspicuously absent from this list is a method for multiplying two functions when each is a superposition of just a few wavelet packets, without re-computing their values at all grid points.

In this article, we consider only wavelet basis functions rather than the more general wavelet packets. Also, we focus our attention on the compactly-supported wavelets of Daubechies and Mallat [5, 13], which are *refinable functions* in the sense that they can be expressed as short linear combinations of dilated and translated versions of themselves. Refinable functions have a cross-scale self-similarity that can be used to compute integrals of their products, and thus to find connection coefficients.

The algebraic properties of refinable functions are well known, and have been heavily exploited in recent papers on wavelets and numerical analysis. Dahmen and Micchelli [4] considered the problem of evaluating integrals of products of refinable functions and their derivatives. Kunoth [10] later implemented the algorithms described in that paper. Latto, Resnikoff, and Tenenbaum [11] also derived a linear system of equations for connection coefficients involving two-scale equations for refinable functions.

Our work supplements theirs and was inspired by a discussion with Ingrid Daubechies. Our goal is to present the algebraic relations satisfied by integrals of refinable functions, to provide a simpler proof of the basic fixed-point theorem behind the iterative numerical algorithm, to implement the algorithm, and then to compute and plot the connection coefficients.

The plots will show rapid decay as the scale or position differences grow large, indicating that relatively few connection coefficients contribute significantly to the product. This suggests multiplying wavelet series will have much lower complexity than that which a coarse estimate from the wavelet's support lengths would indicate.

It should be noted that some recent work by Beylkin on wavelet multiplication provides an alternative method of multiplying short series, relying on the paraproduct formula for wavelet expansions [14] and the observation that certain orthonormal wavelets are good approximations to interpolating functions.

2 Definition of Connection Coefficients

2.1 Abstract Orthonormal Bases

Suppose that $\{e_k : k \in \mathbf{Z}\}$ is an orthonormal basis for $L^2(\mathbf{R})$ consisting of bounded functions. Then the triple product $e_j e_k e_l$ is defined and integrable over $\mathbf{R}$, so we can define the abstract *connection coefficients* of this basis to be the following integrals:

$$\Gamma_{jkl} = \langle e_j, e_k e_l \rangle \stackrel{\text{def}}{=} \int_{\mathbf{R}} \bar{e}_j(t) \, e_k(t) e_l(t) \, dt. \tag{1}$$

These coefficients are used to find the expansion of a product. If $u(t) = \sum_k u_k e_k(t)$ and $v(t) = \sum_l v_l e_l(t)$, then

$$u(t)v(t) = \sum_j \left(\sum_{k,l} \Gamma_{jkl} u_k v_l \right) e_j(t). \tag{2}$$

2.2 Examples

The space $L^2(\mathbf{R})$ is separable and thus has the countable bases needed for the formula in Equations 1 and 2. However, the most useful of these bases require somewhat complicated indexing which can obscure the main ideas. Thus, we will first explore some simpler spaces and their simpler connection coefficients.

Kronecker Basis

The space of sequences ℓ^2 has the *Kronecker basis* $e_k(n) = \delta(n-k)$, where $\delta(x)$ is the Kronecker symbol which is 1 if $x = 0$ and 0 otherwise. The inner product in Equation 1 is a sum rather than an integral, and we see that $\Gamma_{jkl} = \delta(j-k)\delta(j-l)$. The inner summation of Equation 2 simplifies into the pointwise multiplication formula $\sum_{k,l} \Gamma_{jkl} u_k v_l = u_j v_j$.

Fourier Basis

The Fourier basis for $L^2([0,1])$ consists of the functions $e_k(t) = e^{2\pi kt}$, $k \in \mathbf{Z}$. Since this collection of functions is both orthonormal and closed under multiplication, we can easily compute that $\Gamma_{jkl} = \delta(k+l-j)$. We can change variables in the inner summation of Equation 2 to obtain the usual convolution formula $\sum_{k,l} \Gamma_{jkl} u_k v_l = \sum_k u_k v_{j-k}$.

Haar Basis

It has been known since 1910 [9] that there exists a compactly-supported function $\psi = \psi(x)$ having the property that its translates by integers and dilates by powers of two form an orthonormal basis. Namely, the linear span of the following set of orthogonal unit vectors is dense in $L^2(\mathbf{R})$:

$$\{\psi_{sn}(x) \stackrel{\text{def}}{=} 2^{-s/2}\psi(2^{-s}x - n) : s, n \in \mathbf{Z}\}. \quad (3)$$

Haar's "mother" function ψ is simply $\psi(x) = \mathbf{1}(2x) - \mathbf{1}(2x-1)$, where $\mathbf{1}(x)$ is the characteristic or indicator function of the interval $[0, 1)$. It generates a basis of functions that are indexed by a pair of integers, so their connection coefficients require six integer indices:

$$\Gamma^{str}_{nmk} \stackrel{\text{def}}{=} \int_{\mathbf{R}} \psi_{sn}(x)\psi_{tm}(x)\psi_{rk}(x)\,dx \quad (4)$$

$$= 2^{-\frac{s+t+r}{2}} \int_{\mathbf{R}} \psi(2^{-s}x - n)\psi(2^{-t}x - m)\psi(2^{-r}x - k)\,dx$$

For Haar's function, the integral may be evaluated explicitly. The function ψ_{sn} is supported in the interval $2^s[n, n+1)$, and is constant on the left subinterval $2^s[n, n+\frac{1}{2})$, where it is $+2^{-s/2}$, and on the right subinterval $2^s[n+\frac{1}{2}, n+1)$, where it is $-2^{-s/2}$. The product of two Haar functions ψ_{sn} and ψ_{tm} is either zero (if their support intervals are disjoint), or $2^{-s}\mathbf{1}(2^{-s}x - n)$ (if they are equal, i.e., $s = t$ and $n = m$), or $\pm 2^{-\frac{s+t}{2}}\psi_{tm}$ (if their support intervals intersect, and $s > t$). Hence the Haar functions are almost closed under multiplication.

Since ψ is real-valued, the three functions in the connection coefficient integral may be reordered so that $s \geq t \geq r$; in that case,

$$\Gamma^{str}_{nmk} = \begin{cases} 2^{-s/2}, & \text{if } s > t = r \text{ and } m = k \in \mathcal{L}(s-t, n); \\ -2^{-s/2}, & \text{if } s > t = r \text{ and } m = k \in \mathcal{R}(s-t, n); \\ 0, & \text{otherwise, with } s \geq t \geq r. \end{cases} \quad (5)$$

Here it has been useful to define $\mathcal{L}(a, j) = 2^a[j, j+\frac{1}{2})$ and $\mathcal{R}(a, j) = 2^a[j+\frac{1}{2}, j+1)$, the left and right subintervals of the interval $2^a[j, j+1)$, respectively. Another way to write this is $\Gamma^{str}_{nmk} = 2^{-t/2}\delta(t-r)\delta(m-k)\psi_{s-t,n}(m)$.

Accounting for the other orderings of s, t, r, the multiplication formula is thus

$$\sum_{t,m,r,k} \Gamma^{str}_{nmk} u_{tm} v_{rk} =$$

$$2^{-s/2} \sum_{t=-\infty}^{s-1} \left[\sum_{m \in \mathcal{L}(s-t,n)} u_{tm} v_{tm} - \sum_{m \in \mathcal{R}(s-t,n)} u_{tm} v_{tm} \right]$$

$$+ v_{sn} \sum_{t=s+1}^{\infty} 2^{-t/2} \left[\sum_{n \in \mathcal{L}(t-s,m)} u_{tm} - \sum_{n \in \mathcal{R}(t-s,m)} u_{tm} \right]$$

$$+ u_{sn} \sum_{r=s+1}^{\infty} 2^{-r/2} \left[\sum_{n \in \mathcal{L}(r-s,k)} v_{rk} - \sum_{n \in \mathcal{R}(r-s,k)} v_{rk} \right]. \quad (6)$$

Now $n \in \mathcal{L}(a,j) \iff 2^{-a}n - \frac{1}{2} < j \le 2^{-a}n$ and $n \in \mathcal{R}(a,j) \iff 2^{-a}n - 1 < j \le 2^{-a}n - \frac{1}{2}$. For fixed n and $a > 0$ exactly one of these inequalities will have a solution $j \stackrel{\text{def}}{=} \mathcal{M}(a,n)$, and that solution will be unique. Put $\mathcal{S}(a,n) = +1$ if $n \in \mathcal{L}(a, \mathcal{M}(a,n))$ and $\mathcal{S}(a,n) = -1$ if $n \in \mathcal{R}(a, \mathcal{M}(a,n))$; then the second and third lines of Equation 6 simplify as follows:

$$\sum_{t,m,r,k} \Gamma^{str}_{nmk} u_{tm} v_{rk} =$$

$$2^{-s/2} \sum_{t=-\infty}^{s-1} \left[\sum_{m \in \mathcal{L}(s-t,n)} u_{tm} v_{tm} - \sum_{m \in \mathcal{R}(s-t,n)} u_{tm} v_{tm} \right]$$

$$+ \sum_{t=s+1}^{\infty} 2^{-t/2} \mathcal{S}(t-s,n) \left[u_{t,\mathcal{M}(t-s,n)} v_{sn} + u_{sn} v_{t,\mathcal{M}(t-s,n)} \right] (7)$$

Alternatively, one could write $\mathcal{S}(a,n) = \psi(2^a n - \mathcal{M}(a,n))$.

2.3 Discrete Orthonormal Wavelet Basis

The famous generalization of Haar's basis by Daubechies [5] showed that for any positive integer d there exists a compactly-supported function $\psi = \psi(x)$ with d continuous derivatives on $\mathbf{R}$, whose integer translates and power-of-two dilates generate an orthonormal basis for $L^2(\mathbf{R})$ in the same way as the Haar function. The construction begins with the so-called *two-scale equations*:

$$\phi(x) = \sqrt{2} \sum_k h_k \phi(2x - k) \stackrel{\text{def}}{=} H\phi(x); \quad (8)$$

$$\psi(x) = \sqrt{2} \sum_k g_k \phi(2x - k) \stackrel{\text{def}}{=} G\phi(x). \quad (9)$$

Here $\{h_k\}$ and $\{g_k\}$ are two finite sequences of *wavelet filter coefficients* satisfying some orthogonality conditions:

$$\sum_k h_k h_{k+2n} = \delta(n) = \sum_k g_k g_{k+2n}, \quad \sum_k h_k g_{k+2n} = 0, \quad g_k = (-1)^{1-k} h_{1-k}. \tag{10}$$

In addition, we may assume that the coefficients are *conventionally normalized* ([16],pp.158–160) so as to satisfy $\sum_k h_{2k} = \sum_k h_{2k+1} = \sum_k g_{2k} = -\sum_k g_{2k+1} = 1/\sqrt{2}$.

Wavelet bases are composed of translated and rescaled versions of the same function:

$$\psi_{sn}(x) \stackrel{\text{def}}{=} 2^{-s/2} \psi(2^{-s}x - n). \tag{11}$$

The two-scale equations produce the connection coefficients through iteration and filtering.

3 Wavelet Connection Coefficients

As in the Haar case, the connection coefficients are labeled by six indices: Γ^{str}_{nmk}. They may be computed from another set of coefficients, namely those derived from the scaling functions of the wavelet basis:

$$A^{str}_{nmk} \stackrel{\text{def}}{=} \int_{\mathbf{R}} \phi_{sn}(x) \phi_{tm}(x) \phi_{rk}(x)\, dx \tag{12}$$

$$= 2^{-\frac{s+t+r}{2}} \int_{\mathbf{R}} \phi(2^{-s}x - n)\phi(2^{-t}x - m)\phi(2^{-r}x - k)\, dx$$

As before, we may suppose that $s \geq t \geq r$, and then by changing variables we obtain

$$A^{str}_{nmk} = 2^{-\frac{r}{2}} A^{s-r,t-r,0}_{nmk} = 2^{-\frac{r}{2}} A^{s-r,t-r,0}_{0,m-2^{s-t}n,k-2^{s-r}n}, \tag{13}$$

for $n, m, k \in \mathbf{Z}$ and $s \geq t \geq r \in \mathbf{Z}$. Hence in practice, to obtain A^{str}_{nmk} it suffices to compute

$$A^{ij}_{mk} \stackrel{\text{def}}{=} A^{ij0}_{0mk} = 2^{-\frac{i+j}{2}} \int_{\mathbf{R}} \phi(2^{-i}x)\phi(2^{-j}x - m)\phi(x - k)\, dx, \tag{14}$$

for $m, k \in \mathbf{Z}$ and $i \geq j \in \mathbf{Z}$.

Likewise, the complete set of connection coefficients Γ^{str}_{nmk} may be obtained from the following numbers, which must be computed for all $m, k \in \mathbf{Z}$ and for all $i \geq j \in \mathbf{Z}$:

$$\Gamma^{ij}_{mk} \stackrel{\text{def}}{=} 2^{-\frac{i+j}{2}} \int_{\mathbf{R}} \psi(2^{-i}x)\psi(2^{-j}x - m)\psi(x - k)\, dx;$$

$$\Gamma^{str}_{nmk} = 2^{-\frac{r}{2}} \Gamma^{s-r,t-r}_{m-2^{s-t}n, k-2^{s-r}n}. \tag{15}$$

3.1 Obtaining $\Gamma^{ij}_{...}$ from $A^{ij}_{...}$ by Filtering

$$\begin{aligned}
\Gamma^{ij}_{nmk} &= 2^{-\frac{i+j}{2}} \int_{\mathbb{R}} \psi(2^{-i}x - n)\psi(2^{-j}x - m)\psi(x - k)\, dx \\
&= \frac{1}{2} 2^{-\frac{i+j}{2}} \int_{\mathbb{R}} \psi(2^{-i}\frac{x}{2} - n)\psi(2^{-j}\frac{x}{2} - m)\psi(\frac{x}{2} - k)\, dx \\
&= \sqrt{2} \sum_{n',m',k'} g_{n'} g_{m'} g_{k'} \left(2^{-\frac{i+j}{2}} \int_{\mathbb{R}} \phi(2^{-i}x - 2n - n') \right. \\
&\qquad\qquad \left. \times \phi(2^{-j}x - 2m - m')\phi(x - 2k - k')\, dx \right) \\
&= \sqrt{2} \sum_{n',m',k'} g_{n'} g_{m'} g_{k'} A^{ij}_{2n+n',2m+m',2k+k'}.
\end{aligned}$$

This converts to $\Gamma^{ij}_{nmk} = \sqrt{2} G_1 G_2 G_3 A^{ij}_{n,m,k}$ once we define the commuting operators

$$G_1 B(n,m,k) \stackrel{\text{def}}{=} \sum_{n'} g_{n'} B(2n + n', m, k)$$

$$G_2 B(n,m,k) \stackrel{\text{def}}{=} \sum_{m'} g_{n'} B(n, 2m + m', k)$$

$$G_3 B(n,m,k) \stackrel{\text{def}}{=} \sum_{k'} g_{n'} B(n, m, 2k + k').$$

3.2 Obtaining $A^{ij}_{...}$ from $A_{...}$ by Filtering

$$\begin{aligned}
A^{ij}_{nmk} &= 2^{-\frac{i+j}{2}} \int_{\mathbb{R}} \phi(2^{-i}x - n)\phi(2^{-j}x - m)\phi(x - k)\, dx \\
&= 2^{-\frac{i+j}{2}} \int_{\mathbb{R}} \phi(2^{-i+1}\frac{x}{2} - n)\phi(2^{-j}x - m)\phi(x - k)\, dx \\
&= \sum_{n'} h_{n'} \left(2^{-\frac{(i-1)+j}{2}} \int_{\mathbb{R}} \phi(2^{-(i-1)}x - 2n - n') \right. \\
&\qquad\qquad \left. \times \phi(2^{-j}x - m)\phi(x - k)\, dx \right) \\
&= \sum_{n'} h_{n'} A^{i-1,j}_{2n+n',m,k} \stackrel{\text{def}}{=} H_1 A^{i-1,j}_{n,m,k}.
\end{aligned}$$

Iterating i times in the first scale index and j times in the second gives $A^{ij}_{nmk} = H_1^i H_2^j A_{n,m,k}$, using the commuting operators

$$H_1 B(n,m,k) \stackrel{\text{def}}{=} \sum_{n'} h_{n'} B(2n + n', m, k)$$

$$H_2B(n,m,k) \stackrel{\text{def}}{=} \sum_{m'} h_{n'} B(n, 2m+m', k)$$

3.3 Obtaining Γ from A by Filtering

Combining the iterations of H and G yields

$$\Gamma^{ij}_{nmk} = \sqrt{2} G_1 G_2 G_3 H_1^i H_2^j A_{n,m,k}. \tag{16}$$

3.4 Obtaining A as a Fixed Point

Now suppose that $\phi = \phi(t)$ is a normalized fixed point of Equation 8, i.e., $\int \phi(x)\,dx = 1$ and $\|\phi\| = 1$. Then $\{\phi(x-k) : k \in \mathbf{Z}\}$ is a (real) orthonormal system in L^2, though it is not complete, and we can define the following:

$$A_{nmk} \stackrel{\text{def}}{=} A^{000}_{nmk} = \int_{\mathbf{R}} \phi(x-n)\phi(x-m)\phi(x-k)\,dx \tag{17}$$

This quantity will be defined for all triplets of integers, and will vanish whenever some pair of $\{n, m, k\}$ are so different that the supports of the corresponding scaling functions are disjoint.

Now $A_{nmk} = A_{n-k, m-k, 0}$, so it suffices to compute the simpler quantity:

$$A(n,m) \stackrel{\text{def}}{=} A_{nm0} = \int_{\mathbf{R}} \phi(x-n)\phi(x-m)\phi(x)\,dx \tag{18}$$

But this is a matrix which satisfies the following analogue of Equation 8:

Theorem 1 *Suppose that $h_k = 0$ unless $0 \le k < L$. Then $A(n,m) = 0$ unless $-L < n < L$ and $-L < m < L$. Also, A satisfies the homogeneous fixed-point equation*

$$A(n,m) = \sum_{p,q} \alpha(p,q) A(2n-p, 2m-q), \qquad -L < n, m < L,$$

where

$$\alpha(p,q) \stackrel{\text{def}}{=} \sqrt{2} \sum_{k=0}^{L} h_{k-p} h_{k-q} h_k, \qquad -L < p, q < L.$$

Proof: We change variables $x \leftarrow x/2$ in the integral defining $A(n,m)$, use Equation 8, and interchange the finite summation and the integration:

$$A(n,m) = \frac{1}{2}\int_{\mathbf{R}} \phi(\frac{x}{2} - n)\phi(\frac{x}{2} - m)\phi(\frac{x}{2})\,dx$$

$$= \sqrt{2}\int_{\mathbf{R}} \left(\sum_{n'} h_{n'}\phi(x-2n-n')\right)\left(\sum_{m'} h_{m'}\phi(x-2m-m')\right)$$

4. Multiplication of Short Wavelet Series

$$\times \left(\sum_{k'} h_{k'} \phi(x - k') \right) dx$$

$$= \sqrt{2} \sum_{n',m',k'} h_{n'} h_{m'} h_{k'} \int_{\mathbf{R}} \phi(x - 2n - n')$$

$$\times \phi(x - 2m - m') \phi(x - k') \, dx$$

$$= \sqrt{2} \sum_{n',m',k'} h_{n'} h_{m'} h_{k'} \int_{\mathbf{R}} \phi(x + k' - 2n - n')$$

$$\times \phi(x + k' - 2m - m') \phi(x) \, dx$$

$$= \sqrt{2} \sum_{n',m',k'} h_{n'} h_{m'} h_{k'} A(2n + n' - k', 2m + m' - k').$$

$$= \sum_{p,q} \left(\sqrt{2} \sum_{k} h_{k-p} h_{k-q} h_k \right) A(2n - p, 2m - q).$$

At the last step we have substituted $k' \leftarrow k$, $n' \leftarrow k - p$, and $m' \leftarrow k - q$.

To get the support rectangle, note that if h_k vanishes for $k < 0$ or $k \geq L$, then $\phi(x)$ will be supported in $[0, L]$ (see [6], p.167) and $A(n, m)$ will vanish for $|n| \geq L$ or $|m| \geq L$. $\square$

If we put $\hat{f}(\xi, \eta) \stackrel{\text{def}}{=} \sum_{n,m \in \mathbf{Z}} f(n,m) e^{-2\pi i n \xi} e^{-2\pi i m \eta}$ for any double sequence $f = f(n, m)$, then we can write

$$\hat{\alpha}(\xi, \eta) = \sum_{n,m \in \mathbf{Z}} \sqrt{2} \sum_{k=0}^{L-1} h_k h_{k-n} h_{k-m} e^{-2\pi i n \xi} e^{-2\pi i m \eta}$$

$$= \sqrt{2} \left(\sum_{k=0}^{L-1} h_k e^{-2\pi i k (\xi + \eta)} \right) \left(\sum_{p=0}^{L-1} h_p e^{2\pi i p \xi} \right) \left(\sum_{q=0}^{L-1} h_q e^{2\pi i q \eta} \right)$$

$$\stackrel{\text{def}}{=} \sqrt{2} \, m(\xi + \eta) m(-\xi) m(-\eta). \tag{19}$$

The second equality follows from the substitutions $n \leftarrow k-p$ and $m \leftarrow k-q$. The sums in parentheses are trigonometric polynomials, and we adopt the convention of Reference [16], p.176, when defining the filter multiplier m. Note that $\hat{\alpha}(0,0) = 4$, since $m(0) = \sqrt{2}$. Also, $\hat{\alpha}(\xi, \eta) = \hat{\alpha}(\eta, \xi)$ since $\alpha(n, m) = \alpha(m, n)$. Finally, if the filter coefficients are real-valued, then $m(-\xi) = \bar{m}(\xi)$ and $\hat{\alpha}(\xi, -\xi) = \hat{\alpha}(\xi, 0) = \hat{\alpha}(0, \xi) = 2|m(\xi)|^2$.

Remark. Replacing $\{h_k\} \leftarrow \{g_k\}$ and $\phi \leftarrow \psi$ in the arguments above gives the following formula:

$$\Gamma^{00}_{nm} = \sum_{p,q} \beta(p,q) A(2n - p, 2m - q), \qquad -L < n, m < L, \tag{20}$$

where

$$\beta(p,q) \stackrel{\text{def}}{=} \sqrt{2} \sum_{k=0}^{L} g_{k-p} g_{k-q} g_k, \qquad -L < p, q < L. \tag{21}$$

The conjugate quadrature filter relationship between $\{h_k\}$ and $\{g_k\}$ implies that
$$\hat{\beta}(\xi,\eta) = -\sqrt{2}\,\bar{m}(\frac{1}{2}+\xi+\eta)\bar{m}(\frac{1}{2}-\xi)\bar{m}(\frac{1}{2}-\eta). \tag{22}$$

Existence and Uniqueness

¿From the filter coefficients $\{h_k\}$ in the guise of α, we can define an operator T on double sequences:
$$Tf(n,m) \stackrel{\text{def}}{=} \sum_{p,q} \alpha(p,q) f(2n-p, 2m-q) = \sum_{p,q} \alpha(2n-p, 2m-q) f(p,q). \tag{23}$$

The matrix $A = A(n,m)$ is a fixed point of this operator and is completely determined by it except for normalization. Using Lemma 5.13 in Reference [16], p.178, we can rewrite the fixed point equation as follows:

$$\begin{aligned}
\hat{A}(\xi,\eta) \;=\; \widehat{TA}(\xi,\eta) \;=\;& \frac{1}{4}\hat{\alpha}(\frac{\xi}{2},\frac{\eta}{2})\hat{A}(\frac{\xi}{2},\frac{\eta}{2}) \;+\; \\
+\;& \frac{1}{4}\hat{\alpha}(\frac{\xi}{2}+\frac{1}{2},\frac{\eta}{2}+\frac{1}{2})\hat{A}(\frac{\xi}{2}+\frac{1}{2},\frac{\eta}{2}+\frac{1}{2}) \\
+\;& \frac{1}{4}\hat{\alpha}(\frac{\xi}{2}+\frac{1}{2},\frac{\eta}{2})\hat{A}(\frac{\xi}{2}+\frac{1}{2},\frac{\eta}{2}) \;+\; \frac{1}{4}\hat{\alpha}(\frac{\xi}{2},\frac{\eta}{2}+\frac{1}{2})\hat{A}(\frac{\xi}{2},\frac{\eta}{2}+\frac{1}{2}).
\end{aligned}$$

This is rather complicated, and it is easier to compute the action of the adjoint operator:

$$T^*g(p,q) \stackrel{\text{def}}{=} \sum_{n,m} \bar{\alpha}(2n-p, 2m-q) g(n,m); \quad \widehat{T^*g}(\xi,\eta) = \overline{\hat{\alpha}(\xi,\eta)}\hat{g}(2\xi, 2\eta). \tag{24}$$

Theorem 2 *The equation $A = TA$ with normalization $\sum_{n,m} A(n,m) = 1$ has a unique solution.*

Proof: We first show that the normalization is preserved by T:

$$\begin{aligned}
\sum_{n,m} Tf(n,m) &= \sum_{n,m}\sum_{p,q} \alpha(2n-p, 2m-q) f(p,q) \\
&= \sum_{p,q} \left(\sqrt{2} \sum_{n,m,k} h_{k-2n+p} h_{k-2m+q} h_k \right) f(p,q).
\end{aligned}$$

Now $\sum_n h_{2n} = \sum_n h_{2n+1} = 1/\sqrt{2}$, so the sum over n,m,k inside the parentheses equals 1 for all p,q. Hence the right hand side simplifies to $\sum_{p,q} f(p,q)$.

To show the existence of $A(n,m)$, start with any fixed $f = f(n,m)$ satisfying $\sum_{n,m} f(n,m) = 1$, define $A_k = T^k f$ for $k = 1, 2, ...$, and show that $A_k \to A$ as $k \to \infty$. Then A will solve $A = TA$ and $\sum_{n,m} A(n,m) = 1$.

4. Multiplication of Short Wavelet Series

But if $g = g(n,m)$ is any double sequence, then $\langle g, T^k f \rangle = \langle T^{*k} g, f \rangle$ which by Plancherel's theorem is equal to the inner product between the Fourier transforms: $\langle \widehat{T^{*k} g}, \hat{f} \rangle$ is approximated by

$$\sum_{p=0}^{2^k-1} \sum_{q=0}^{2^k-1} \int_0^1 \int_0^1 \hat{\phi}(\xi + p + \eta + q) \hat{\phi}(-\eta - p) \hat{\phi}(-\xi - q) \overline{\hat{g}(\xi, \eta)} \, d\xi d\eta$$

$$\to \int_0^1 \int_0^1 \hat{\phi}_1(\xi + \eta) \hat{\phi}_1(-\eta) \hat{\phi}_1(-\xi) \overline{\hat{g}(\xi, \eta)} \, d\xi d\eta, \quad \text{as } k \to \infty,$$

Now for each fixed (ξ, η), $\frac{1}{2^{2k}} \prod_{j=1}^k \hat{\alpha}(\frac{\xi}{2^j}, \frac{\eta}{2^j}) \to \hat{\phi}(\xi + \eta) \hat{\phi}(-\eta) \hat{\phi}(-\xi)$ and $\hat{f}(\frac{\xi}{2^k}, \frac{\eta}{2^k}) \to 1$ as $k \to \infty$. This convergence is uniform on compact subsets of $\mathbf{R}^2$, and since the filter $\{h_k\}$ determines a regular scaling function ϕ, $\hat{\phi}$ vanishes to sufficiently high order at ∞ for the integrand to converge in $L(\mathbf{R}^2)$ as well. Then, since $\hat{g} = \hat{g}(\xi, \eta)$ is 1-periodic in both ξ and η, the integration breaks up over the $2^k \times 2^k$ periods: $\langle \widehat{T^{*k} g}, \hat{f} \rangle$ is approximated by

$$\sum_{p=0}^{2^k-1} \sum_{q=0}^{2^k-1} \int_0^1 \int_0^1 \hat{\phi}(\xi + p + \eta + q) \hat{\phi}(-\eta - p) \hat{\phi}(-\xi - q) \overline{\hat{g}(\xi, \eta)} \, d\xi d\eta$$

$$\to \int_0^1 \int_0^1 \hat{\phi}_1(\xi + \eta) \hat{\phi}_1(-\eta) \hat{\phi}_1(-\xi) \overline{\hat{g}(\xi, \eta)} \, d\xi d\eta, \quad \text{as } k \to \infty,$$

where $\hat{\phi}_1(\xi) \stackrel{\text{def}}{=} \sum_{j \in \mathbf{Z}} \hat{\phi}(\xi + j)$ is the 1-periodization of $\hat{\phi}$. Thus $A_k \to A$ in $\ell^2(\mathbf{Z}^2)$ and the limit sequence satisfies $\hat{A}(\xi, \eta) = \hat{\phi}_1(\xi + \eta) \hat{\phi}_1(-\eta) \hat{\phi}_1(-\xi)$.

Finally, note that the solution A is unique, since if $A' = TA'$ with $\sum_{n,m} A'(n,m) = 1$, then starting with $f = A'$ in the iteration above gives $A' = T^n A' \to A$ as $n \to \infty$. □

Example: Haar Wavelet Basis

For this wavelet basis, we have $h_0 = h_1 = \frac{1}{\sqrt{2}}$ and $h_k = 0$ if $k \notin \{0,1\}$. Thus $L = 2$, and

$$\alpha = \frac{1}{2} \left\{ \begin{array}{ccc} 0 & 1 & 1 \\ 1 & 2 & 1 \\ 1 & 1 & 0 \end{array} \right\}; \quad A(n,m) = \delta(n)\delta(m) \qquad (25)$$

Example: Coifman 12 Wavelet Basis

For this wavelet basis, we have $L = 12$ and $\{h_k : 0 \leq k < 12\}$ is the following table of values:

$$\begin{array}{ll}
\{1.6387336463179785 \times 10^{-2}, & -4.1464936781966485 \times 10^{-2}, \\
-6.7372554722299874 \times 10^{-2}, & 3.8611006682309290 \times 10^{-1}, \\
8.1272363544960613 \times 10^{-1}, & 4.1700518442377760 \times 10^{-1}, \\
-7.6488599078264594 \times 10^{-2}, & -5.9434418646471240 \times 10^{-2}, \\
2.3680171946876750 \times 10^{-2}, & 5.6114348193659885 \times 10^{-3}, \\
-1.8232088709100992 \times 10^{-3}, & -7.2054944536811512 \times 10^{-4}\}
\end{array}$$

These filter coefficients define the α shown in Figure 1. The entries are multiplied by 1000 and truncated to integers for display purposes, the origin $m = k = 0$ of $\alpha(m,k)$ is at the center, m increases downwards, and k increases to the right as in the convention for matrices.

The corresponding fixed point A is plotted below in Figure 4.

4 Numerical Examples

4.1 Procedure

We start with a filter sequence $\{h_k\}$ taken from the list of filters in Reference [16], which are known to have smooth scaling function limits. We restrict our attention to the Daubechies filters of lengths 4, 6, 10, and 12, and the Coifman filters of lengths 6 and 12. We compute the kernel α of the fixed-point problem and then iterate from the elementary double sequence $A(m,k) = 1 \iff m = k = 0$ until the maximum change per iteration in a coefficient of A falls below 10^{-6}.

To get the other scaling and connection coefficients, we apply the filter operators G_1, G_2, G_3, H_1, and H_2 as needed.

4.2 Space and Time Requirements

Once the low-pass filter $\{h_n : n = 0, 1, ..., L - 1\}$ is chosen, it is known that $A(m,k)$ will vanish for $|m| \geq L$ or $|k| \geq L$, so it is only necessary to allocate $(2L - 1) \times (2L - 1)$ memory locations to hold A. Likewise, $\alpha(m,k)$ fits into a $(2L + 1) \times (2L + 1)$ array.

Each application of an operator G_1, G_2, G_3, H_1, or H_2 costs L operations per output coefficient, since we sum over the L non-zero coefficients of either $\{h_n\}$ or $\{g_n\}$. The number of coefficients we need to compute grows, however. For filters supported on $\{0, 1, ..., L-1\}$ and fixed $i, j \geq 0$, the matrices $A^{ij}(m,k)$ and $\Gamma^{ij}(m,k)$ will vanish outside $-L < m < 2^{i-j}L$ and $-L < k < 2^i L$. However, as the plots below will show, many of the coefficients with indices in this range are negligible.

4. Multiplication of Short Wavelet Series   89

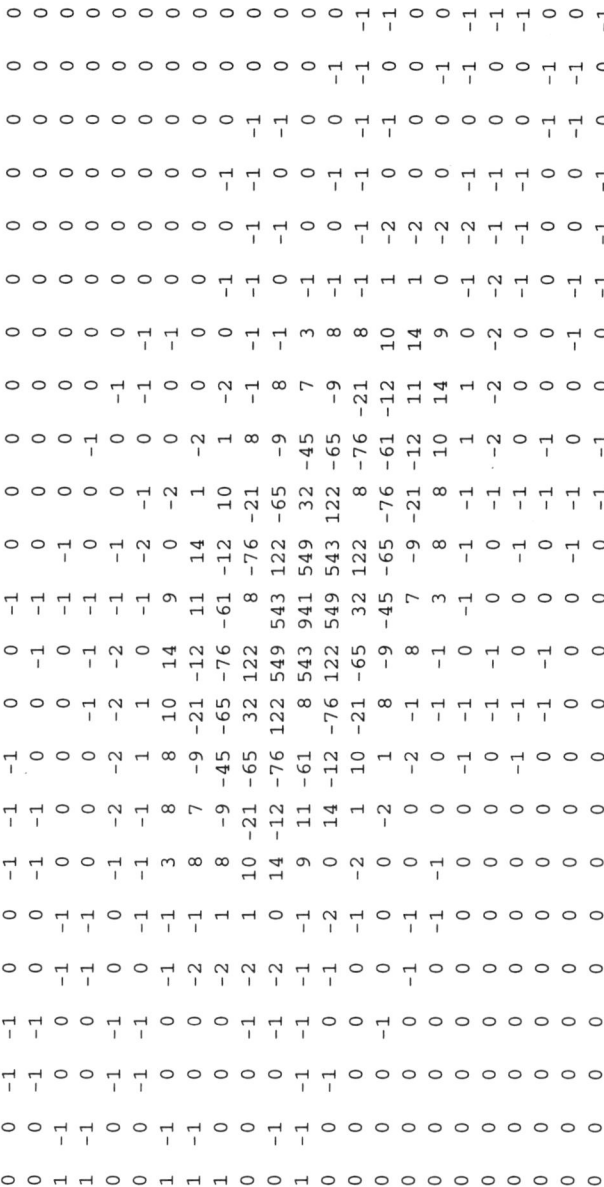

FIGURE 1. Matrix α for Coifman 12

4.3 Graphs

In the following figures, we plot level lines of the logarithm of the absolute value of scaling and connection coefficients. The origin $m = k = 0$ is always at the center of the square. The graphs are oriented such that m increases to the right and k increases upwards as in the convention for xy plots in the first quadrant.

Graphs of $A(m,k) = \int \varphi(x)\varphi(x-m)\varphi(x-k)\,dx$ for filters D4, D6, D10, D12, C6, and C12

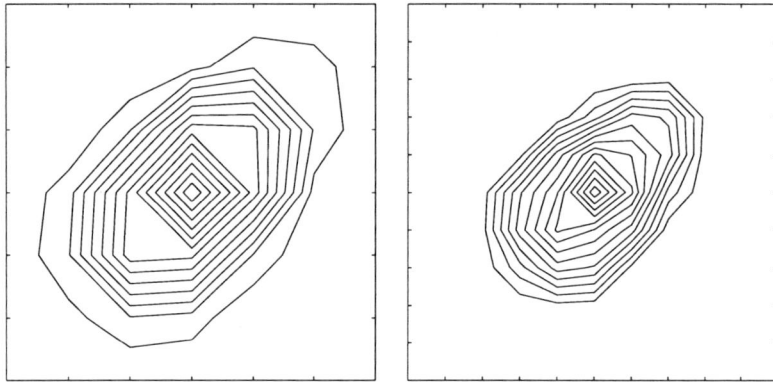

FIGURE 2. $A(m,k)$ for Daubechies scaling functions of support 4 and 6

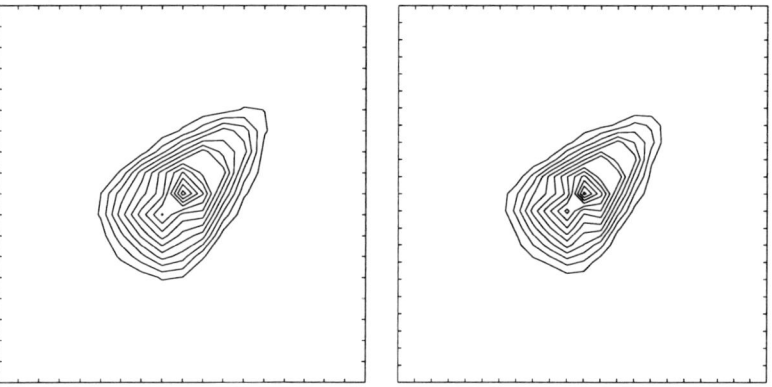

FIGURE 3. $A(m,k)$ for Daubechies scaling functions of support 10 and 12

4. Multiplication of Short Wavelet Series   91

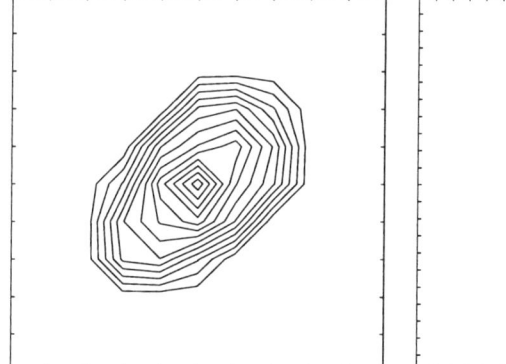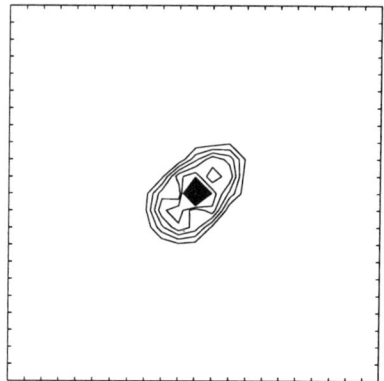

FIGURE 4. $A(m,k)$ for Coifman scaling functions of support 6 and 12

Graphs of $\Gamma(m,k) = \int \psi(x)\psi(x-m)\psi(x-k)\,dx$ for filters D4, D6, D10, D12, C6, and C12

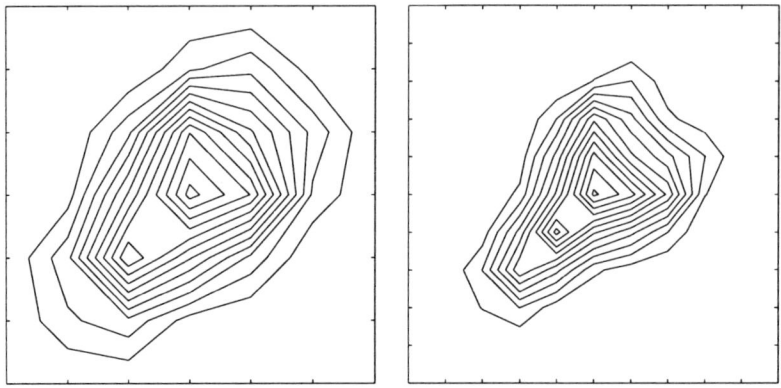

FIGURE 5. $\Gamma(m,k)$ for Daubechies wavelets of support 4 and 6

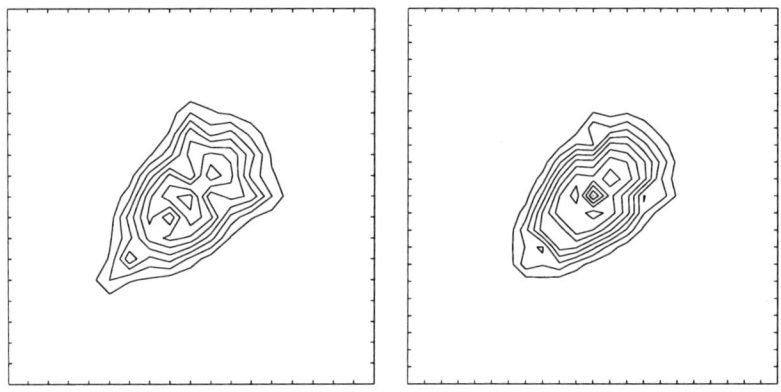

FIGURE 6. $\Gamma(m,k)$ for Daubechies wavelets of support 10 and 12

4. Multiplication of Short Wavelet Series   93

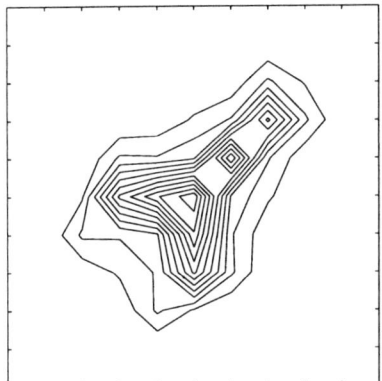 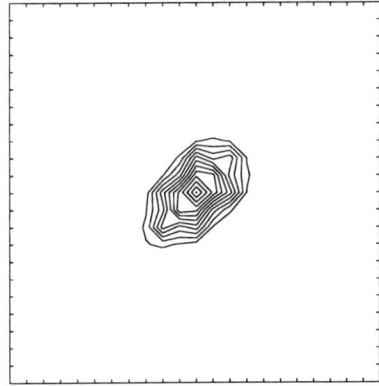

FIGURE 7. $\Gamma(m,k)$ for Coifman wavelets of support 6 and 12

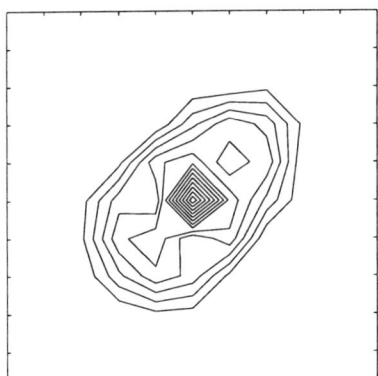 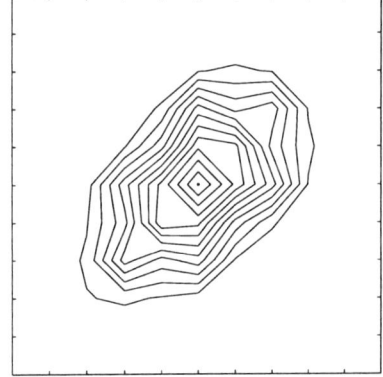

FIGURE 8. Close-up of $\Gamma(m,k)$ for Coifman scaling function and wavelet of order 12

Graphs of $A^{i,j}(m,k) = 2^{-\frac{i+j}{2}} \int \varphi(\frac{x}{2^i})\varphi(\frac{x}{2^j} - m)\varphi(x - k)\, dx$ for filters D6, C6, and C12

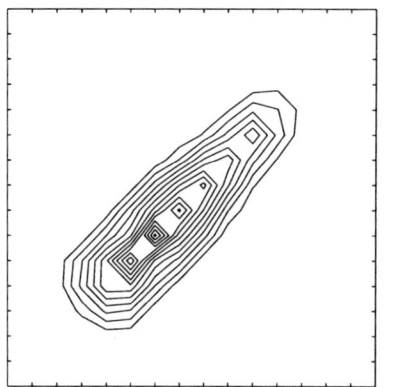 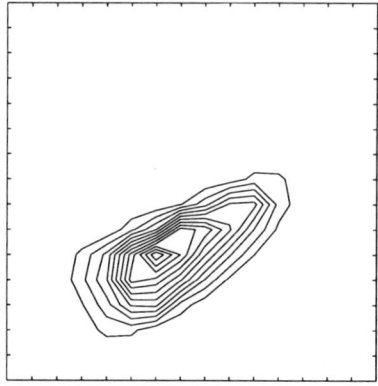

FIGURE 9. $A^{1,0}(m,k)$ and $A^{1,1}(m,k)$ for Daubechies 6

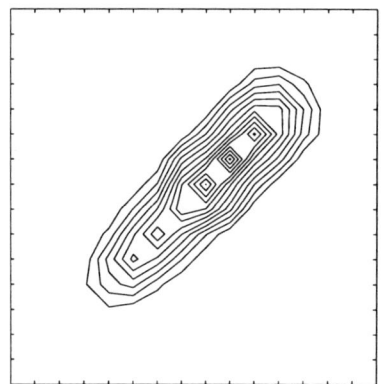 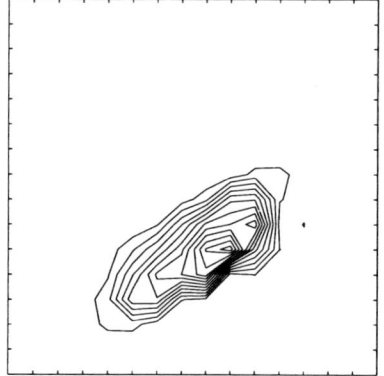

FIGURE 10. $A^{1,0}(m,k)$ and $A^{1,1}(m,k)$ for Coifman 6

4. Multiplication of Short Wavelet Series   95

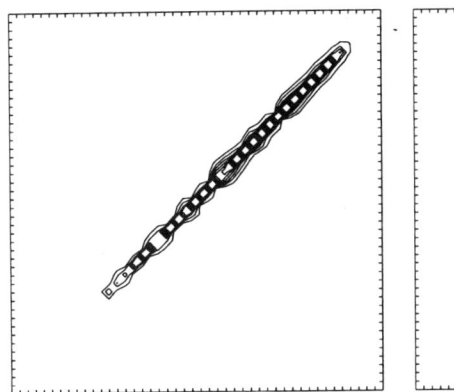

FIGURE 11. $A^{3,0}(m,k)$ and $A^{3,1}(m,k)$ for Coifman 6

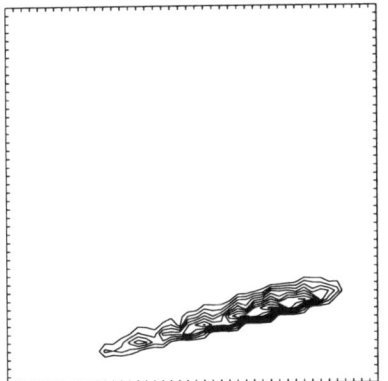

FIGURE 12. $A^{3,2}(m,k)$ and $A^{3,3}(m,k)$ for Coifman 6

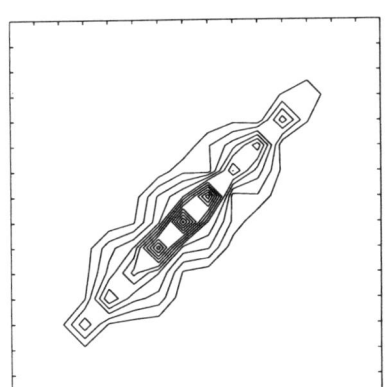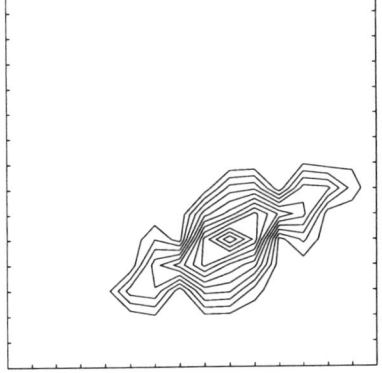

FIGURE 13. Close-ups of $A^{1,0}(m,k)$ and $A^{1,1}(m,k)$ for Coifman 12

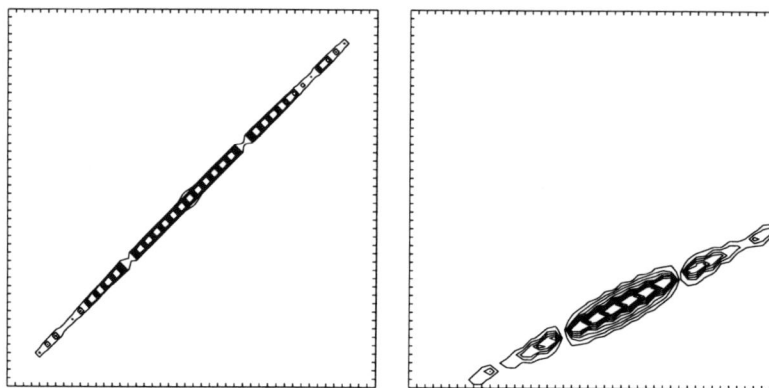

FIGURE 14. Close-ups of $A^{3,0}(m,k)$ and $A^{3,1}(m,k)$ for Coifman 12

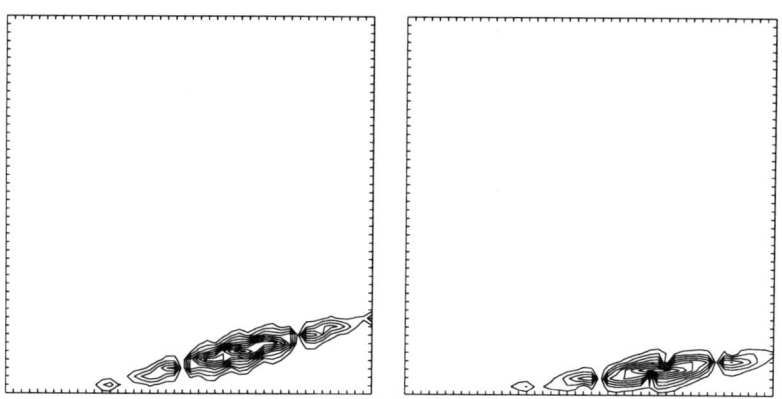

FIGURE 15. Close-ups of $A^{3,2}(m,k)$ and $A^{3,3}(m,k)$ for Coifman 12

4. Multiplication of Short Wavelet Series   97

Graphs of $\Gamma^{i,j}(m,k) = 2^{-\frac{i+j}{2}} \int \psi(\frac{x}{2^i})\psi(\frac{x}{2^j} - m)\psi(x - k)\, dx$ for filters D6, C6, and C12.

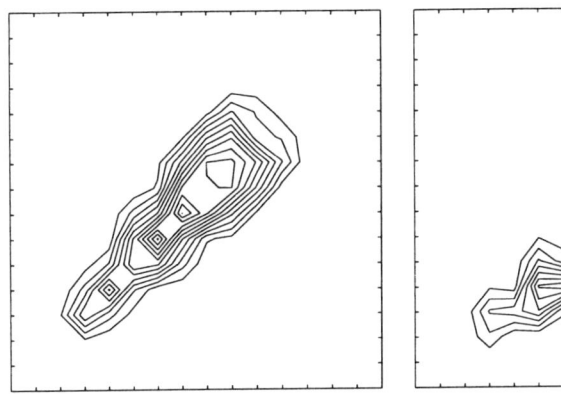
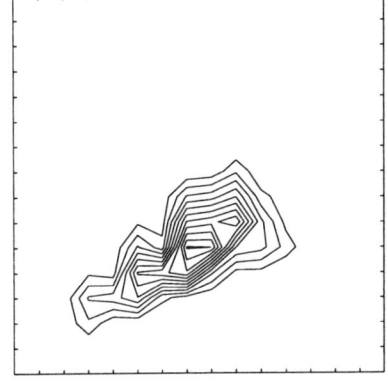

FIGURE 16. $\Gamma^{1,0}(m,k)$ and $\Gamma^{1,1}(m,k)$ for Daubechies 6

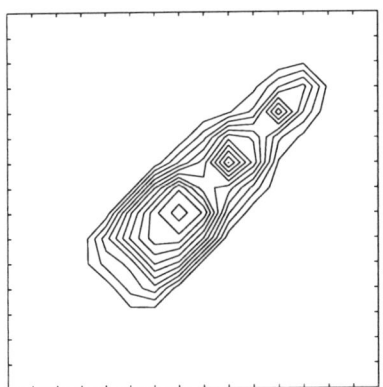
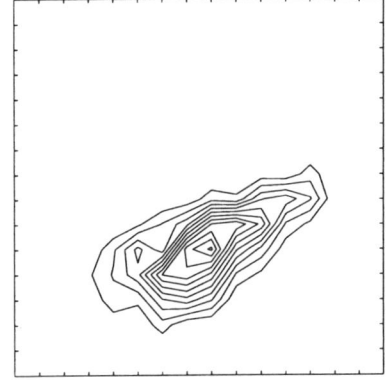

FIGURE 17. $\Gamma^{1,0}(m,k)$ and $\Gamma^{1,1}(m,k)$ for Coifman 6

 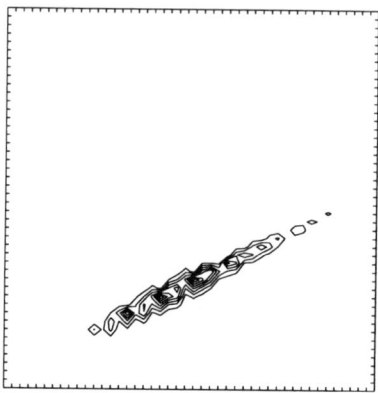

FIGURE 18. $\Gamma^{3,0}(m,k)$ and $\Gamma^{3,1}(m,k)$ for Coifman 6

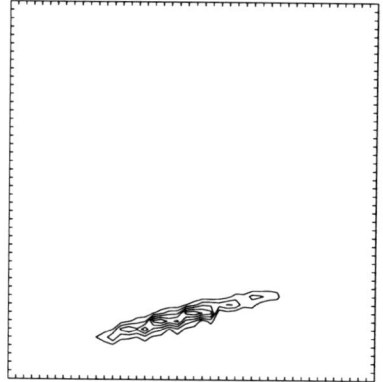 

FIGURE 19. $\Gamma^{3,2}(m,k)$ and $\Gamma^{3,3}(m,k)$ for Coifman 6

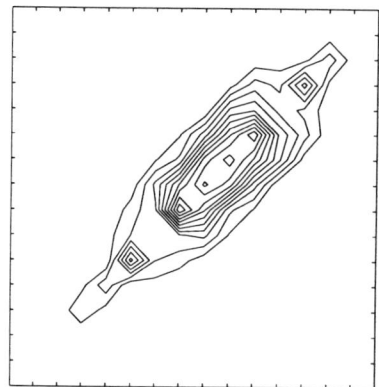 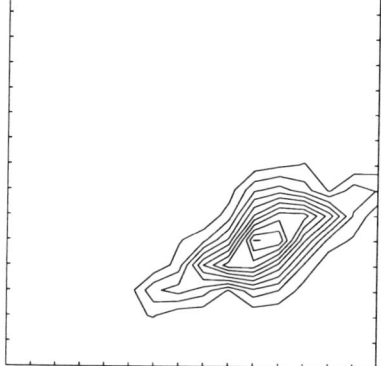

FIGURE 20. Close-ups of $\Gamma^{1,0}(m,k)$ and $\Gamma^{1,1}(m,k)$ for Coifman 12

4. Multiplication of Short Wavelet Series   99

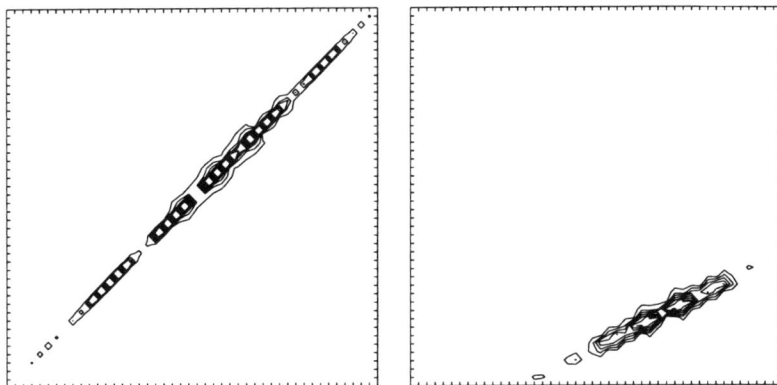

FIGURE 21. Close-ups of $\Gamma^{3,0}(m,k)$ and $\Gamma^{3,1}(m,k)$ for Coifman 12

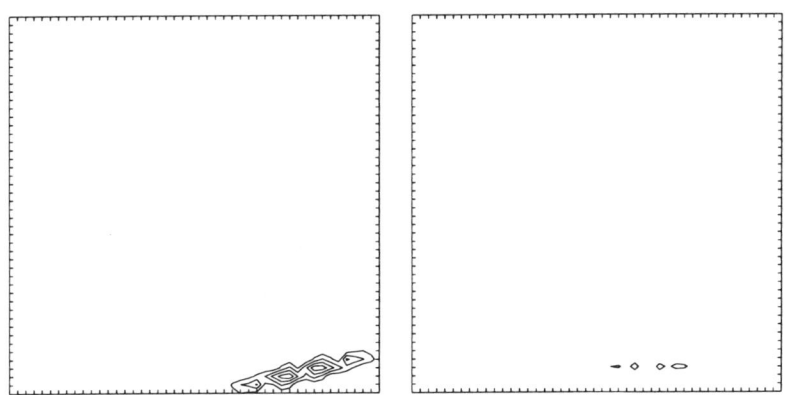

FIGURE 22. Close-ups of $\Gamma^{3,2}(m,k)$ et $\Gamma^{3,3}(m,k)$ for Coifman 12

5 References

[1] Gregory Beylkin. On the representation of operators in bases of compactly supported wavelets. *SIAM Journal of Numerical Analysis*, 6-6:1716–1740, 1992.

[2] Gregory Beylkin, Ronald R. Coifman, and Vladimir Rokhlin. Fast wavelet transforms and numerical algorithms I. *Communications on Pure and Applied Mathematics*, XLIV:141–183, 1991.

[3] Ronald R. Coifman and Mladen Victor Wickerhauser. Entropy based algorithms for best basis selection. *IEEE Transactions on Information Theory*, 32:712–718, March 1992.

[4] Wolfgang Dahmen and Charles A. Micchelli. Using the refinement equation for evaluating integrals of wavelets. *SIAM Journal of Numerical Analysis*, 30(2):507–537, April 1993.

[5] Ingrid Daubechies. Orthonormal bases of compactly supported wavelets. *Communications on Pure and Applied Mathematics*, XLI:909–996, 1988.

[6] Ingrid Daubechies. *Ten Lectures on Wavelets*, volume 61 of *CBMS-NSF Regional Conference Series in Applied Mathematics*. SIAM Press, Philadelphia, 1992.

[7] Marie Farge, Eric Goirand, Yves Meyer, Frédéric Pascal, and Mladen Victor Wickerhauser. Improved predictability of two-dimensional turbulent flows using wavelet packet compression. *Fluid Dynamics Research*, 10:229–250, 1992.

[8] Eric Goirand, Mladen Victor Wickerhauser, and Marie Farge. A parallel two dimensional wavelet packet transform and its application to matrix-vector multiplication. In Rodolphe L. Motard and Babu Joseph, editors, *Wavelet Applications in Chemical Engineering*, pages 275–319. Kluwer Academic Publishers, Norwell, Massachusetts, 1994.

[9] A. Haar. Zur theorie der orthogonalen funktionensysteme. *Mathematische Annalen*, 69:331–371, 1910.

[10] Angela Kunoth. Computing refinable integrals, version 1.1. Available by anonymous FTP from ftp.igpm.rwth-aachen.de, 1995. C++ source code and documentation.

[11] Andy Latto, Howard L. Resnikoff, and Eric Tenenbaum. The evaluation of connection coefficients of compactly support wavelets. Preprint, Aware, Inc., Cambridge, Massachusetts, 1991.

[12] J. Liandrat, Valerie Perrier, and Philippe Tchamitchian. Numerical resolution of nonlinear partial differential equations using the wavelet approach. In Mary Beth Ruskai, Gregory Beylkin, Ronald Coifman, Ingrid Daubechies, Stéphane Mallat, Yves Meyer, and Louise Raphael, editors, *Wavelets and Their Applications*, pages 227–238. Jones and Bartlett, Boston, 1992.

[13] Stéphane G. Mallat. A theory for multiresolution signal decomposition: The wavelet decomposition. *IEEE Transactions on Pattern Analysis and Machine Intelligence*, 11:674–693, 1989.

[14] Yves Meyer. *Ondelettes et Opérateurs*, volume I: Ondelettes. Hermann, Paris, 1990.

[15] Mladen Victor Wickerhauser. An adapted waveform functional calculus. In Moody Chu, Robert Plemmons, David Brown, and Donald Ellison, editors, *Proceedings of the Cornelius Lanczos Centenary, Raleigh, North Carolina, 12–17 December 1993*, pages 418–421, Philadelphia, 1994. SIAM, SIAM Press.

[16] Mladen Victor Wickerhauser. *Adapted Wavelet Analysis from Theory to Software*. AK Peters, Ltd., Wellesley, Massachusetts, 9 May 1994.

[17] Mladen Victor Wickerhauser. Large-rank approximate principal component analysis with wavelets for signal feature discrimination and the inversion of complicated maps. *Journal of Chemical Information and Computer Science*, 34(5):1036–1046, September/October 1994. Proceedings of Math-Chem-Comp 1993, Rovinj, Croatia.

[18] Mladen Victor Wickerhauser. Time localization techniques for wavelet transforms. *Croatica Chemica Acta*, 68(1):1–27, April 1995. Proceedings of the Ninth Dubrovnik International Course and Math-Chem-Comp 1994.

[19] Mladen Victor Wickerhauser, Marie Farge, Eric Goirand, Eva Wesfreid, and Echeyde Cubillo. Efficiency comparison of wavelet packet and adapted local cosine bases for compression of a two-dimensional turbulent flow. In Charles K. Chui, Laura Montefusco, and Luigia Puccio, editors, *Wavelets: Theory, Algorithms, and Applications*, Proceedings of the International Conference in Taormina, Sicily, 14–20 October 1993, pages 509–531. Academic Press, San Diego, California, 1994.

5
Conjugate Quadrature Filters

Wayne M. Lawton[1]

ABSTRACT Conjugate quadrature filters (CQF's) have applications to wavelet construction and signal processing. We show that any continuous *frequency response* (Fourier transform) P of a CQF p can be uniformly approximated by the frequency response Q of CQF q having finite length. Our first proof uses Hardy space theory and the parametrization of CQF's by the infinite dimensional Lie group of *paraunitary* matrices. Our second proof uses a *differential equation method* to approximate CQF's by CQF's having factorial decay, followed by a *phase retreival method* to approximate CQF's having factorial decay by CQF's having finite support.

1 Introduction

We let $\mathbf{Z}$, $\mathbf{R}$, $\mathbf{C}$, and $\mathbf{T}$ denote the integers, reals, complex numbers, and unit circle. A *conjugate quadrature filter* (CQF) is a sequence p whose Fourier transform P is a measurable function that satisfies

$$|P(w)|^2 + |P(-w)|^2 = 1, \ w \in \mathbf{T}. \tag{1.1}$$

These filters, having either finite or infinite support, have applications to wavelet construction and signal processing. In this paper we discuss the mathematics of approximating CQF's. The results we obtain include the following:

Theorem 1.1 *If p is a CQF whose Fourier transform P, defined by*

$$P(w) := \sum_{k \in \mathbf{Z}} p(k) w^k, \ w \in \mathbf{T},$$

is continuous, then for any positive number ϵ there exists a CQF q having finite support whose Fourier Q transform satisfies the following inequality:

$$|P(w) - Q(w)| < \epsilon, \ w \in \mathbf{T}. \tag{1.2}$$

---

[1]Department of Mathematics, National University of Singapore, 10 Kent Ridge Crescent, Singapore 119260, Republic of Singapore

Section 2 describes how to parametrize CQF's by *paraunitary matrices*, and Section 3 uses this parametrization to prove Theorem 1.1. Section 4 describes a *differential equation method*, to approximate a CQF by a CQF having *factorial decay*. Section 5 uses complex analysis methods to obtain two results, used in Section 6, that describe how the roots of a polynomial change when its coefficients are perturbed. Section 6 describes a *phase retrieval method* to approximate a CQF having factorial decay by a CQF having finite support. Section 7 provides a brief review of the development of CQF's.

2 Paraunitary Parametrization

We let $||F||_\mathbf{T}$ denote the *maximum norm* of a function F on $\mathbf{T}$, and we let $\mathbf{A}$ denote an *algebra* of continuous functions on $\mathbf{T}$ with the topology induced by this norm. We let $\mathrm{CQF}(\mathbf{A})$ denote the corresponding *topological space* of CQF's whose Fourier transforms are in $\mathbf{A}$. In particular we consider the algebra $\mathbf{A}_1$ of continuous functions, and the algebra $\mathbf{A}_5$ of Fourier transforms of finitely supported sequences (isomorphic to the algebra of *Laurent polynomials*). If $P \in \mathbf{A}_5$ is the Fourier transform of a finitely supported sequence p, we define the *length* $\ell(P)$ as follows:

$$\ell(P) := \max\{\,|m - n| : m, n \in \mathbf{Z},\, p(m) \neq 0,\, p(n) \neq 0\,\} \qquad (2.1)$$

We let $\mathbf{G}_5$ denote the *topological closure* of $\mathrm{CQF}(\mathbf{A}_5)$ in $\mathrm{CQF}(\mathbf{A}_1)$, and we observe that Theorem 1.1 is equivalent to the identity $\mathbf{G}_5 = \mathrm{CQF}(\mathbf{A}_1)$. We let $\mathbf{A}_2$ denote the algebra of functions F on $\mathbf{T}$ such that the function h on $\mathbf{R}$ defined by $h(t) := F(e^{it})$ is continuously differentiable, and we let $\dot{F}$ denote the function on $\mathbf{T}$ defined by the equation $\dot{F}(w) := \frac{dh}{dt}(t_0)$, where $w = e^{it_0}$. We let $\mathbf{A}_3$ denote the algebra of functions F on $\mathbf{T}$ that extend to analytic functions in an open annular region containing $\mathbf{T}$. A standard result shows that $\mathbf{A}_3$ coincides with the set of Fourier transforms of sequences that have *exponential decay*. We define a set of sequences that decay at a faster rate below:

Definition 2.1 *A sequence p has factorial decay if there exist positive numbers α, β, and γ such that*

$$|p(n)| \leq \alpha \beta^{|n|} (|n|!)^{-\gamma}, \quad n \in \mathbf{Z}. \qquad (2.2)$$

These sequences are closed under *convolution*, hence the set of functions defined by Fourier transforms of sequences with factorial decay forms an algebra. We let $\mathbf{A}_4$ denote this algebra. We define three functions $\lambda_1, \lambda_2,$ and λ_3 on $\mathbf{T}$ as follows: $\lambda_1(w) := w$, $\lambda_2(w) := -w$, and $\lambda_3(w) := w^2$.

Definition 2.2 *An algebra* **A** *is admissible if it contains* λ_1, *it is closed under complex conjugation, and for any function F on* **T**, *the following conditions are equivalent:* $F \in \mathbf{A}$, $F \circ \lambda_2 \in \mathbf{A}$, $F \circ \lambda_3 \in \mathbf{A}$.

We observe that the algebras $\mathbf{A}_1, \ldots, \mathbf{A}_5$ form a strictly decreasing sequence of algebras, and that the Weierstrass approximation theorem implies $\mathbf{A}_5$ is dense in $\mathbf{A}_1$. If B and C are functions on **T**, we let (B, C) denote the 2×2 matrix

$$(B, C) := \begin{bmatrix} B & C \\ -\overline{C} & \overline{B} \end{bmatrix}. \tag{2.3}$$

We call B and C the *members* of (B, C).

Definition 2.3 *A paraunitary matrix over an algebra* **A** *is a matrix having the form (B, C), whose members are in* **A** *and satisfy the following equation:*

$$|B|^2 + |C|^2 = 1. \tag{2.4}$$

We observe that if an algebra **A** of functions on **T** is closed under complex conjugation, then the set of paraunitary matrices over **A** forms an *infinite dimensional Lie group*. We call this group the *special unitary Lie group* over **A** and we denote it by $SU(2, \mathbf{A})$. The corresponding Lie algebra consists of matrices having the form (B, C) whose members are in **A** and satisfy the following equation

$$\overline{B} = -B. \tag{2.5}$$

We call this the *special unitary Lie algebra* over **A** and we denote it by $\mathbf{su}(2, \mathbf{A})$. We observe that an element in $SU(2, \mathbf{A})$ corresponds to a function that maps **T** into the three dimensional Lie group $SU(2, \mathbf{C})$, and an element in $\mathbf{su}(2, \mathbf{A})$ corresponds to a function that maps **T** into the three dimensional Lie algebra $\mathbf{su}(2, \mathbf{C})$. The following result describes a useful relationship between these Lie groups and Lie algebras:

Lemma 2.1 *Let $\widetilde{\mathbf{A}}$ be an algebra that contains an algebra* **A** *and that contains $\dot{B}$ for all $B \in \mathbf{A}$. Then the function Φ, defined by the following equation*

$$\Phi(g) := g^{-1}\dot{g}, \tag{2.6}$$

maps $SU(2, \mathbf{A})$ into $\mathbf{su}(2, \widetilde{\mathbf{A}})$, hence g is uniquely determined by $\Phi(g)$ and by the initial value of g at 1, as the solution of the following differential equation:

$$\dot{g} = g\,\Phi(g). \tag{2.7}$$

Proof If $g = (B, C)$ then $\Phi(g) = (X, Y)$ where $X = \overline{B}\dot{B} + \overline{C}\dot{C}$. Differentiating the expression in equation (2.4) yields $\overline{X} + X = 0$. □

We observe that if g_1 and g_2 are in $SU(2, \mathbf{A})$, then

$$\Phi(g_1 g_2) = g_2^{-1} \Phi(g_1) g_2 + \Phi(g_2), \quad g_1, g_2 \in SU(2, \mathbf{A}). \tag{2.8}$$

The following result parametrizes CQF's by paraunitary matrices:

Lemma 2.2 *Let $\mathbf{A}$ be an admissible algebra. Then a sequence p is in $CQF(\mathbf{A})$ iff its Fourier transform P is related to an element (B, C) in $SU(2, \mathbf{A})$ by the following equation:*

$$P(w) = \frac{1}{\sqrt{2}} B(w^2) + \frac{1}{\sqrt{2}} w C(w^2), \quad w \in \mathbf{T}. \tag{2.9}$$

Proof Let (B, C) be an element in $SU(2, \mathbf{A})$ and define P by equation (2.9). The fact that (B, C) satisfies equation (2.4) implies that P satisfies equation (1.1). Moreover, P is in $\mathbf{A}$ since B and C are in $\mathbf{A}$ and $\mathbf{A}$ is admissible. This shows that p is in $CQF(\mathbf{A})$. To show the converse, let p be in $CQF(\mathbf{A})$, let P denote the Fourier transform of p, and then construct functions B_1 and C_1 on $\mathbf{T}$ by the following equations:

$$B_1(w) := \frac{P(w) + P(-w)}{\sqrt{2}}, \quad C_1(w) := \frac{P(w) - P(-w)}{w\sqrt{2}}, \quad w \in \mathbf{T}. \tag{2.10}$$

We observe that B_1 and C_1 are in $\mathbf{A}$, since $\mathbf{A}$ is admissible, and that they satisfy the following equations:

$$B_1(-w) := B_1(w), \quad C_1(-w) := C_1(w), \quad w \in \mathbf{T}. \tag{2.11}$$

Therefore, there exist functions B and C on $\mathbf{T}$ that satisfy the following equations:

$$B_1(w) = B(w^2), \quad C_1(w) := C(w^2), \quad w \in \mathbf{T}. \tag{2.12}$$

Finally, we observe that B and C are in $\mathbf{A}$, since $\mathbf{A}$ is admissible, and that B and C satisfy equation (2.4), since P satisfies equation (1.1). □

3 Approximation by CQF's with Finite Support

Lemma 2.2 justifies identifying the topological space $CQF(\mathbf{A})$ by the topological group $SU(2, \mathbf{A})$. For $j = 2, \ldots, 5$ we let $\mathbf{G}_j$ denote the topological closure of the subgroup $SU(2, \mathbf{A}_j)$ in $SU(2, \mathbf{A}_1)$, and observe it is a closed subgroup. We observe that it suffices, to prove Theorem 1.1, to show that $\mathbf{G}_5 = SU(2, \mathbf{A}_1)$. We let $SU(2, \mathbf{A}_3)_+$ denote the subset of elements in $SU(2, \mathbf{A}_3)$ whose members have positive modulus.

Lemma 3.1 *The closure of* $SU(2, \mathbf{A}_3)_+$ *equals* $SU(2, \mathbf{A}_1)$.

Proof For any element $(B_1, C_1) \in SU(2, \mathbf{A}_1)$, the Weierstrass approximation theorem implies that there exists B_5, C_5 in $\mathbf{A}_5$ whose modulus is positive and that approximates B_1, C_1, respectively. Then $(B_5 D^{-1}, C_5 D^{-1})$, where $D := (|B_5|^2 + |C_5|^2)^{1/2}$, is in $SU(2, \mathbf{A}_3)_+$ and it approximates (B_1, C_1). □

We now complete the proof of Theorem 1.1 by constructing a subset S of $\mathbf{G}_5$ such that every element in $SU(2, \mathbf{A}_3)_+$ can be expressed as a product of elements in S. We let $L^2(\mathbf{T})$ denote the set of measurable functions on $\mathbf{T}$ that have square integrable modulus, and we let $H^2(\mathbf{T})$ denote the Hardy space of functions in $L^2(\mathbf{T})$ that are Fourier transforms of sequences supported on the nonnegative integers.

Definition 3.1 *An outer function is a function F in $H^2(\mathbf{T})$ defined by the following equation:*

$$F(w) := \lim_{r \uparrow 1} \exp\left[\frac{1}{2\pi} \int_0^{2\pi} \frac{e^{it} + rw}{e^{it} - rw} \log G(t)\, dt\right], \quad (3.1)$$

where G is a nonnegative real–valued function in $L^2(\mathbf{T})$ that satisfies the following inequality

$$\int_0^{2\pi} \log G(e^{it})\, dt > -\infty. \quad (3.2)$$

We observe that the exponent on the right side of equation (3.1) equals $\log G$ + the Hilbert transform of $\log G$. See ([9], page 193) and ([34], page 370) for a proof that the limit in equation (3.1) exists and defines a function in F in H^2 that satisfies the equation

$$|F| = G. \quad (3.3)$$

We observe that if G equals the square root of a nonnegative function P in $\mathbf{A}_5$, then the outer function F defined by equation (3.1) is also in $\mathbf{A}_5$. Moreover, F is the restriction to $\mathbf{T}$ of the unique polynomial that has a positive constant term, has all roots on or outside $\mathbf{T}$, and whose squared modulus restricted to $\mathbf{T}$ equals P. This polynomial is the *minimal phase spectral factor* of P given by the Riesz–Fejer theorem.

We construct a set S as the union of two subsets S_1 and S_2 of $SU(2, \mathbf{A}_3)_+$. The subset S_1 consists of elements whose members have the form $\lambda_1^m F$, where m is an integer, $\lambda_1(w) := w$, and F is an outer function. The subset S_2 consists of elements having the form $(B, 0)$, where necessarily $|B| = 1$. The next two results complete the proof of Theorem 1.1.

Lemma 3.2 $S \subset \mathbf{G}_5$.

Proof We first show that $S_1 \subset \mathbf{G}_5$. We observe that if the members of any element in $SU(2, \mathbf{A}_5)$ are multiplied by monomials having modulus one, then the resulting element is in $SU(2, \mathbf{A}_5)$. Therefore it suffices to construct, for any element $(B_4, C_4) \in S_1$ whose members are outer functions in $\mathbf{A}_3$ with positive moduli, a sequence $\{(P_j, Q_j)\} \in SU(2, \mathbf{A}_5)$ that converges uniformly to (B_4, C_4). For $j \geq 1$ we construct positive functions D_j, E_j in $\mathbf{A}_5$ by convolving $|B_4|^2$, $|C_4|^2$, respectively, with the Fejer kernel $K_j \in \mathbf{A}_1$ that has length $2j$. Then we construct the *minimal phase spectral factor* P_j, $Q_j \in \mathbf{A}_5$, of D_j, E_j, respectively. We observe that

$$|P_j|^2 + |Q_j|^2 = K_j * (|B_4|^2 + |C_4|^2) = K_j * 1 = 1,$$

where $*$ denotes convolution on $\mathbf{T}$. Therefore $(P_j, Q_j) \in SU(2, \mathbf{A}_5)$. Furthermore, we observe that P_j, Q_j is determined from $|P_j|$, $|Q_j|$, respectively, by equation (3.1), and also that B_4, C_4 is determined from $|B_4|$, $|C_4|$, respectively, by equation (3.1). We further observe that the sequence $|P_j|$, $|Q_j|$ consists of positive functions in $\mathbf{A}_3$ that converge uniformly to the positive function $|B_4|$, $|C_4|$, respectively. Therefore the Hilbert transform of $\log |P_j|$, $\log |Q_j|$, and therefore P_j, Q_j, converges uniformly to B_4, C_4, respectively. This shows that $S_1 \subset \mathbf{G}_5$. We now show that $S_2 \subset \mathbf{G}_5$. We let $(B_4, 0) \in S_2$ and we construct a sequence P_j in $\mathbf{A}_5$ such that $|P_j| \leq 1$ and P_j converges uniformly to B_4. Then we define $Q_j \in \mathbf{A}_5$ to be the minimal phase factor of $1 - |P_j|^2$. We observe that Q_j converges uniformly to 0, and that $|P_j|^2 + |Q_j|^2 = 1$, hence (P_j, Q_j) is a sequence of elements in $SU(2, \mathbf{A}_5)$ that converges uniformly to $(B_4, 0)$, hence $S_1 \subset \mathbf{G}_5$. □

Definition 3.2 *The winding number, denoted by $\omega(P)$, of an element P in $\mathbf{A}_2$ whose modulus is positive, is defined by the following equation:*

$$\omega(P) := \frac{1}{2\pi} \int_0^{2\pi} \frac{\dot P(e^{it})}{P(e^{it})} \, dt \tag{3.4}$$

We observe that the winding number is always an integer, that if P and Q are elements in $\mathbf{A}_2$ having positive moduli, then $\omega(PQ) = \omega(P) + \omega(Q)$, and that $\omega(P)$ is even iff there exists $R \in \mathbf{A}_2$ such that $P = R^2$.

Lemma 3.3 *Every element in $SU(2, \mathbf{A}_3)_+$ can be expressed in the form $h_1 g h_2^{-1}$ where $g \in S_1$ and $h_1, h_2 \in S_2$.*

Proof For any element $(B, C) \in SU(2, \mathbf{A}_3)_+$, Lemma 3.2 implies there exists an element $g = (B_4, C_4) \in S_1$ such that B_4, C_4, has the same modulus and the same winding number as B, C, respectively. We next observe that for any $h_1 = (P_1, 0)$ and $h_2 = (P_2, 0)$ in S_2,

$$h_1 g h_2^{-1} = (P_1 P_2 B_4, P_1 \overline{P_2} C_4).$$

It suffices to choose P_1 and P_2 to satisfy the following equations:

$$P_1 P_2 B_4 = B, \quad P_1 \overline{P_2} C_4 = C.$$

These equations are equivalent to the following equations:

$$P_1^2 = \frac{BC}{B_4 C_4}, \quad P_2^2 = \frac{B\overline{C}}{B_4 \overline{C}_4}.$$

Observe that the functions on the right sides of these equations have modulus one, and that their winding numbers are zero. Therefore, the square roots of these functions exist and yield P_1 and P_2. □

4 Differential Equation Method

In the remainder of this paper we present a second proof of Theorem 1.1. Our proof consists of the following result together with Theorem 6.1

Theorem 4.1 *The subset* $\Phi^{-1}(su(2, \mathbf{A}_5))$ *of* $SU(2, \mathbf{A}_1)$, *where*

$$\Phi : SU(2, \mathbf{A}_2) \to su(2, \mathbf{A}_1)$$

is defined in Lemma 2.1 by choosing $\mathbf{A} = \mathbf{A}_2$ *and* $\widetilde{\mathbf{A}} = \mathbf{A}_1$, *is a subset of* $SU(2, \mathbf{A}_4)$ *and its closure equals* $SU(2, \mathbf{A}_1)$.

The proof follows directly from the next two lemmas.

Lemma 4.1 *If* $g \in SU(2, \mathbf{A}_2)$ *and* $\Phi(g) \in su(2, \mathbf{A}_5)$, *then* $g \in SU(2, \mathbf{A}_4)$.

Proof We observe that g and $\Phi(g)$ can be expanded in Fourier series to obtain

$$g(w) = \sum_{n \in \mathbf{Z}} b(n) w^n,$$

and

$$\Phi(g)(w) = \sum_{k=-L}^{L} c(k) w^k,$$

where L is a positive integer, and where b and c are sequences having values in the set of 2×2 matrices over $\mathbf{C}$. Then equation (2.7) implies

$$b(n) = \frac{-i}{n} \sum_{k=-L}^{L} b(n-k) c(k), \, n \in \mathbf{Z}, \, n \neq 0. \tag{4.1}$$

We define a function M on the positive integers by the equation

$$M(n) := \max\{ |b(k)| : |k| \geq n \}, \, n \in Z, \, n \geq 1. \tag{4.2}$$

Equation (4.1) implies that M satisfies

$$M(n) \leq \frac{1}{n} \left(\sum_{k=-L}^{L} |c(k)| \right) M(n-L), \, n \geq L.$$

We let p be the largest integer such that $pL \leq n$, and obtain

$$M(n) \leq M(n - pL) \left(\sum_{k=-L}^{L} |c(k)| \right)^p \frac{1}{n(n-L)\cdots(n-pL)}. \tag{4.3}$$

We define the following constants:

$$\alpha := M(n - pL), \quad \gamma := \frac{1}{L}, \quad \beta := \left(\sum_{k=-L}^{L} |c(k)| \right)^\gamma,$$

and substitute into inequality (4.3) to obtain inequality (2.2). □

We recall the standard fact that the *exponential mapping*

$$\exp : su(2, \mathbf{C}) \to SU(2, \mathbf{C}) \tag{4.4}$$

is a local homeomorphism ([1], page 23), and furthermore, that it is surjective since $SU(2, \mathbf{C})$ is compact ([1], page 165).

Lemma 4.2 *The closure of $\Phi^{-1}(su(2, \mathbf{A}_5))$ equals $SU(2, \mathbf{A}_1)$.*

Proof It suffices to approximate an element g_2 in $SU(2, \mathbf{A}_2)$ by an element g in $SU(2, \mathbf{A}_2)$ that satisfies $\Phi(g) \in su(2, \mathbf{A}_5)$. We first construct the function $h_2 : \mathbf{R} \to SU(2, \mathbf{C})$ by the equation $h_2(t) := g_2(e^{it})$. For any pair (a, Δ), where $a \in su(2, \mathbf{A}_1)$ and $\Delta \in su(2, \mathbf{C})$, we observe that $a + \Delta \in su(2, \mathbf{A}_1)$ and we define the function

$$h_{a,\Delta} : \mathbf{R} \to SU(2, \mathbf{C})$$

as the unique solution of the differential equation

$$\dot{h}_{a,\Delta} = h_{a,\Delta}\,(a + \Delta),$$

that satisfies $h_{a,\Delta}(0) = h_2(0)$, and we define the function

$$\Psi_a : su(2, \mathbf{C}) \to SU(2, \mathbf{C})$$

as follows:

$$\Psi_a(\Delta) := h_{a,\Delta}(2\pi) \tag{4.5}$$

We observe that

$$h_{\Phi(g_2),(0,0)}(2\pi) = h_2(2\pi) = g_2(1),$$

and consequently that

$$\Psi_{\Phi(g_2)}((0,0)) = h_2(2\pi).$$

Furthermore, we employ a transversality argument to show that there exists an open neighborhood U of $(0,0)$ such that $\Psi_{\Phi(g_2)}$ maps U onto a neighboorhood of $g_2(1)$ in $SU(2,\mathbf{C})$. Then we invoke the Weierstrass approximation theorem to choose an element $a_5 \in su(2,\mathbf{A}_5)$ that approximates $\Phi(g_2)$ with sufficient accuracy to ensure that Ψ_{a_5} maps U onto a neighborhood of $g_2(1)$, and we choose a point $\Delta_5 \in U$ such that

$$\Psi_{a_5}(\Delta_5) = g_2(1).$$

We define the function $g : \mathbf{T} \to SU(2,\mathbf{C})$ by the equation

$$g(e^{it}) := h_{a_5,\Delta_5}(t), \ t \in [0, 2\pi),$$

and we observe that $g \in SU(2,\mathbf{A}_2)$ since $h(2\pi) = h(0)$, and that g approximates g_2 since h_{a_5,Δ_5} approximates h. Clearly $\Phi(g) = a_5 + \Delta_5 \in su(2,\mathbf{A}_5)$ and the proof is complete. □

We observe that if $\Phi(g_2)$ is constant, then for any $\Delta \in su(2,\mathbf{C})$,

$$h_{\Phi(g_2),\Delta}(t) = h_2(0) \exp(t(a + \Delta)),$$

and the fact that $\Psi_{\Phi(g_2)}$ maps some open neighborhood U of $(0,0)$ in $su(2,\mathbf{C})$ onto a neighborhood of $g_2(1)$ in $SU(2,\mathbf{C})$ can be derived using properties of the exponential mapping and the local expansions used in the derivation of the Baker–Campbell-Hausdorff formula [40]. We also observe that a_5 can be computed from a using standard methods for approximation by trigonometric polynomials, while Δ_5 can be computed using the *shooting method* [3]. The parameters in equation (2.2) that describe the factorial decay of the Fourier coefficients of g can be obtained, using Theorem 4.1, from $a_5 + \Delta_5$.

5 Roots of Polynomials

We derive some properties about roots of polynomials required in the next section. For any positive integer n, and number $r \geq 1$, define

$$f(r) := \max\{1, r\} - 1,$$

$$h(n,r) := \sqrt{1 + (1+r)^2 + \cdots + (1 + r + \cdots + r^{n-1})^2}.$$

Theorem 5.1 *Let ρ and ϵ be positive numbers, let P be a monic polynomial having degree n all of whose roots have modulus $\leq \rho$, and let Q be a monic polynomial having degree n that approximates P on $\mathbf{T}$ to satisfy the inequality*

$$||P - Q||_\mathbf{T} < \frac{\left(\frac{\epsilon}{2n}\right)^n}{1 + f(\rho + \frac{\epsilon}{2n})h(n, \rho + \frac{\epsilon}{2n})}. \tag{5.1}$$

Then the roots of P and Q are ϵ close.

Proof We define $\Delta := \frac{\epsilon}{2n}$, and we let λ_i, $i = 1, \ldots, n$ denote the roots of P. For $i = 1, \ldots, n$ we let D_i denote the closed disk having radius Δ and center λ_i, and construct the closed subset D of $\mathbf{C}$ by

$$D := D_1 \cup \cdots \cup D_n.$$

Then we use the fact that every topological space can be decomposed into pairwise disjoint connected components to express D as follows:

$$D = E_1 \cup \cdots \cup E_m,$$

where $m \leq n$, and for $j = 1, \ldots, m$, the set E_j is connected and disjoint from E_k whenever $k \neq j$. For $j = 1, \ldots, m$ we let Γ_j denote the boundary of E_j and define

$$\Gamma = \Gamma_1 \cup \cdots \cup \Gamma_m.$$

We observe that each root of P has distance $\geq \Delta$ from Γ, therefore

$$|P(z)| \geq \Delta^n, \ z \in \Gamma.$$

Furthermore, we observe that

$$P(z) - Q(z) = \frac{1}{2\pi} \int_0^{2\pi} \frac{(e^{-i\theta}z)^{n+1} - 1}{e^{-i\theta}z - 1} [P(e^{i\theta}) - Q(e^{i\theta})] \, d\theta, \ z \in \mathbf{C},$$

therefore

$$|P(z) - Q(z)| \leq \|P - Q\|_\mathbf{T} \left[1 + f(\rho + \Delta) h(n, \rho + \Delta)\right], \ z \in \Gamma.$$

Hence if $\|P - Q\|_\mathbf{T}$ satisfies inequality (5.1), then $|P - Q| < |P|$ on Γ and Rouche's theorem ([34], page 242) implies Q must have the same number of zeros as P inside each region E_j, $j = 1, \ldots, m$. The proof is complete since the diameter of each region E_j does not exceed $2n\Delta = \epsilon$. □

Theorem 5.2 *Let $R > 0$ and let n be a positive integer. If $\mu_1, \ldots, \mu_n$ are nonzero complex numbers that satisfy the inequalities*

$$|T_j| \leq R^j, \ j = 1, \ldots, n,$$

where

$$T_j := \sum_{i=1}^n \mu_i^j, \ j = 1, \ldots, n,$$

then their moduli $|\mu_1|, \ldots, |\mu_n|$ satisfy the inequalities

$$|\mu_j| < R\left(1 + e + \sqrt{e(1+e)}\right) \approx 6.8975\, R.$$

Proof We use the numbers $\mu_1, \ldots, \mu_n$ to construct the polynomial

$$P(z) := \prod_{i=1}^{n}(1 - \mu_i z), \quad z \in \mathbf{C},$$

and the entire function

$$F(z) := \exp\left(-\sum_{k=1}^{n} \frac{1}{k} T_k z^k\right), \quad z \in \mathbf{C}.$$

Since the zeros of P are $\mu_1^{-1}, \ldots, \mu_n^{-1}$, it suffices to show P has no zeros on the closed disk of radius sR^{-1} where s is the constant defined by

$$s := (1 + e + \sqrt{e(1+e)})^{-1}.$$

We observe that the first $n+1$ terms of the Taylor expansions of F and P at 0 coincide. This fact implies that

$$P(z) = 1 + p(1)z + \cdots + p(n)z^n,$$

where for $k = 2, \ldots, n$,

$$p(k) = \frac{1}{2\pi i} \int_\Gamma z^{-k-1} F(z) dz,$$

and Γ is any positively oriented simple contour whose interior contains 0. For any $t \in (0, 1)$ the inequalities for $\mu_1, \ldots, \mu_n$ imply

$$|F(z)| \leq \frac{1}{1-t}, \quad |z| \leq tR^{-1},$$

and therefore, choosing Γ to be a circular contour having radius tR^{-1} and minimizing over $t \in (0, 1)$ yields

$$|p(k)| \leq \min_{t \in (0,1)} \frac{R^k}{(1-t)t^k} = R^k (1+k)\left(1+\frac{1}{k}\right)^k.$$

Since $s \approx 0.1450 < 1$, we complete the proof by observing that the following inequality

$$|P(z) - 1| \leq \sum_{k=1}^{\infty} (1+k)\left(1+\frac{1}{k}\right)^k s^k < \frac{es(2-s)}{(1-s)^2} = 1, \quad |z| \leq sR^{-1},$$

implies P has no zeros in the disk of radius sR^{-1}. □

6 Phase Retrieval Approximation Method

In Section 4 we described a method to approximate a CQF having a continuous Fourier transform by a CQF having factorial decay. We propose and discuss a method, based on *phase retrieval* [12], [21, 22], for approximating a CQF having factorial decay by a CQF having finite support. We recall that a CQF having factorial decay can be represented by an element (B, C) in $SU(2, \mathbf{A}_4)$. Our method approximates (B, C) by a sequence of elements $(B_N, C_N) \in SU(2, \mathbf{A}_5)$ as follows:

1. for each integer N, truncate B, C to obtain Laurent polynomials E_N, F_N, respectively, whose coefficient sequences are supported on the set $\{-N, \ldots, N\}$,

2. compute
$$B_N = \frac{E_N}{\max\{1, \|E_N\|_{\mathbf{T}}\}},$$

3. compute C_N to be the spectral factor, of the nonnegative Laurent polynomial $1 - |B_N|^2$, that minimizes $\|C_N - F_N\|_{\mathbf{T}}$,

We observe that $(B_N, C_N) \in SU(2, \mathbf{A}_5)$. The following result justifies our proposed method:

Theorem 6.1 *If $(B, C) \in SU(2, \mathbf{A}_4)$, then the sequence (B_N, C_N) constructed above converges to (B, C).*

Proof We observe that for all N, the matrix (B_N, C_N) is in $SU(2, \mathbf{A}_5)$, and that the sequence B_N converges to B. For all N, let H_N denote the minimal phase spectral factor of $|F_N|^2$ and let G_N denote the minimal phase spectral factor of $1 - |B_N|^2$. Then we can factor H_N, F_N and G_N as follows:
$$H_N(z) = a_N\, z^N (z^{-1} - \lambda_1) \cdots (z^{-1} - \lambda_{2N}),$$
$$F_N(z) = b_N\, z^{p-N} (z^{-1} - \lambda_1) \cdots (z^{-1} - \lambda_p)(z - \overline{\lambda_{p+1}}) \cdots (z - \overline{\lambda_{2N}}),$$
$$G_N(z) = c_N\, z^{-N}(z^{-1} - \mu_1) \cdots (z^{-1} - \mu_{2N}),$$
where λ_n^{-1}, $n = 1, \ldots, 2N$ are the roots of H_N, where λ_n^{-1}, $n = 1, \ldots, p$ and $\overline{\lambda_n}$, $n = p+1, \ldots, N$ are the roots of F_N, where μ_n^{-1}, $n = 1, \ldots, 2N$ are the roots of G_N, and where a_N, b_N, c_N are the geometric means of H_N, F_N, G_N, respectively (note that the roots depend on N). We observe that $a_N = b_N$ and that the sequences $\|F_N - C\|_{\mathbf{T}}$, $\|H_N - G_N\|_{\mathbf{T}}$, and $|a_N - c_N|$ have factorial decay. For each N we apply Theorem 5.1 to the polynomials P and Q defined by
$$P(z) := z^N H_N(z^{-1})$$
and
$$Q(z) := z^N H_N(z^{-1}),$$

to obtain upper bounds on the distance between the sets $\{\lambda_n\}$ and $\{\mu_n\}$, and conclude these upper bounds converge to zeros very rapidly as N increases. This implies that $||D_N - F_N||_\mathbf{T}$ converges to zeor where D_N is the Laurent polynomial defined below:

$$D_N(z) := c_N\, z^{p-N}(z^{-1} - \mu_1)\ldots(z^{-1} - \mu_p)(z - \overline{\mu_{p+1}})\ldots(z - \overline{\mu_{2N}}).$$

Clearly, D_N is a spectral factor of $1-|B_N|^2$, and since $||F_N - C||_\mathbf{T}$ converges to zero, we observe that

$$||C_N - C||_\mathbf{T} \leq ||C_N - F_N||_\mathbf{T} + ||F_N - C||_\mathbf{T} \leq ||D_N - F_N||_\mathbf{T} + ||F_N - C||_\mathbf{T},$$

therefore $||C_N - C||_\mathbf{T}$ converges to zero. □

We observe that Theorem 5.2 can be used to derive a lower bound on the required length of a CQF to approximate a CQF having the form $(B, 0)$, in terms of the phase of B.

7 The Development of Conjugate Quadrature Filters

Vaidyanathan [38, 39] describes the origin of the concepts discussed in Section 2 from their analog versions in the early work of Brune [2] and Darlington [4] to the construction of finitely supported CQF's for perfect reconstruction paraunitary filter banks by Mintzer [30], Smith and Barnwell [35, 36], and Vetterli [41]. Parks and McClellan [31] developed a method, based on Chebyshev approximation, that can be used to construct finitely supported CQF's whose modulus approximates the modulus of a specified CQF. Vaidyanathan and Hoang [37] described an efficient *lattice* filterbank structure that is equivalent to factoring elements in $SU(2, \mathbf{A}_5)$ into products of factors $[z, 0]$ and $[\beta, \gamma]$, $\beta, \gamma \in \mathbf{C}$. Daubechies [7, 8] used certain finitely supported CQF's, obtained as minimal phase spectral factors of filters constructed by Herrmann [11], to construct a family of orthonormal bases consisting of compactly supported wavelets that extend the Haar basis [10]. She showed these wavelets were continuous and that their regularity increased with the filter length. The author [15] showed that every finitely supported CQF p whose Fourier transform satisfies $P(1) = 1$ yields a family of wavelets that form a *tight frame* for $L^2(\mathbf{R})$. Cohen [5, 6] and the author [16, 17] independently developed necessary and sufficient conditions for the tight frame to be an orthonormal basis. Cohen's conditions are expressed in terms of the zeros of Fourier transform of the CQF while the author's are expressed in terms of the eigenvalues of a *transition operator* constructed from the CQF. The author, Lee and Shen extended these and related results to multivariate CQF's [25, 26]. The author [18] also showed that there exist non-minimal phase spectral factors, of Herrmann's filters, that have lengths $4n + 2$, $n \geq 2$ and are complex valued and symmetric, and he demonstrated their utility for handling boundaries in subband signal coding. Pollen [32] showed that the set of finite CQF's that yield tight frames form a group under a modified *twisted* product structure on $SU(2, \mathbf{A}_5)$, and he developed a corresponding factorization theorem [33]. The author and Pollen [13] observed that the *metric* on $SU(2, \mathbf{A}_5)$ defined by $||(B, C)||^2 := ||B||_2^2 + ||C||_2^2$ is translation invariant. Micchelli [28, 29] showed that given any n negative numbers, there exists a CQF p having length $2n - 1$ with real coefficients whose Fourier transform P (considered as a Laurent polynomial) vanishes at these prescribed negative numbers. He also applied this result to interpolatory subdivision by sums of exponentials and to wavelet construction in Sobolev spaces. The author and Micchelli [23, 24, 19, 20] extended this result by showing that for any finite subset $\Lambda \subset \mathbf{C} \setminus \{0\}$, there exists a finitely supported CQF p such that set of zeros of the Laurent polynomial P contains Λ iff

$$(\Lambda \cup \overline{\Lambda}^{-1}) \cap -(\Lambda \cup \overline{\Lambda}^{-1}) = \phi. \tag{7.1}$$

They also discussed applications to the analysis of vibration signals and extended univariate results to multivariate polynomials.

Acknowledgment Research supported in part by the NUS Wavelets Program funded by the National Science and Technology Board and the Ministry of Education, Republic of Singapore.

8 References

[1] T. Bröcker and T. Dieck, Representations of Compact Lie Groups, Springer, New York, 1985.

[2] O. Brune, *Synthesis of a finite two terminal network whose driving point impedance is a prescribed function of frequency*, J. Mathematics and Physics, Volume 10, pages 191-235, 1931.

[3] G. Dahlquist and A. Björck, Numerical Methods, Prentice–Hall, Englewood Cliffs, New Jersey, 1974.

[4] S. Darlington, *Synthesis of reactance four–poles*, J. Mathematical Physics, Volume 18, pages 257–353, September 1939.

[5] A. Cohen, *Ondelettes, analyses multirésolutions et filtres miroir en quadrature*, Ann. Inst. H. Poincaré, Anal. nonlinéaire, Volume 7, pages 439-459, 1990.

[6] A. Cohen, *Ondelettes, analyses multirésolutions et traitement numérique du signal*, PhD. Thesis, Université Paris, Dauphine, 1990.

[7] I. Daubechies, *Orthonormal bases of compactly supported wavelets*, Communications on Pure and Applied Mathematics, Volume 41, pages 909-996, 1988

[8] I. Daubechies, Ten Lectures on Wavelets, SIAM, Philadelphia, 1992.

[9] H. Dym and H. P. McKean, Fourier Series and Integrals, Academic Press, New York, 1972.

[10] A. Haar, *Zur Theorie der orthogonalen Funktionen-Systeme*, Mathematsche Annallen, Volume 69, pages 331-371, 1910.

[11] O. Herrmann, *On the approximation problem in nonrecursive digital filter design*, IEEE Transactions on Circuit Theory, CT-18, pages 411-413, 1971.

[12] N. Hurt, Phase Retrieval and Zero Crossings: Mathematical Methods in Image Reconstruction, Kluwer Academic Publishers, Boston, 1989.

[13] W. Lawton and D. Pollen, *Group structures and invariant metrics for quadrature mirror filters and their Aware angular parameterization*, Aware, Inc. Technical Report, Cambridge, Massachussetts, 1988.

[14] W. Lawton, *Approximating wavelet conjugate quadrature filters using spectral factorization and lattice decomposition*, Aware, Inc. Technical Report, Cambridge, Massachussetts, 1991.

[15] W. Lawton, *Tight frames of compactly supported affine wavelets*, J. Mathematical Physics, Volume 31, Number 8, pages 1898-1901, August 1990.

[16] W. Lawton, *Necessary and sufficient conditions for constructing orthonormal wavelet bases*, J. Mathematical Physics, Volume 32, Number 1, pages 57-61, January 1991.

[17] W. Lawton, *Multilevel properties of the wavelet-Galerkin operator*, J. Mathematical Physics, Volume 32, pages 1440–1443, 1991.

[18] W. Lawton, *Application of complex-valued wavelet transforms to subband decomposition*, IEEE Transactions on Signal Processing, Volume 41, Number 12, pages 3566-3568, December 1993.

[19] W. Lawton, *Rational wavelet design for molecular vibration analysis*, address to the Workshop on Wavelets and their Applications, Chinese University of Hong Kong, Hong Kong, May 5-8, 1997.

[20] W. Lawton, *Bezout's identity with inequality constraints*, address to the Guangzhou International Symposium on Computational Mathematics, Guangzhou, China, August 11-15, 1997.

[21] W. Lawton, *Uniqueness results for the phase retrieval problem for radial functions*, Journal of the Optical Society of America, Vol. 71, Number 12, pages 1519-1522, December 1981.

[22] W. Lawton, *Mathematical results for the phase retrieval problem for bandlimited functions of several variables*, Proceedings of the Topical Meeting of the Optical Society of America on Signal Recovery and Synthesis with Incomplete Information and Partial Constraints, Incline Village, Nevada, January 12-14, 1983.

[23] W. Lawton and C. A. Micchelli, *Construction of conjugate quadrature filters with specified zeros*, Numerical Algorithms, Volume 14, Number 4, pages 383-399, 1997.

[24] W. Lawton and C. A. Micchelli, *Design of conjugate quadrature filters having specified zeros*, Proceedings of ICASSP97, held at Munich, Germany, April 21-24, 1997.

[25] W. Lawton, S. L. Lee and Z. Shen, *An algorithm for matrix extension and wavelet construction*, Mathematics of Computation, Volume 65, pages 723–737, 1996.

[26] W. Lawton, S. L. Lee and Z. Shen, *Stability and orthonormality of multivariate refinable functions*, SIAM J. Mathematical Analysis, Volume 28, pages 999–1014, 1997.

[27] W. Lawton, S. L. Lee and Z. Shen, *Convergence of multidimensional cascade algorithm*, to appear in Numerische Mathematik.

[28] C. A. Micchelli, *Interpolatory subdivision schemes and wavelets*, J. Approximation Theory, Volume 86, pages 41-71, 1996.

[29] C. A. Micchelli, *On a family of filters arising in wavelet construction*, Applied and Computational Harmonic Analysis, Volume 4, pages 38-50, 1997.

[30] F. Mintzer, *Filters for distortion–free two–band multirate filter banks*, IEEE Transactions on Acoustics, Speech, and Signal Processing, Volume ASSP-33, Number 3, pages 626-630, June 1985.

[31] T. W. Parks and J. H. McClellan, *Chebyshev approximation for nonrecursive digital filters with linear phase*, IEEE Transactions on Circuit Theory, Volume CT-19, pages 189-194, March 1972.

[32] D. Pollen, *The unique factorization for the topological group of co-efficient vectors for one-dimensional, multiplier-two scaling function, wavelet systems*, Aware, Inc. Technical Report, Cambridge, Massachussetts, July 1988.

[33] D. Pollen, $SU_I(2,F[z,1/z])$ *for F a subfield of C*, J. American Mathematical Society, Volume 3, page 611, 1990.

[34] W. Rudin, Real and Complex Analysis, McGraw–Hill, New York, 1966.

[35] M. J. Smith and T. P. Barnwell, *A procedure for designing exact reconstruction filter banks for tree structured sub-band coders*, Proceedings of IEEE International Conference on Acoustics, Speech and Signal Processing, San Diego, March 1986.

[36] M. J. Smith and T. P. Barnwell, *Exact reconstruction techniques for tree structured subbandcoders*, IEEE Transactions on Acoustics, Speech, and Signal Processing, Volume ASSP-34, pages 434-441, 1986.

[37] P. P. Vaidyanathan and P. Q. Hoang, *Lattice structures for optimal design and robust implementation of two-channel perfect reconstruction QMF banks*, IEEE Transactions on Acoustics, Speech, and Signal Processing, Volume ASSP-36, pages 81-94, January 1988.

[38] P. P. Vaidyanathan, *Multirate digital filters, filterbanks, polyphase networks, and applications: a tutorial*, Proceedings IEEE, Volume 78, 1990.

[39] P. P. Vaidyanathan, *Multirate Systems and Filterbanks*, Prentice-Hall, Englewood Cliffs, New Jersey, 1993.

[40] V. S. Varadarajan, Lie Groups, Lie Algebras, and Their Representations, Springer, New York, 1984.

[41] M. Vetterli, *Splitting a signal into subsampled channels allowing perfect reconstruction*, Proceedings IASTED Conference on Applications of Signal Processing and Digital Filtering, Paris, France, June 1985.

6
Polynomial Reproduction by Refinable Functions

Carlos A. Cabrelli [1]   Christopher Heil [2] and Ursula M. Molter[3]

> ABSTRACT In this paper we give an expository review of the problem of reproducing polynomials from integer translates of a compactly supported refinable function vector. This property, which is called the accuracy of the function vector, is important for the construction of wavelet bases and is related to order of approximation, smoothness and other properties.

1 Introduction

In this paper, we will study the question of when a compactly supported function can exactly reproduce polynomials as linear combinations of its integer translates. We will show that to each compactly supported function f we can associate a maximum non-negative integer n such that f reproduces all the polynomials of degree less than n. This number n is the *accuracy* of f.

Accuracy has played an important role in both approximation theory and in wavelet theory. In approximation theory, it is closely related to the approximation properties of shift invariant spaces, often generated by splines or finite elements. In wavelet theory, one of the most successful and systematic ways of constructing smooth, compactly supported, orthonormal wavelet bases for $\mathcal{L}^2(\mathbf{R})$ is based on the factorization of a 2π-periodic *symbol* which determines a *scaling function* [Dau92]. This factorization of

---

[1]Departamento de Matemática, Facultad de Ciencias Exactas y Naturales, Universidad de Buenos Aires, Ciudad Universitaria,Pabellón I, 1428 Buenos Aires, Argentina (e-mail: ccabrell@dm.uba.ar)
This research was supported in part by Grants EX048 (UBA) and PIA 646/96 (CONICET)

[2]School of Mathematics, Georgia Institute of Technology, Atlanta, Georgia, 30332-0160 (e-mail: heil@math.gatech.edu)
This research was supported in part by NSF Grant DMS-9401340.

[3]same address as the first author (e-mail: umolter@dm.uba.ar)

the symbol is closely related to the accuracy of the scaling function. Each scaling function satisfies a *dilation equation* of the type

$$f(x) = \sum_k c_k\, f(2x - k). \tag{1}$$

If the scaling function has accuracy n, then the corresponding wavelet will have n zero moments. This implies that the wavelet transform of the smooth part of a signal will yield only small coefficients, which leads to good compression ratios in applications involving signal compression. It is also the key to characterizing spaces of smooth functions via wavelet transforms.

Equation (1) also plays a key role in the context of subdivision schemes [CDM91], where it is known as a *refinement equation*, and the solution f is a *refinable function*. Accuracy is necessary for a refinable function to be smooth, although it is not sufficient.

Generalizations of the refinement equation (1) allow functions with domain $\mathbf{R}^d$ and a general dilation matrix A in place of the dilation factor 2, or allow multiple functions $f_1,\ldots,f_r$ to each be written as linear combinations of translated and dilated versions of all of the f_i. Refinable functions or their generalizations to higher dimensions or multiple functions may also be viewed as a particular case of self-similar functions as studied in fractal geometry, e.g., [Baj57], [Dub85], [Bar86], [CM98].

The problem of determining when a solution to equation (1) exists and of determining the smoothness properties of the resulting solution has been studied by many researchers. Usually, the goal is to characterize these properties in terms of conditions on the coefficients c_k. A short and incomplete list of references, might include [CH94], [Dau88], [DL91], [DL92], [Dub86], [DGL91], [Eir92], [LW95], [MP89], [Rio92], [Vil92], [Wan95], and others.

Our objective in this paper is to give an accessible and self-contained account of the characterization of the accuracy of refinable functions in a way that introduces techniques used in the study of generalized forms of the refinement equation. We do not intend to give a complete survey of all results on accuracy; rather, we focus on that part of accuracy that is most relevant to wavelet theory. The history of this problem is long and convoluted; we will mention only a few papers that bear directly on our discussion, and provide in the references some of the additional papers that are related to this problem.

The classical study of the order of approximation by integer translates of a single function begins with the *Strang–Fix conditions*, first derived in [Sch46] and [SF73]. For a refinable function satisfying some extra hypotheses, the Strang–Fix conditions reduce to a finite set of equations on the coefficients c_k, called the *sum rules*. In approximation theory, it is important to study order of approximation assuming only very weak extra conditions on f; we refer to [dBVR94a] and related references for this type of approach. In wavelet theory, it is natural to assume much stronger conditions on f, such as compact support, and linear independence or even orthogonality of the translates $\{f(x+k)\}$. We therefore restrict our attention in this paper to the case of compactly supported functions with linearly independent translates. These hypotheses also allow us in some cases to present sharper results or simpler proofs, and allow us to avoid convergence problems. However, many of the results are still valid without the compactness constraint if we impose instead a certain decay of the refinable function.

One generalization of the refinement equation is to allow multiple functions. In wavelet theory, this leads to *multiwavelet bases*, which can simultaneously combine several desirable properties that cannot be simultaneously realized by classical wavelet bases, such as symmetry and compact support [DGHM96]. Plonka [Plo97] and Heil, Strang, and Strela [SS94], [HSS96], independently characterized the conditions for accuracy in the one-dimensional, multi-function setting, with later insights by Jia, Riemenschneider, and Zhou [JRZ96].

In one dimension with a single function, the sum rules are equivalent to a factorization of the 2π-periodic symbol defined by $m_0(\omega) = \sum_{k=0}^{N} c_k e^{-i\omega k}$. However, such factorizations need not exist in higher dimensions. Instead, in the construction of smooth wavelets or multiwavelets in higher dimensions, it may be preferable to start directly from the sum rules to construct an appropriate choice of coefficients $\{c_k\}$ which lead to a scaling function which is both sufficiently smooth and is orthogonal to its lattice translates. For example, Belogay and Wang [BW97] used such an approach to induce a partial factorization the symbol resulting from certain choices of coefficients, thereby leading to the construction of two-dimensional, non-separable, orthogonal wavelets using a special dilation matrix of determinant 2. On the other hand, many time-domain conditions for accuracy do carry over to the higher-dimensional, single-function case, although with much greater technical difficulties. The papers [CHM96], [CHM97] extend the characterization of accuracy to the case of higher dimensions combined with multiple functions. This work was applied to construct non-separable multiwavelets in $\mathbf{R}^2$ [CHM98].

For clarity, we have chosen to describe in complete detail only the simplest case of one function in one variable. This appears in Section 2. Many of the statements of the results translate easily to the generalized cases, but the proofs are usually much more technical. We therefore have often presented proofs of the one-dimensional, single-function results which illustrate the ideas behind the proofs of the more general cases without the obscuring technical details. However, since some results are not valid in the general case, or need extra hypothesis, or are simply unknown, we summarize the general cases in Sections 3 and 4, stating the results but omitting the proofs, referring instead to [CHM96], [CHM97] for complete details. Specifically, in Section 3 we discuss some of the difficulties that appear when considering an arbitrary dilation matrix in higher dimensions, and in Section 4 we present the statements of these results in the general context of higher dimensions and multiple functions. This setting of course includes most of the previous results as particular cases. Finally, in Section 5 we briefly discuss some of the relationships that appear between accuracy and order of approximation or smoothness.

We emphasize that, because of the philosophy of the paper, we have often combined into one statement results that summarize the work of many independent authors. Moreover, many of the ideas presented in this paper have arisen in multiple contexts and have several independent proofs by numerous authors. Thus it is nearly impossible to give complete individual credit for the results. Instead, we have attempted to give a simple, self-contained introduction to the problem of accuracy. We present our own personal viewpoint and our own personal approaches to the proofs, but include in our bibliography numerous references that include related results.

2 One Dimension with a Single Function

Throughout this section we will only consider functions that are defined on the real line $\mathbf{R}$ and take values in the complex plane $\mathbf{C}$. Sequences and series with unspecified limits are understood to be indexed by the set of integers $\mathbf{Z}$.

Definition 1 Let f be a compactly supported function. Then a function g is *reproducible by integer translates of f* (or simply *reproducible by f*) if there exist complex scalars $\{\alpha_k\}_{k \in \mathbf{Z}}$ such that

$$g(x) = \sum_k \alpha_k f(x+k). \tag{2}$$

If we define the *shift-invariant space generated by f* to be

$$S(f) = \left\{ g: \mathbf{R} \to \mathbf{C} \ : \ \exists \alpha_k \in \mathbf{C} \text{ such that } g(x) = \sum_k \alpha_k f(x+k) \right\},$$

then g is reproducible by f if and only if $g \in S(f)$.

We say that translates of f are *linearly independent* if for every choice of scalars $\alpha_k \in \mathbf{C}$ we have

$$\sum_k \alpha_k f(x+k) = 0 \quad \Longleftrightarrow \quad \alpha_k = 0 \text{ for every } k.$$

Throughout this paper, we will usually work with functions which satisfy the following "standard hypotheses":

a) f is compactly supported, and
b) translates of f are linearly independent. (3)

We remark that if $f \in \mathcal{L}^2(\mathbf{R})$ is compactly supported and has independent translates, then the translates of f are a Riesz basis for the subspace of $\mathcal{L}^2(\mathbf{R})$ that they span [JW93], [Jia95]. As a consequence, there exist constants $A, B > 0$ such that:

$$\forall \omega \in \mathbf{R}, \quad A \leq \sum_k |\hat{f}(\omega + 2k\pi)|^2 \leq B, \tag{4}$$

where $\hat{f}(\omega) = \int f(x) e^{-ix\omega} dx$ is the Fourier transform of f.

2.1 Accuracy

In this section we study the question of the reproducibility of polynomials from integer translates of a general compactly supported (but not necessarily refinable) function f. That is, given f, we seek to determine what conditions must be imposed on f in order that all polynomials up to a given degree can be exactly reproduced from integer translates of f.

Let Π_n be the space of all polynomials of degree less or equal than n, i.e.,

$$\Pi_n = \left\{ p(x) = \sum_{k=0}^n a_k x^k : a_k \in \mathbf{C} \right\}.$$

We show in Proposition 1 below that if a polynomial p is reproducible by f, then every polynomial of smaller degree is also reproducible by f. Therefore, a natural question is whether there is a maximum number n such that $\Pi_n \subset \mathcal{S}(f)$, or whether there exists an f that reproduces every polynomial. We show in Proposition 4 that for compactly supported functions there is always a maximum n, and that this n depends on the diameter of the support of f.

These remarks motivate the following definition.

Definition 2 Let f be a compactly supported function. Then f has *accuracy* n if n is the maximum integer such that $\Pi_{n-1} \subset \mathcal{S}(f)$.

We will also discuss an interesting property related to the scalars used to reproduce a given polynomial p from translates of f. Suppose that f satisfies (3). If a polynomial p can be written as $p(x) = \sum \alpha_k f(x+k)$, then we show in Proposition 2 that $\alpha_k = u_p(k)$ where u_p is a polynomial of the same degree as p. Hence, $p(x) = \sum u_p(k) f(x+k)$. Furthermore, we show in Proposition 3 that $\frac{dp}{dx}(x) = \sum \frac{du_p}{dx}(k) f(x+k)$.

As a consequence of these remarks, if f has accuracy n then the map $\mathbf{u}: \Pi_{n-1} \to \Pi_{n-1}$ defined by $\mathbf{u}(p) = u_p$ satisfies:

i) **u** is linear,

ii) **u** preserves degree, and

iii) **u** commutes with $\frac{d}{dx}$.

We now present the proofs of the above remarks.

Proposition 1 *Let f be a compactly supported function. If $p \in S(f)$ is a polynomial of degree m, then $\Pi_m \subset S(f)$.*

Proof:

Since polynomials of different degrees are linearly independent and since $S(f)$ is a linear space, it is enough to prove the following statement:

If q is a polynomial of degree $s > 0$ such that $q \in S(f)$, then there exists a polynomial u of degree $s-1$ such that $u \in S(f)$.

Let q be any polynomial of degree $s > 0$ such that $q \in S(f)$. Without loss of generality, assume that $q(x) = x^s + \sum_{r=0}^{s-1} b_r x^r$. Since q is reproducible by f, there exists scalars α_k^s such that

$$q(x) = \sum_k \alpha_k^s f(x+k).$$

Then we have that

$$q(x+1) = \sum_k \alpha_k^s f(x+1+k) = \sum_k \alpha_{k-1}^s f(x+k),$$

and also that

$$q(x+1) = (x+1)^s + \sum_{r=0}^{s-1} b_r (x+1)^r$$

$$= x^s + (s + b_{s-1})x^{s-1} + \sum_{r=0}^{s-2} (b_r + b'_r)x^r,$$

where

$$b'_r = \binom{s}{r} + \sum_{t=r+1}^{s-1} b_t \binom{t}{r}.$$

Therefore, if we set

$$u(x) = q(x+1) - q(x) = sx^{s-1} + \sum_{r=1}^{s-2} b'_r x^r,$$

then u is a polynomial of degree $s-1$ which satisfies

$$u(x) = \sum_k (\alpha_{k-1}^s - \alpha_k^s) f(x+k).$$

Hence $u \in S(f)$, which completes the proof. □

The preceeding result plays an important role in the proof of the following proposition.

Proposition 2 *Assume that f satisfies the standard hypotheses (3). Let $p \in S(f)$ be a polynomial of degree m, and write $p(x) = \sum \alpha_k f(x+k)$. Then there exists a polynomial $u_p(x)$ of degree m such that $\alpha_k = u_p(k)$.*

Proof:
Suppose first that $p(x) = x^m$ and $p \in S(f)$. Then by Proposition 1, we must have $\Pi_m \subset S(f)$. Hence, for each $0 \le r \le m$, there exist scalars α_k^r such that
$$x^r = \sum_k \alpha_k^r f(x+k). \tag{5}$$

Then, for each $\ell \in \mathbf{Z}$ we have by the binomial theorem that

$$\sum_k \alpha_{k+\ell}^n f(x+k) = (x-\ell)^m$$
$$= \sum_{r=0}^m \binom{m}{r} (-\ell)^{m-r} x^r$$
$$= \sum_{r=0}^m \binom{m}{r} (-\ell)^{m-r} \sum_k \alpha_k^r f(x+k)$$
$$= \sum_k \left(\sum_{r=0}^m \binom{m}{r} (-\ell)^{m-r} \alpha_k^r \right) f(x+k).$$

Since the translates of f are independent, it follows that

$$\forall k \in \mathbf{Z}, \quad \alpha_{k+\ell}^m = \sum_{r=0}^m \binom{m}{r} (-\ell)^{m-r} \alpha_k^r.$$

Define $u_p(\ell)$ to be the $k=0$ case of this equation, i.e.,

$$u_p(\ell) = \alpha_\ell^m = \sum_{r=0}^m \binom{m}{r} (-\ell)^{m-r} \alpha_0^r \tag{6}$$

Then u_p is a polynomial in ℓ, and u_p has degree m if and only if $\alpha_0^0 \ne 0$. Now, by (6) for the case $m=0$, we have $\alpha_\ell^0 = \alpha_0^0$ for every ℓ, so $1 = \alpha_0^0 \sum f(x+k)$ by (5). The independence of the translates of f therefore implies that $\alpha_0^0 \ne 0$, and hence that u_p has degree m.

This proves the proposition for the case that $p(x) = x^m$. Now suppose that p is any polynomial in $S(f)$ with degree m. Then by Proposition 1, we have $\Pi_m \subset S(f)$. The result then follows by writing $p(x) = \sum_{s=0}^m a_s x^s$ and applying the results above to each individual term x^s. □

Corollary 1 *Assume that f satisfies the standard hypotheses (3). If f has accuracy n, then the map $\mathbf{u}\colon \Pi_{n-1} \to \Pi_{n-1}$ defined by $\mathbf{u}(p) = u_p$ is a linear bijection of Π_{n-1} onto itself.*

Proposition 3 *Assume that f is compactly supported, let $s \geq 0$, and suppose that there is a polynomial u such that*

$$x^s = \sum_k u(k)\, f(x+k).$$

Then for each $0 \leq t \leq s$, we have

$$x^t = C_t \sum_k u^{(s-t)}(k)\, f(x+k),$$

where

$$C_t = (-1)^{s-t} \frac{t!}{s!}$$

and $u^{(s-t)}$ is the $(s-t)$th derivative of u.

Proof:
Note first that $(x+\ell)^s = \sum_k u(k-\ell)\, f(x+k)$ for each $\ell \in \mathbf{Z}$. For each fixed x, define

$$g_x(y) = (x+y)^s \qquad \text{and} \qquad h_x(y) = \sum_k u(k-y)\, f(x+k).$$

Then g_x and h_x are both polynomials in the variable y. Moreover, $g_x(\ell) = h_x(\ell)$ for every integer ℓ, so we must have $g_x(y) = h_x(y)$ for every $y \in \mathbf{R}$. Thus, for every x and y we have

$$(x+y)^s = \sum_k u(k-y)\, f(x+k). \tag{7}$$

By taking the derivative with respect to y on both sides of (7) and then setting $y = 0$, we find that

$$sx^{s-1} = -\sum_k u'(k)\, f(x+k).$$

The proof then follows by repetition of this argument. □

Now let $\delta(K)$ denote the diameter of a compact set K, let $[a]$ denote the integer part of a real number a, and let Π denote the set of all polynomials with complex coefficients. We then have the following proposition.

6. Polynomial Reproduction 129

Proposition 4 *Let f be a compactly supported function. Then the set of polynomials reproducible by f is a finite-dimensional subspace of Π, and*

$$\dim(\Pi \cap \mathcal{S}(f)) \leq [\delta(\mathrm{supp}(f))] + 1.$$

Proof:

If $f(x) = 0$ a.e., then the result is trivial. Otherwise, let I be the minimal interval containing the support of f, and let m be any natural number such that $|I| < m$. First we show that no polynomial of degree m can be reproduced by f.

We proceed by contradiction. Assume that there exists a polynomial of degree m which is reproducible by f. Then, by Proposition 1, the m linearly independent polynomials $(x - x_0), \ldots, (x - x_0)^m$ can all be reproduced by f. That is, there exist scalars α_k^s such that

$$(x - x_0)^s = \sum_k \alpha_k^s f(x + k), \qquad 1 \leq s \leq m. \tag{8}$$

Since $|I| < m$, there exists an $x_0 \in I$ and an integer ℓ_0 such that $f(x_0 + \ell)$ can be nonzero for at most $\ell \in \{\ell_0 + 1, \ldots, \ell_0 + m\}$. Moreover, we can find a ball $B(x_0, r)$ such that if $x \in B(x_0, r)$ then $f(x + \ell)$ can be nonzero for at most $\ell \in \{\ell_0 + 1, \ldots, \ell_0 + m\}$. Hence, from (8),

$$\forall x \in B(x_0, r), \quad (x - x_0)^s = \sum_{k=1}^m \alpha_k^s f(x + \ell_0 + k).$$

Define vectors $V = (f(x_0 + \ell_0 + 1), \ldots, f(x_0 + \ell_0 + m)) \in \mathbf{C}^m$ and $\alpha^s = (\overline{\alpha_1^s}, \ldots, \overline{\alpha_m^s}) \in \mathbf{C}^m$. Then the dot product of V and α^s is

$$\langle V, \alpha^s \rangle = \sum_{k=1}^m \alpha_k^s f(x_0 + \ell_0 + m) = (x_0 - x_0)^s = 0, \qquad 1 \leq s \leq m.$$

Since $V \neq 0$, the m vectors $\{\alpha^1, \ldots, \alpha^m\}$ must therefore be linearly dependent in $\mathbf{C}^m$. Hence, there exist scalars λ_s, not all zero, such that $\sum_{s=1}^m \overline{\lambda_s} \alpha^s = 0$. Therefore, if we define a polynomial $P(x)$ by

$$P(x) = \sum_{s=1}^m \lambda_s (x - x_0)^s,$$

then for $x \in B(x_0, r)$ we have

$$P(x) = \sum_{s=1}^m \lambda_s \sum_{k=1}^m \alpha_k^s f(x + \ell_0 + k)$$

$$= \sum_{k=1}^m \left(\sum_{s=1}^m \lambda_s \alpha_k^s \right) f(x + \ell_0 + k) = 0.$$

Hence we must have $P \equiv 0$, from which it follows that $(x-x_0), \ldots, (x-x_0)^m$ are linearly dependent polynomials, which is a contradiction.

Now, if we choose m such that $m - 1 \leq |I| < m$, then

$$\dim(\Pi \cap \mathcal{S}(f)) \leq m = [|I|] + 1 = [\delta(\mathrm{supp}(f))] + 1.$$

□

2.2 A Fourier Characterization of Accuracy

In this section, we will study how polynomial reproducibility is reflected in the Fourier transform of f. The resulting well-known conditions for accuracy are usually called the *Strang–Fix* conditions, first discussed in [Sch46] and [SF73].

Definition 3 A compactly supported function $f \in \mathcal{L}^2(\mathbf{R})$ satisfies the *Strang–Fix conditions of order n* if

$$\hat{f}(0) \neq 0 \quad \text{and} \quad \hat{f}^{(s)}(2k\pi) = 0, \ \forall k \in \mathbf{Z} - \{0\}, \ 0 \leq s \leq n-1. \quad (9)$$

The next theorem shows the equivalence between the Strang–Fix conditions and accuracy. The link between the time and frequency domains is provided by the Poisson Summation Formula. The proof here is adapted from [SF73] and is included because of its clarity and elegance.

Theorem 1 *Assume that $f \in C^1(\mathbf{R})$ satisfies the standard hypotheses (3). Then the following statements are equivalent:*

i) f satisfies the Strang–Fix conditions of order n.

ii) f has accuracy n.

Proof: We will apply the Poisson Summation Formula in the form

$$\sum_k g_s(k) = \sum_k \hat{g}_s(2k\pi) \quad (10)$$

to the functions $g_s(x) = x^s f(t+x)$ for $0 \leq s \leq n-1$. Note that

$$g_s(k) = k^s f(t+k),$$

and that

$$\hat{g}_s(\omega) = i^s D^s(e^{i\omega t} \hat{f}(\omega))$$
$$= i^s \sum_{r=0}^{s} \binom{s}{r} i^r t^r \hat{f}^{(s-r)}(\omega) e^{i\omega t}. \quad (11)$$

6. Polynomial Reproduction   131

i) ⇒ ii). Assume that f satisfies the Strang–Fix conditions of order n. Then we see from (11) that $\hat{g}_s(2k\pi) = 0$ for $k \neq 0$. On the other hand, for $k = 0$ we have

$$\hat{g}_s(0) = \sum_{r=0}^{s} \left[\binom{s}{r} i^{s+r} \hat{f}^{(s-r)}(0)\right] t^r.$$

This is a polynomial in t whose leading coefficient is $(-1)^s \hat{f}(0) \neq 0$. Hence, by (10),

$$\sum_k k^s f(t+k) = \sum_k g_s(k) = \sum_k \hat{g}_s(2k\pi) = \hat{g}_s(0)$$

is a polynomial in t of degree s. Hence $\mathcal{S}(f)$ contains at least one polynomial of degree $s = n-1$, and therefore by Proposition 1 contains all polynomials of degree less than n. Hence f has accuracy n.

ii) ⇒ i). Assume that f has accuracy $n > 0$. We will use induction on s to show that (9) holds.

Consider first the case $s = 0$. By Proposition 2, we know that the constant polynomial can be reproduced using coefficients which are themselves constant. Hence we must have $\sum f(x+k) = c$ for some constant c, and this constant must be nonzero since translates of f are independent. Therefore, by (10) for the case $s = 0$,

$$\sum_k e^{i2k\pi t} \hat{f}(2k\pi) = \sum_k \hat{g}_0(2k\pi) = \sum_k g_0(k) = \sum_k f(t+k) = c.$$

This is only possible if $\hat{f}(0) = c$ and $\hat{f}(2k\pi) = 0$ for $k \neq 0$.

For the inductive step, assume that for some $0 \leq s \leq n - 1$ we have

$$\hat{f}(0) \neq 0 \quad \text{and} \quad \hat{f}^{(r)}(2k\pi) = 0, \ \forall k \in \mathbf{Z} - \{0\},\ 0 \leq r \leq s - 1.$$

Then by (11),

$$\sum_k k^s f(t+k) = \sum_k g_s(k)$$

$$= \sum_k \hat{g}_s(2k\pi)$$

$$= i^s \sum_{k \neq 0} \hat{f}^{(s)}(2k\pi) e^{i2k\pi t} + \sum_{r=1}^{s} \left[\binom{s}{r} i^{s+r} \hat{f}^{(s-r)}(0)\right] t^r.$$

However, by Corollary 1 and our assumption of accuracy, we know that $\sum_k k^s f(t+k)$ is a polynomial in t of degree s. This is only possible if $\sum_{k \neq 0} \hat{f}^{(s)}(2k\pi) e^{i2k\pi t} \equiv 0$, which implies that $\hat{f}^{(s)}(2k\pi) = 0$ for $k \neq 0$. □

2.3 Refinable Functions

Definition 4 A compactly supported function f is *refinable* if there exists a sequence of complex numbers $\{c_k\}_{k \in \mathbf{Z}}$ such that f satisfies the *refinement equation*

$$f(x) = \sum_k c_k\, f(2x - k).$$

We will restrict our attention to the case where only finitely many c_k are nonzero. By translating f if necessary, we may without loss of generality assume that there exists a positive integer N such that $c_k = 0$ when $k \neq 0, \ldots, N$. That is, we can assume that the refinement equation has the form

$$f(x) = \sum_{k=0}^{N} c_k\, f(2x - k). \tag{12}$$

The *symbol* of this refinement equation is the 2π-periodic function

$$m_0(\omega) = \sum_{k=0}^{N} c_k\, e^{-i\omega k}.$$

If $f \in \mathcal{L}^2(\mathbf{R})$ is refinable, then $\hat{f}$ must satisfy

$$\hat{f}(2\omega) = \frac{1}{2} m_0(\omega)\, \hat{f}(\omega). \tag{13}$$

The motivation from wavelet theory for studying the accuracy of refinable functions is that wavelets associated with a multiresolution analysis are constructed from a refinable function called the *scaling function*. The wavelet inherits most of its properties from the scaling function. For example, if the scaling function has high accuracy then the multiresolution analysis will have good approximation properties.

The accuracy or other properties of refinable functions can be characterized in terms of the properties of the mask $\{c_k\}$ or in terms of properties of the symbol m_0. We will first present in Section 2.4 a Fourier characterization of accuracy for refinable functions, and then in Section 2.5 present a time-domain approach.

2.4 Strang–Fix Conditions for Refinable Functions

The next theorem shows that the Strang–Fix conditions for a refinable function are equivalent to the requirement that the symbol have zeros at π, i.e., that $m_0^{(s)}(\pi) = 0$ for $s = 0, \ldots, n-1$. Equivalently, this states that $\left(\frac{1+e^{i\omega}}{2}\right)^{n-1}$ is a factor of m_0.

Theorem 2 Assume that $f \in \mathcal{L}^2(\mathbf{R})$ satisfies (3). Then the following statements are equivalent:

i) f satisfies the Strang–Fix conditions of order n.

ii) π is a zero of order n of m_0, i.e., $m_0^{(s)}(\pi) = 0$ for $0 \leq s \leq n-1$.

Proof:
i) $\Rightarrow$ ii). Assume that f satisfies the Strang–Fix conditions of order n. We will show that ii) holds by using induction on s.

Consider first the case $s = 0$. Since $\hat{f}(0) \neq 0$, it follows from (13) that $m_0(0) = 2$. Therefore, using (13) again and the fact that m_0 is 2π-periodic, we have that

$$\sum_k |\hat{f}(2k\pi)|^2 = \sum_k \frac{1}{4}|m_0(k\pi)|^2 |\hat{f}(k\pi)|^2$$

$$= \frac{1}{4}|m_0(0)|^2 \sum_k |\hat{f}(2k\pi)|^2 + \frac{1}{4}|m_0(\pi)|^2 \sum_k |\hat{f}(\pi + 2k\pi)|^2$$

$$= \sum_k |\hat{f}(2k\pi)|^2 + \frac{1}{4}|m_0(\pi)|^2 \sum_k |\hat{f}(\pi + 2k\pi)|^2. \tag{14}$$

Now, since f has independent translates, we have from equation (4) that $\sum |\hat{f}(\pi + 2k\pi)|^2 > 0$. It therefore follows from (14) that $m_0(\pi) = 0$.

For the inductive step, assume that for some $0 \leq s \leq n-1$ we have

$$m_0^{(r)}(\pi) = 0, \qquad 0 \leq r \leq s-1.$$

Taking the sth derivative of (13), evaluating at $\omega = \pi + 2k\pi$ and using the facts that $\hat{f}^{(r)}(2k\pi) = 0$ and $m_0(\pi) = 0$, we have that

$$\forall k \in \mathbf{Z}, \quad m_0^{(s)}(\pi)\, \hat{f}(\pi + 2k\pi) = 0.$$

Since $\sum |\hat{f}(\pi + 2k\pi)|^2 > 0$, we must have $\hat{f}(\pi + 2k\pi) \neq 0$ for some k, and therefore $m_0^{(s)}(\pi) = 0$.

ii) $\Rightarrow$ i). Assume that $m_0^{(s)}(\pi) = 0$ for $0 \leq s \leq n-1$. We will use induction on s to prove that the Strang–Fix conditions in (9) hold.

Consider first the case $s = 0$. Choose any $k \neq 0$, and write $k = 2^j \ell$ with $j \geq 0$ and ℓ odd. Iterating (13), we have

$$\hat{f}(2k\pi) = \hat{f}(2^{j+1}\pi\ell) = \frac{1}{2^{j+1}} m_0(2^j \pi\ell) \cdots m_0(2\pi\ell) m_0(\pi\ell) \hat{f}(\pi\ell).$$

However, m_0 is 2π-periodic, so $m_0(\pi\ell) = m_0(\pi) = 0$. Hence $\hat{f}(2k\pi) = 0$ for $k \neq 0$. On the other hand, we know that $\sum |\hat{f}(2k\pi)|^2 > 0$, so we must have $\hat{f}(0) \neq 0$.

For the inductive step, assume that for some $0 \leq s \leq n-1$ we have $\hat{f}^{(r)}(2k\pi) = 0$ for all $k \neq 0$ and $0 \leq r \leq s-1$. Taking the derivative of

(13), we find that

$$2^s \hat{f}^{(s)}(2k\pi) = \frac{1}{2} \sum_{r=0}^{s} \binom{s}{r} m_0^{(r)}(k\pi) \hat{f}^{(s-r)}(k\pi). \tag{15}$$

Now, if k is odd, then we have by hypothesis ii) that $m_0^{(r)}(k\pi) = m_0^{(r)}(\pi) = 0$ for $0 \le r \le s$, and therefore $\hat{f}^{(s)}(2k\pi) = 0$ by (15). On the other hand, if $k \ne 0$ is even then we have by the inductive hypothesis that $\hat{f}^{(s-r)}(k\pi) = 0$ for $0 < r \le s$. Hence (15) reduces for this case to

$$2^s \hat{f}^{(s)}(2k\pi) = \frac{1}{2} m_0(k\pi) \hat{f}^{(s)}(k\pi). \tag{16}$$

Therefore, if we write $k = 2^j \ell$ with $j > 0$ and ℓ odd, we can iterate (16) to obtain

$$(2^s)^{j+1} \hat{f}^{(s)}(2k\pi) = \frac{1}{2} m_0(2^j \ell \pi) \cdots m_0(\ell \pi) \hat{f}^{(s)}(\ell \pi) = 0,$$

since $m_0(\ell \pi) = m_0(\pi) = 0$. □

2.5 A Time-Domain Approach for Refinable Functions

In this section we again consider the accuracy of refinable functions, but we approach the question in the time domain rather than the frequency domain. We introduce a convenient matrix notation, which leads to the definition of a fundamental operator associated with the refinement equation. This operator is a bi-infinite matrix L whose entries are coefficients of the refinement equation, specifically,

$$L = [c_{2i-j}]_{i,j \in \mathbf{Z}} = \begin{bmatrix} \ddots & & & & & & \\ \cdots & c_3 & c_2 & c_1 & c_0 & & \\ & \cdots & c_3 & c_2 & c_1 & c_0 & \\ & & \cdots & c_3 & c_2 & c_1 & c_0 \\ & & & & & & \ddots \end{bmatrix}.$$

Note that only finitely many entries of any given row or column of L are nonzero. Moreover, there is a double shift between the rows of L; thus L is a "downsampled Toeplitz operator" or a "two-slanted matrix."

For each $x \in \mathbf{R}$, let $F(x)$ be the infinite column vector with components $f(x+k)$, i.e.,

$$F(x) = [f(x+k)]_{k \in \mathbf{Z}} = \begin{bmatrix} \vdots \\ f(x-1) \\ f(x) \\ f(x+1) \\ \vdots \end{bmatrix}.$$

Note that for each given x, only finitely many components $f(x+k)$ can be nonzero, since f has compact support.

If f satisfies the refinement equation (12), then

$$\begin{aligned}
LF(2x) &= [c_{2i-j}]_{i,j\in\mathbf{Z}}\,[f(2x+j)]_{j\in\mathbf{Z}} \\
&= \left[\sum_j c_{2i-j}\,f(2x+j)\right]_{i\in\mathbf{Z}} \\
&= \left[\sum_k c_k\,f(2x+2i-k)\right]_{i\in\mathbf{Z}} = [f(x+i)]_{i\in\mathbf{Z}} = F(x).
\end{aligned}$$

The converse is also true, so the refinement equation (12) can be rewritten in the compact matrix form

$$LF(2x) = F(x). \tag{17}$$

In order to state a result giving necessary and/or sufficient conditions for a refinable function to have accuracy n, we need to introduce some notation. Given a finite list of scalars $\{v_0, \ldots, v_{n-1}\}$, we will associate the polynomials

$$y_{[s]}(x) = \sum_{r=0}^{s} \binom{s}{r} (-x)^{s-r} v_r, \qquad 0 \le s \le n-1. \tag{18}$$

Note that $y_{[s]}$ has degree s if and only if $v_0 \ne 0$, and that $y_{[0]}(x) \equiv v_0$. By applying the binomial theorem, we obtain the following useful property of these polynomials:

$$y_{[s]}(x+y) = \sum_{t=0}^{s} \binom{s}{t} (-y)^{s-t}\, y_{[t]}(x). \tag{19}$$

Next, we place the evaluations of the polynomial $y_{[s]}$ at integers into an infinite row vector that we call $Y_{[s]}$, i.e.,

$$Y_{[s]} = (y_{[s]}(k))_{k\in\mathbf{Z}} = (\ldots,\, y_{[s]}(-1),\, y_{[s]}(0),\, y_{[s]}(1),\, \ldots). \tag{20}$$

With this notation, our time-domain characterization of accuracy is as follows.

Theorem 3 *Assume that $f \in \mathcal{L}^1(\mathbf{R})$ satisfies the standard hypotheses (3). If f is refinable and if $\hat{f}(0) \ne 0$, then the following statements are equivalent:*

a) f has accuracy n.

b) There exist scalars $\{v_0, \ldots, v_{n-1}\}$ such that $v_0 \neq 0$ and such that the infinite row vector $Y_{[n-1]}$ defined by (18) and (20) is a left eigenvector for L for the eigenvalue $2^{-(n-1)}$, i.e.,

$$Y_{[n-1]} = 2^{n-1} Y_{[n-1]} L, \qquad (21)$$

or equivalently,

$$y_{[n-1]}(\ell) = 2^{n-1} \sum_k y_{[n-1]}(k) c_{2k-\ell}, \qquad \ell \in \mathbf{Z}. \qquad (22)$$

c) There exist scalars $\{v_0, \ldots, v_{n-1}\}$ such that $v_0 \neq 0$ and such that the infinite row vectors $Y_{[s]}$ defined by (18) and (20) are left eigenvectors for L for the eigenvalues 2^{-s}, i.e.,

$$Y_{[s]} = 2^s Y_{[s]} L, \qquad 0 \leq s \leq n-1, \qquad (23)$$

or equivalently,

$$y_{[s]}(\ell) = 2^s \sum_k y_{[s]}(k) c_{2k-\ell}, \qquad 0 \leq s \leq n-1 \text{ and } \ell \in \mathbf{Z}. \qquad (24)$$

d) There exist scalars $\{v_0, \ldots, v_{n-1}\}$ such that $v_0 \neq 0$ and such that

$$\begin{aligned} v_s &= \sum_k \sum_{t=0}^{s} (-1)^{s-t} \binom{s}{t} (2k)^{s-t} 2^t v_t c_{2k} \\ &= \sum_k \sum_{t=0}^{s} (-1)^{s-t} \binom{s}{t} (2k+1)^{s-t} 2^t v_t c_{2k+1}. \end{aligned} \qquad (25)$$

e) $\sum_k c_k = 2$ and $\sum_k (-1)^k k^s c_k = 0$ for $0 \leq s \leq n-1$.

Remark 1 The equations in statement e) of Theorem 3 are the *sum rules*.

Remark 2 Note that condition c) of Theorem 3 implies that in order for f to provide accuracy n, it is necessary that $1, \frac{1}{2}, \ldots, \frac{1}{2^{n-1}}$ be eigenvalues of L. However, this eigenvalue condition alone is not sufficient for accuracy. The extra requirement needed is that the left eigenvectors $Y_{[s]}$ have the special polynomial structure described in (18) and (20). Moreover, it will be clear from the proof that if the numbers $\{v_0, \ldots, v_{n-1}\}$ are scaled by the nonzero factor $v_0 \hat{f}(0)$, then $x^s = \sum y_{[s]}(k) f(x+k)$ for each $0 \leq s \leq n-1$. Thus, the components of the left eigenvectors $Y_{[s]}$ are precisely the scalars that are used to reproduce x^s from f.

6. Polynomial Reproduction   137

Remark 3 Note that condition d) of Theorem 3 implies that the scalars $\{v_0, \ldots, v_{n-1}\}$ are entirely determined from the coefficients c_k by a finite set of finite linear equations. Further, these equations have a triangular form, i.e., once $v_0, \ldots, v_{s-1}$ have been found, the equations can be solved for v_s.

Proof of Theorem 3:

a) $\Rightarrow$ b). Assume that f has accuracy n. Then, by Proposition 2, there exist polynomials u_s of degree s such that

$$x^s = \sum_k u_s(k) f(x+k), \qquad 0 \le s \le n-1. \tag{26}$$

Set $v_s = u_s(0)$. Then the proof of Proposition 2, and equation (6) in particular, implies that

$$u_{n-1}(x) = \sum_{r=0}^{n-1} \binom{n-1}{r} (-x)^{n-1-r} v_r = y_{[n-1]}(x).$$

Since $y_{[n-1]} = u_{n-1}$ has degree $n-1$, we must therefore have $v_0 \ne 0$. Moreover, we can rewrite (26) for $s = n-1$ as

$$x^{n-1} = \sum_k y_{[n-1]}(k) f(x+k) = Y_{[n-1]} F(x). \tag{27}$$

Therefore, by using the refinement equation in the form $F(x) = L F(2x)$, we see that

$$\begin{aligned} Y_{[n-1]} F(2x) &= (2x)^{n-1} & \text{by (27)} \\ &= 2^{n-1} x^{n-1} \\ &= 2^{n-1} Y_{[n-1]} F(x) & \text{by (27)} \\ &= 2^{n-1} Y_{[n-1]} L F(2x) & \text{by (17)}. \end{aligned}$$

Now, both $Y_{[n-1]} F(2x)$ and $2^{n-1} Y_{[n-1]} L F(2x)$ are linear combinations of the translates $\{f(2x+k)\}_{k \in \mathbb{Z}}$. For example,

$$Y_{[n-1]} F(2x) = \sum y_{[n-1]}(k) f(2x+k).$$

Replacing x by $x/2$ and considering our assumption that the integer translates of f are independent, this implies that the coefficients of the linear combinations $Y_{[n-1]} F(x)$ and $2^{n-1} Y_{[n-1]} L F(x)$ must be equal, i.e., that $Y_{[n-1]} = 2^{n-1} Y_{[n-1]} L$.

b) $\Rightarrow$ c). Assume that there exist scalars $\{v_0, \ldots, v_{n-1}\}$ such that $v_0 \ne 0$ and $Y_{[n-1]} = 2^{n-1} Y_{[n-1]} L$. We must show that the equations in (24) are also satisfied.

Choose any $j, \ell \in \mathbf{Z}$. Then:

$$\sum_{s=0}^{n-1} \binom{n-1}{s} (-2j)^{n-1-s} \left(2^s \sum_k y_{[s]}(k) \, c_{2k-\ell} \right)$$

$$= \sum_{s=0}^{n-1} 2^{n-1} \binom{n-1}{s} (-j)^{n-1-s} \sum_k y_{[s]}(k) \, c_{2k-\ell}$$

$$= 2^{n-1} \sum_k \left(\sum_{s=0}^{n-1} \binom{n-1}{s} (-j)^{n-1-s} y_{[s]}(k) \right) c_{2k-\ell}$$

$$= 2^{n-1} \sum_k y_{[n-1]}(j+k) \, c_{2k-\ell} \qquad \text{by (19)}$$

$$= 2^{n-1} \sum_k y_{[n-1]}(k) \, c_{2k-(2j+\ell)}$$

$$= y_{[n-1]}(2j+\ell) \qquad \text{by hypothesis b)}$$

$$= \sum_{s=0}^{n-1} \binom{n-1}{s} (-2j)^{n-1-s} \, y_{[s]}(\ell) \qquad \text{by (19)}.$$

Since both the first and last lines of the formula above are polynomials in j, we must therefore have that $y_{[s]}(\ell) = 2^s \sum_k y_{[s]}(k) \, c_{2k-\ell}$ for $0 \leq s \leq n-1$ and $\ell \in \mathbf{Z}$. Thus c) holds.

c) $\Rightarrow$ a). Assume that there exist scalars $\{v_0, \ldots, v_{n-1}\}$ such that $v_0 \neq 0$ and $Y_{[s]} = 2^s Y_{[s]} L$ for $0 \leq s \leq n-1$. We must show that f has accuracy n. For each $0 \leq s \leq n-1$, define a function $G_{[s]}(x)$ by

$$G_{[s]}(x) = \sum_k y_{[s]}(k) \, f(x+k) = Y_{[s]} F(x). \qquad (28)$$

Note that for each fixed x, only finitely many terms of the series in (28) are nonzero. Using the hypothesis $Y_{[s]} = 2^s Y_{[s]} L$ and the refinement equation in the form $L F(2x) = F(x)$, we have that

$$\begin{aligned} G_{[s]}(2x) = Y_{[s]} F(2x) &= 2^s Y_{[s]} L F(2x) \\ &= 2^s Y_{[s]} F(x) \\ &= 2^s G_{[s]}(x). \end{aligned} \qquad (29)$$

Therefore $G_{[s]}(x)$ behaves similarly to x^s when dilated by 2. We will show that, in fact, there is a constant C independent of s such that $G_{[s]}(x) = C x^s$ for $0 \leq s \leq n-1$.

We proceed by induction on s. Consider first the case $s = 0$. Since $y_{[0]}(k) = v_0$ for every k, the function $G_{[0]}$ is defined by the formula

$$G_{[0]}(x) = v_0 \sum_k f(x+k).$$

Hence, $G_{[0]}(x)$ is 1-periodic. Further, by equation (29) we have $G_{[0]}(2x) = G_{[0]}(x)$. Thus $G_{[0]}(x)$ satisfies

$$G_{[0]}(2x) = G_{[0]}(x) \quad \text{and} \quad G_{[0]}(x - \ell) = G_{[0]}(x), \quad \ell \in \mathbf{Z}. \quad (30)$$

Therefore, if we define a map $\tau \colon [0, 1) \to [0, 1)$ by

$$\tau(x) = 2x \pmod{1} = \begin{cases} 2x, & 0 \leq x < 1/2, \\ 2x - 1, & 1/2 \leq x < 1, \end{cases}$$

then (30) implies that $G_{[0]}(\tau(x)) = G_{[0]}(x)$ for each $x \in [0, 1)$. However, τ is ergodic mapping of $[0, 1)$ into itself. It therefore follows from the Ergodic Theorem [Wal82, Theorem 1.6] that $G_{[0]}$ is constant a.e. on $[0, 1)$. By periodicity, we therefore have $G_{[0]}(x) = C$ a.e. on $\mathbf{R}$. Moreover, we can evaluate this constant explicitly:

$$\begin{aligned} C &= \int_0^1 G_{[0]}(x)\,dx = v_0 \sum_k \int_0^1 f(x + k)\,dx \\ &= v_0 \int_{\mathbf{R}} f(x)\,dx = v_0 \hat{f}(0) \neq 0. \end{aligned}$$

For the inductive step, assume that for some $0 \leq s \leq n - 1$ we have $G_{[t]}(x) = Cx^t$ a.e. for $0 \leq t \leq s - 1$. Then:

$$\begin{aligned} G_{[s]}(x - \ell) &= \sum_k y_{[s]}(k)\,f(x - \ell + k) \\ &= \sum_k y_{[s]}(k + \ell)\,f(x + k) \\ &= \sum_k \sum_{t=0}^s \binom{s}{t} (-\ell)^{s-t} y_{[t]}(k)\,f(x + k) \quad \text{by (19)} \\ &= \sum_{t=0}^s \binom{s}{t} (-\ell)^{s-t} G_{[t]}(x) \\ &= G_{[s]}(x) + \sum_{t=0}^{s-1} \binom{s}{t} (-\ell)^{s-t} Cx^t \quad \text{by induction} \\ &= G_{[s]}(x) + C \sum_{t=0}^s \binom{s}{t} (-\ell)^{s-t} x^t - Cx^s \\ &= G_{[s]}(x) + C(x - \ell)^s - Cx^s \quad \text{binomial theorem.} \end{aligned}$$

Therefore, if we define

$$H_{[s]}(x) = G_{[s]}(x) - Cx^s,$$

then we have shown that

$$H_{[s]}(x - \ell) = H_{[s]}(x), \quad \ell \in \mathbf{Z},$$

i.e., $H_{[s]}(x)$ is 1-periodic. Further, it follows from (29) and the statement $(2x)^s = 2^s x^s$ that

$$H_{[s]}(2x) = 2^s H_{[s]}(x).$$

The combination of the two preceeding equations implies that

$$H_{[s]}(\tau(x)) = 2^s H_{[s]}(x), \qquad x \in [0,1). \tag{31}$$

We will show that this implies that $H_{[s]}(x) \equiv 0$, which consequently implies the desired fact that $G_{[s]}(x) = Cx^s$ a.e.

Let $E \subset [0,1)$ be a set of positive measure on which $H_{[s]}$ is bounded, say $|H_{[s]}(x)| \leq M$ for $x \in E$. Since τ is ergodic, we know from the Ergodic Theorem [Wal82, p. 35], that for almost every $x \in [0,1)$,

$$\lim_{n \to \infty} \frac{\#\{n \geq k > 0 : \tau^k(x) \in E\}}{n} = |E| > 0. \tag{32}$$

Fix any $x \in [0,1)$ such that (32) holds. Then there exists an increasing sequence $\{n_j\}_{j=1}^{\infty}$ of positive integers such that $\tau^{n_j}(x) \in E$ for each j. Hence for each j we have by (31) and the boundedness of $H_{[s]}$ that

$$M \geq |H_{[s]}(\tau^{n_j}(x))| = |(2^s)^{n_j} H_{[s]}(x)|.$$

Therefore we must have $H_{[s]}(x) = 0$ a.e. on $[0,1)$, and since $H_{[s]}$ is 1-periodic, it must therefore vanish a.e. on **R**. Hence $G_{[s]}(x) = Cx^s$ a.e., which completes the proof.

a), b), c) $\Leftrightarrow$ d). The proof of these equivalences involves techniques that are essentially similar to those used in the arguments above, and therefore will be omitted.

c) $\Rightarrow$ e). Assume that there exist scalars $\{v_0, \ldots, v_{n-1}\}$ such that $v_0 \neq 0$ and $Y_{[s]} = 2^s Y_{[s]} L$ for $0 \leq s \leq n-1$. To show that e) holds, we will proceed by induction on s.

For the case $s = 0$, note first that $y_{[0]}(x) = v_0$ for every x. Therefore, equation (24) for the case $s = 0$ implies that for each $\ell \in \mathbf{Z}$ we have

$$v_0 = y_{[0]}(\ell) = \sum_k y_{[0]}(k) c_{2k-\ell} = v_0 \sum_k c_{2k-\ell}. \tag{33}$$

Since $v_0 \neq 0$, it follows by setting $\ell = 0$ and $\ell = 1$ in (33) that $\sum c_{2k} = 1 = \sum c_{2k-1}$. Hence we have that $\sum c_k = 2$ and that $\sum (-1)^k c_k = 0$.

Assume now, inductively, that for some $0 \leq s \leq n-1$ we have

$$\sum_k (2k)^r c_{2k} = \sum_k (2k-1)^r c_{2k-1}, \qquad 0 \leq r \leq s-1.$$

We will compute $v_s = y_{[s]}(0)$ in two different ways. First,

$$\begin{aligned}
v_s &= y_{[s]}(0) \\
&= 2^s \sum_k y_{[s]}(k)\, c_{2k} && \text{by (24)} \\
&= 2^s \sum_k \sum_{t=0}^{s} \binom{s}{t} (-k)^{s-t} v_t\, c_{2k} && \text{by (18)} \\
&= \sum_k \sum_{t=0}^{s} \binom{s}{t} (-1)^{s-t} (2k)^{s-t} 2^t\, v_t\, c_{2k} \\
&= (-1)^s v_0 \sum_k (2k)^s c_{2k} + \sum_{t=1}^{s} \binom{s}{t} (-1)^{s-t} 2^t v_t \sum_k (2k)^{s-t} c_{2k}.
\end{aligned}$$

Second, making use of the identity $\binom{s}{t}\binom{t}{r} = \binom{s}{r}\binom{s-r}{t-r}$, we have

$$\begin{aligned}
v_s &= y_{[s]}(1-1) \\
&= \sum_{t=0}^{s} \binom{s}{t} y_{[t]}(1) && \text{by (19)} \\
&= \sum_{t=0}^{s} \binom{s}{t} 2^t \sum_k y_{[t]}(k)\, c_{2k-1} && \text{by (24)} \\
&= \sum_k \sum_{t=0}^{s} \binom{s}{t} 2^t \sum_{r=0}^{t} \binom{t}{r} (-k)^{t-r} v_r\, c_{2k-1} && \text{by (18)} \\
&= \sum_k \sum_{t=0}^{s} \sum_{r=0}^{t} \binom{s}{r}\binom{s-r}{t-r} (-2k)^{t-r} 2^r v_r\, c_{2k-1} \\
&= \sum_k \sum_{r=0}^{s} \binom{s}{r} \sum_{t=r}^{s} \binom{s-r}{t-r} (-2k)^{t-r} 2^r v_r\, c_{2k-1} && \text{interchange sums} \\
&= \sum_k \sum_{r=0}^{s} \binom{s}{r} (1-2k)^{s-r} 2^r v_r\, c_{2k-1} && \text{binomial theorem} \\
&= (-1)^s v_0 \sum_k (2k-1)^s c_{2k-1} + \\
&\quad \sum_{r=1}^{s} \binom{s}{r} (-1)^{s-r} 2^r v_r \sum_k (2k-1)^{s-r} c_{2k-1}.
\end{aligned}$$

The second terms in the last line of both of these calculations are equal by the inductive hypothesis. Further, v_0 is a nonzero scalar, so this implies that

$$\sum_k (2k)^s c_{2k} = \sum_k (2k-1)^s c_{2k-1},$$

which completes the induction.

e) $\Rightarrow$ c). Assume that e) holds. We will inductively define scalars v_s for $s = 0, \ldots, n-1$ so that the polynomials $y_{[s]}$ satisfy (24).

Begin by setting $v_0 = 1$. Note that by hypothesis e), the numbers

$$m_{s,t} = (-1)^{s-t} \binom{s}{t} \sum_k (2k-\ell)^{s-t} c_{2k-\ell}$$

are independent of $\ell \in \mathbf{Z}$ when $0 \le s-t \le n-1$. Therefore, once $v_0, \ldots, v_{s-1}$ have been defined, we can define v_s by the equation

$$v_s = 2^s v_s + \sum_{t=0}^{s-1} m_{s,t} 2^t v_t.$$

With this definition, we have for each $0 \le s \le n-1$ that

$$\begin{aligned}
v_s &= 2^s v_s \sum_k c_{2k-\ell} + \sum_{t=0}^{s-1} (-1)^{s-t} \binom{s}{t} \sum_k (2k-\ell)^{s-t} c_{2k-\ell} \\
&= \sum_k \sum_{t=0}^{s} (-1)^{s-t} \binom{s}{t} (2k-\ell)^{s-t} 2^t v_t c_{2k-\ell}.
\end{aligned}$$

Therefore,

$$\begin{aligned}
2^s \sum_k y_{[s]}(k) c_{2k-\ell} \\
&= 2^s \sum_k \sum_{r=0}^{s} \binom{s}{r} (-k)^{s-r} v_r c_{2k-\ell} \qquad \text{by (18)} \\
&= \sum_k \sum_{r=0}^{s} \binom{s}{r} (-1)^{s-r} (2k-\ell+\ell)^{s-r} 2^r v_r c_{2k-\ell} \\
&= \sum_k \sum_{r=0}^{s} \binom{s}{r} (-1)^{s-r} \sum_{t=r}^{s} \binom{s-r}{t-r} (2k-\ell)^{t-r} \ell^{s-t} 2^r v_r c_{2k-\ell} \\
&= \sum_k \sum_{t=0}^{s} \sum_{r=0}^{t} (-1)^{s-r} \binom{s}{t} \binom{t}{r} (2k-\ell)^{t-r} \ell^{s-t} 2^r v_r c_{2k-\ell} \\
&= \sum_{t=0}^{s} (-1)^{s-t} \binom{s}{t} \ell^{s-t} \sum_k \sum_{r=0}^{t} (-1)^{t-r} \binom{t}{r} (2k-\ell)^{t-r} 2^r v_r c_{2k-\ell} \\
&= \sum_{t=0}^{s} \binom{s}{t} (-\ell)^{s-t} v_t \\
&= y_{[s]}(\ell),
\end{aligned}$$

so (24) holds for $0 \le s \le n-1$. $\square$

2.6 Accuracy and Orthogonal Wavelets

In this section we will study the relationship between accuracy and properties of orthogonal wavelets. We briefly recall the construction of such wavelets, referring to [Dau92] for a detailed description.

The construction of an orthogonal wavelet basis begins by choosing a refinable function $\varphi \in \mathcal{L}^2(\mathbf{R})$ satisfying the refinement equation

$$\varphi(x) = \sum_{k=0}^{N} c_k \, \varphi(2x - k), \tag{34}$$

which has the further property that its integer translates $\{\varphi(x+k)\}_{k \in \mathbf{Z}}$ are orthonormal. A necessary condition, which is only "rarely" insufficient, for φ to have orthonormal translates is that $\sum_k c_k \bar{c}_{k-2j} = 2\delta_{0j}$. Once a scaling function with orthonormal translates is chosen, the corresponding wavelet is

$$\psi(x) = \sum_k (-1)^{k-1} \bar{c}_{-k-1} \, \varphi(2x - k). \tag{35}$$

Equivalently, ψ is defined by the equation

$$\hat{\psi}(2\omega) = \frac{1}{2} m_1(\omega) \, \hat{\varphi}(\omega), \tag{36}$$

where

$$m_1(\omega) = \frac{1}{2} \sum_k (-1)^{k-1} \bar{c}_{-k-1} \, e^{-i\omega k} = e^{i\omega} \, \overline{m_0(\omega + \pi)}. \tag{37}$$

This wavelet has the remarkable property that $\{2^{n/2} \psi(2^n x + k)\}_{n,k \in \mathbf{Z}}$ forms an orthonormal basis for $\mathcal{L}^2(\mathbf{R})$. Further, $V_0 = \text{span}\{\varphi(x+k)\}$ and $W_0 = \text{span}\{\psi(x+k)\}$ are orthogonal complements in $\mathcal{L}^2(\mathbf{R})$.

High accuracy of the scaling function is a desirable feature for an orthogonal wavelet basis. For example, it leads to good approximation properties for the subspaces $V_j = \text{span}\{2^{j/2} \varphi(2^j x + k)\}$ which define the associated multiresolution analysis. Further, the smoothness of the scaling function and wavelet is limited by the accuracy of the scaling function (see Section 5). The following result lists some implications of accuracy for scaling functions. We will make use of the fact that if a scaling function φ has orthonormal translates, then we must have $\hat{\varphi}(0) = 1$.

Theorem 4 *Assume that $\varphi \in \mathcal{L}^2(\mathbf{R})$ satisfies the refinement equation (34) and has orthonormal translates. Let ψ be the associated wavelet, defined by (35). Then the following are equivalent.*

i) $\int x^s \psi(x) \, dx = 0$ *for* $s = 0, \ldots, n-1$.

ii) $m_0^{(s)}(\pi) = 0$ *for* $s = 0, \ldots, n-1$.

iii) $\sum_k (2k)^s c_{2k} = \sum_k (2k+1)^s c_{2k+1}$ for $s = 0, \ldots, n-1$.

iv) φ has accuracy n.

Proof:
The equivalence of ii), iii), and iv) follows upon combining Theorems 2, 3, and 4.

i) $\Rightarrow$ ii) Assume that $\int x^s \psi(x)\, dx = 0$ for $s = 0, \ldots, n-1$. Note that both $\hat{\varphi}$ and $\hat{\psi}$ are continuous functions since φ and ψ have compact support. In light of (37), it therefore suffices to show that $m_1^{(s)}(0) = 0$ for $s = 0, \ldots, n-1$. We proceed by induction on s. For the case $s = 0$, we have $0 = \hat{\psi}(0) = \frac{1}{2} m_1(0) \hat{\varphi}(0) = \frac{1}{2} m_1(0)$.

Assume now that for some $0 \le s \le n-1$ we have $m_1^{(r)}(0) = 0$ for $r = 0, \ldots, s-1$. Differentiating both sides of (36), we have that

$$2^s \hat{\psi}^{(s)}(2\omega) = \frac{1}{2} \sum_{r=0}^{s} \binom{s}{r} m_1^{(r)}(\omega)\, \hat{\varphi}^{(s-r)}(\omega).$$

Therefore, by the inductive hypothesis and the fact that $\hat{\varphi}(0) = 1$, we have

$$0 = 2^s \hat{\psi}^{(s)}(0) = \frac{1}{2} \sum_{r=0}^{s} \binom{s}{r} m_1^{(r)}(0)\, \hat{\varphi}^{(s-r)}(0) = \frac{1}{2} m_1^{(s)}(0).$$

iv) $\Rightarrow$ i) Assume that φ has accuracy n. Fix any s with $0 \le s \le n-1$. Then there exist scalars α_k such that $x^s = \sum \alpha_k \varphi(x+k)$. Since φ and ψ are both compactly supported, we can interchange the integral and the sum in the following calculation:

$$\int x^s \psi(x)\, dx = \int \sum_k \alpha_k \varphi(x+k)\, \psi(x)\, dx$$
$$= \sum_k \bar{\alpha}_k \int \psi(x) \overline{\varphi(x+k)}\, dx = 0,$$

since φ and ψ have orthogonal translates. □

3 Higher Dimensions, One Function - Sum Rules

In this section we present the statements of results which generalize some of the theorems of earlier sections to refinement equations in higher dimensions. We try to indicate how the techniques used previously can be extended to this more general settings. Refinable functions with domain $\mathbf{R}^2$ in particular play important roles in applications such as image processing.

We will omit the details and proofs of most results; these can be found in [CHM96] and [CHM97]. In Section 4 we will further extend these results to the case of multiple refinable functions.

We will use the standard multi-index notation $x^\alpha = x_1^{\alpha_1} \cdots x_d^{\alpha_d}$, where $x = (x_1, \ldots, x_d)^t \in \mathbf{R}^d$ and $\alpha = (\alpha_1, \ldots, \alpha_d)$ is a vector of nonnegative integers. The degree of x^α is $|\alpha| = \alpha_1 + \cdots + \alpha_d$. The number of multi-indices α of degree s is $d_s = \binom{s+d-1}{d-1}$. We write $\beta \leq \alpha$ if $\beta_i \leq \alpha_i$ for $i = 1, \ldots, d$.

One way to generalize the refinement equation (1) to higher dimensions is simply to retain the equation as written, but to allow x to be an element of $\mathbf{R}^d$ instead of $\mathbf{R}$. In this case many results carry over with little change. Such refinement equations have been studied in detail, for example in [CDM91]. However, the uniform dilation $2x$ appearing in (1) is often too limiting, and therefore we would like to allow dilations which involve rotations, shears, etc. To do this, we replace the factor 2 by a *dilation matrix* A. Such a matrix must satisfy:

a) $A(\mathbf{Z}^d) \subset \mathbf{Z}^d$, and

b) A is *expansive*, i.e., $|\lambda| > 1$ for all eigenvalues λ of A.

In this case $m = |\det(A)|$ is necessarily an integer, and therefore the quotient group $\mathbf{Z}^d/A(\mathbf{Z}^d)$ has order m. We will say that a *full set of digits* $d_1, \ldots, d_m \in \mathbf{Z}^d$ is a complete set of representatives of $\mathbf{Z}^d/A(\mathbf{Z}^d)$. In this case, $\mathbf{Z}^d$ is partitioned into the disjoint cosets

$$\Gamma_i = A(\mathbf{Z}^d) - d_i = \{Ak - d_i : k \in \mathbf{Z}^d\}, \qquad i = 1, \ldots, m.$$

For example, in the one-dimensional case $d = 1$ with $A = m$, the numbers $0, \ldots, m-1$ are a full set of digits. We remark that the lattice $\mathbf{Z}^d$ is chosen for convenience only; any full-rank lattice $\Gamma \subset \mathbf{R}^d$ could be used instead with appropriate modifications of the results. However, such a general lattice can always be reduced to the lattice $\mathbf{Z}^d$ by a change of basis.

Using the above notation, the refinement equation that we will study in this section has the form

$$f(x) = \sum_{k \in \Lambda} c_k f(Ax - k), \qquad x \in \mathbf{R}^d, \qquad (38)$$

where Λ is a finite subset of $\mathbf{Z}^d$, the c_k are complex scalars, and f maps $\mathbf{R}^d$ into $\mathbf{C}$. We say that f has accuracy n if every multivariate polynomial $q(x) = q(x_1, \ldots, x_d)$ of degree strictly less than n can be written exactly as an infinite linear combination of the translates $\{f(x+k)\}_{k \in \mathbf{Z}^d}$. We remark that although there remain close connections between accuracy and order of approximation, some of the implications valid in one dimension do not carry over to higher dimensions.

There are two simple but key properties that recur in the one-dimensional proofs in Section 2, and especially in the proof of Theorem 3. These are the homogeneity of dilation and the binomial theorem, i.e., the facts that for $x, y \in \mathbf{R}$ we have

$$(2x)^s = 2^s x^s \quad \text{and} \quad (x-y)^s = \sum_{t=0}^{s} \binom{s}{t} (-y)^{s-t} x^t. \tag{39}$$

To illustrate the difficulties that occur in higher dimensions with a general dilation matrix, let us consider a specific two-dimensional example. With $d = 2$, take A to be the *quincunx matrix* $A = \begin{bmatrix} 1 & -1 \\ 1 & 1 \end{bmatrix}$. The construction of wavelets using this matrix has received special attention, e.g., [KV92], [GM92], [CD93], [Vil94]. Note that $m = \det(A) = 2$ for this matrix.

Consider now the dilation by A of a given monomial, say of $x^\alpha = x_1^2 x_2$ where $\alpha = (2,1)$. Unfortunately, $(Ax)^\alpha$ is no longer itself a monomial; indeed, since

$$Ax = \begin{bmatrix} 1 & -1 \\ 1 & 1 \end{bmatrix} \begin{bmatrix} x_1 \\ x_2 \end{bmatrix} = \begin{bmatrix} x_1 - x_2 \\ x_1 + x_2 \end{bmatrix},$$

we have

$$(Ax)^\alpha = \begin{bmatrix} x_1 - x_2 \\ x_1 + x_2 \end{bmatrix}^\alpha = (x_1 - x_2)^2 (x_1 + x_2)$$
$$= x_1^3 - x_1^2 x_2 - x_1 x_2^2 + x_2^3$$
$$= x^{(3,0)} - x^{(2,1)} - x^{(1,2)} + x^{(0,3)}.$$

Hence to express $(Ax)^\alpha$ we require terms x^β with all $|\beta| \leq |\alpha|$.

To overcome this difficulty, let us consider *all* the monomials x^α of a given degree together, by collecting them into a vector. For example, for $|\alpha| = 3$, define the vector of all monomials of degree 3 to be

$$X_{[3]}(x) = \begin{bmatrix} x_1^3 \\ x_1^2 x_2 \\ x_1 x_2^2 \\ x_2^3 \end{bmatrix}.$$

Then,

$$X_{[3]}(Ax) = \begin{bmatrix} (x_1 - x_2)^3 \\ (x_1 - x_2)^2(x_1 + x_2) \\ (x_1 - x_2)(x_1 + x_2)^2 \\ (x_1 + x_2)^3 \end{bmatrix} = \begin{bmatrix} 1 & -3 & 3 & -1 \\ 1 & -1 & -1 & 1 \\ 1 & 1 & -1 & -1 \\ 1 & 3 & 3 & 1 \end{bmatrix} \begin{bmatrix} x_1^3 \\ x_1^2 x_2 \\ x_1 x_2^2 \\ x_2^3 \end{bmatrix}.$$

Hence, if we define $A_{[3]}$ to be the 4×4 matrix which appears in the line above, then we have that

$$X_{[3]}(Ax) = A_{[3]} X_{[3]}(x).$$

Except that we must keep in mind that $A_{[3]}$ is a matrix instead of a scalar, this equation is formally analogous to the one-dimensional equation $(2x)^3 = 2^3 x^3$, and allows us to proceed with proofs that are similar in structure to the one-dimensional versions.

It is now clear how to define an analogous matrix $A_{[s]}$ corresponding to an arbitrary dilation matrix A and a given degree s. We first define $X_{[s]}(x) = [x^\alpha]_{|\alpha|=s}$ to be the vector of all monomials of degree s. The ordering of these monomials is unimportant, as long as the same ordering is used consistently. Recalling that there are $d_s = \binom{s+d-1}{d-1}$ multi-indices α of degree s, we then let $A_{[s]}$ be the $d_s \times d_s$ matrix which has the property that $X_{[s]}(Ax) = A_{[s]} X_{[s]}(x)$. It is easy to see that such a matrix will always exist. The following result lists some of the remarkable properties of these matrices $A_{[s]}$.

Lemma 1 *Let A, B be arbitrary $d \times d$ matrices.*

a) *If $d = 1$ (so A is a scalar), then $A_{[s]} = A^s$.*

b) *$A_{[0]}$ is the scalar 1, and $A_{[1]} = A$.*

c) *$(AB)_{[s]} = A_{[s]} B_{[s]}$. Hence, if A is invertible then so is $A_{[s]}$, and $(A_{[s]})^{-1} = (A^{-1})_{[s]}$.*

d) *Let $\lambda = (\lambda_1, \ldots, \lambda_d)^t$ be the vector whose entries are the eigenvalues of A. Then the eigenvalues of $A_{[s]}$ are $[\lambda^\alpha]_{|\alpha|=s}$. Hence, if A is expansive and $s > 0$, then $A_{[s]}$ is expansive.*

The above remarks provide one way to generalize the first half of (39) to higher dimensions. The next problem is to generalize the second half, i.e., to generalize the binomial theorem. Using our vectors $X_{[s]}(x)$ containing all monomials of degree s, it is now clear how to proceed. We simply must understand the behavior of $X_{[s]}(x)$ under translation by an element $y \in \mathbf{R}^d$. First, define the following extension of binomial coefficients to higher dimensions:

$$\binom{\alpha}{\beta} = \begin{cases} \binom{\alpha_1}{\beta_1} \cdots \binom{\alpha_d}{\beta_d}, & \text{if } \beta_i \leq \alpha_i \text{ for every } i, \\ 0, & \text{if } \beta_i > \alpha_i \text{ for some } i. \end{cases}$$

Then for any multi-index α of degree $|\alpha| = s$ we have:

$$(x-y)^\alpha$$
$$= (x_1 - y_1)^{\alpha_1} \cdots (x_d - y_d)^{\alpha_d}$$
$$= \prod_{i=1}^{d} \sum_{\beta_i=0}^{\alpha_i} \binom{\alpha_i}{\beta_i} (-y_i)^{\alpha_i - \beta_i} x_i^{\beta_i}$$

$$= \sum_{\beta_1=0}^{\alpha_1} \cdots \sum_{\beta_d=0}^{\alpha_d} \binom{\alpha_1}{\beta_1} \cdots \binom{\alpha_d}{\beta_d} (-y_1)^{\alpha_1-\beta_1} \cdots (-y_d)^{\alpha_d-\beta_d} x_1^{\beta_1} \cdots x_d^{\beta_d}$$

$$= \sum_{|\beta| \le s} (-1)^{|\alpha|-|\beta|} \binom{\alpha}{\beta} y^{\alpha-\beta} x^\beta. \tag{40}$$

Now, for each integer $0 \le t \le s$ and each $y \in \mathbf{R}^d$, define a $d_s \times d_t$ matrix

$$Q_{[s,t]}(y) = (-1)^{s-t} \left[\binom{\alpha}{\beta} y^{\alpha-\beta} \right]_{|\alpha|=s, |\beta|=t}.$$

Note that we can view $Q_{[s,t]}$ as a matrix of polynomials, each entry of which is either 0 or is a monomial of degree $s - t$ in y. The utility of this matrix of polynomials is that, by (40),

$$X_{[s]}(x-y) = [(x-y)^\alpha]_{|\alpha|=s}$$

$$= \left[\sum_{t=0}^{s} \sum_{|\beta|=t} (-1)^{s-t} \binom{\alpha}{\beta} y^{\alpha-\beta} x^\beta \right]_{|\alpha|=s}$$

$$= \sum_{t=0}^{s} Q_{[s,t]}(y) X_{[t]}(x).$$

This equation plays the same role for higher dimensions that the binomial theorem plays for one dimension.

The following lemma lists basic properties of the matrix of polynomials $Q_{[s,t]}$, and its interaction with the matrices $A_{[s]}$ defined above.

Lemma 2 *a)* $Q_{[s,t]}(x+y) = \sum_{u=t}^{s} Q_{[s,u]}(y) Q_{[u,t]}(x)$.

b) $Q_{[s,t]}(Ay) = A_{[s]} Q_{[s,t]}(y) A_{[t]}^{-1}$.

Using this machinery, many of the results for one dimension can be extended to higher dimensions with arbitrary dilation matrices. Since in the next section we will consider the additional generalization to multiple functions, we include here the statement of only one result on the characterization of accuracy for a single multivariate refinable function. This result generalizes the one-dimensional sum rules of Theorem 3, part e).

Theorem 5 (Sum Rules) *Assume that $f \colon \mathbf{R}^d \to \mathbf{C}$ satisfies the refinement equation (38), that f is integrable and compactly supported, and that translates of f along $\mathbf{Z}^d$ are independent. Let $m = |\det(A)|$, and let $d_1, \ldots, d_m \in \mathbf{Z}^d$ be a full set of digits. Set $\Gamma_i = A(\mathbf{Z}^d) - d_i$. Then the following statements are equivalent.*

  i) f has accuracy n.

  ii) $\displaystyle\sum_{k \in \mathbf{Z}^d} c_k = m$ and $\displaystyle\sum_{k \in \Gamma_1} k^\alpha c_k = \cdots = \sum_{k \in \Gamma_m} k^\alpha c_k$ *for* $0 \le |\alpha| \le n-1$.

4 Higher Dimensions with Multiple Functions

In this section we state the generalizations of previous results to the case of multiple refinable functions $f_1, \ldots, f_r$. Since we have already discussed the transition to higher dimensions, we will assume that each f_i has domain $\mathbf{R}^d$. We will omit the proofs of these results, referring mainly to [CHM96] for details. In order to transmit the main ideas, we have not presented the results in their weakest form.

A finite collection of functions $f_1, \ldots, f_r$ are refinable if each function f_i can be rewritten as a finite linear combination of the rescaled and translated functions $f_j(Ax - k)$. That is, there exist scalars $c_{i,j,k}$ such that

$$f_i(x) = \sum_{j=1}^{r} \sum_{k \in \Lambda} c_{i,j,k} f_j(Ax - k), \qquad i = 1, \ldots, r, \qquad (41)$$

where Λ is some finite subset of $\mathbf{Z}^d$. We can write this more compactly if we define a vector-valued function $f: \mathbf{R}^d \to \mathbf{C}^r$ by $f(x) = (f_1(x), \ldots, f_r(x))^t$. Then we can rewrite (41) as

$$f(x) = \sum_{k \in \Lambda} c_k f(Ax - k), \qquad (42)$$

where the c_k are now some appropriate $r \times r$ *matrices*. Note that this equation is the same as (38) except that f is now vector-valued and the c_k are matrices.

The *accuracy* of the collection of functions $f_1, \ldots, f_r$, or simply the accuracy of f, is the the largest integer n such that all multivariate polynomials $p(x) = p(x_1, \ldots, x_d)$ with $\deg(p) < n$ lie in the shift-invariant space

$$\mathcal{S}(f) = \left\{ \sum_{k \in \mathbf{Z}^d} \sum_{i=1}^{r} \alpha_{k,i} f_i(x + k) : \alpha_{k,i} \in \mathbf{C} \right\}.$$

We can write this more compactly by using row vectors and column vectors. We let $\mathbf{C}^r$ be the space of all column vectors of length r; for example, $f(x) = (f_1(x), \ldots, f_r(x))^t \in \mathbf{C}^r$ for each $x \in \mathbf{R}^d$. We let $\mathbf{C}^{1 \times r}$ be the space of all row vectors of length r; for example, we can set $\alpha_k = (\alpha_{k,1}, \ldots, \alpha_{k,r}) \in \mathbf{C}^{1 \times r}$. With this notation, $\mathcal{S}(f)$ can be rewritten as

$$\mathcal{S}(f) = \left\{ \sum_{k \in \mathbf{Z}^d} \alpha_k f(x + k) : \alpha_k \in \mathbf{C}^{1 \times r} \right\}. \qquad (43)$$

We will use this type of vector notation throughout this section, and we will speak interchangeably of the function $f: \mathbf{R}^d \to \mathbf{C}^r$ and the collection of functions $f_1, \ldots, f_r$. For example, we say that translates of $f_1, \ldots, f_r$ are independent, or simply that translates of f are independent, if $\sum b_k f(x + k) = 0$ with $b_k \in \mathbf{C}^{1 \times r}$ implies $b_k = 0$ for every k.

A further notational convenience is to allow "vectors" that are indexed by $\mathbf{Z}^d$, and to regard these as behaving like row vectors or column vectors. For example, given the column vector $f(x) = (f_1(x), \ldots, f_r(x))^t \in \mathbf{C}^r$, we define an "infinite column vector" $F(x)$ by the formula

$$F(x) = [f(x+k)]_{k \in \mathbf{Z}^d}.$$

Next, we need to generalize the notation that we introduced immediately preceeding the statement of Theorem 3. We shall often be given a finite list of row vectors $\{v_\alpha \in \mathbf{C}^{1 \times r} : 0 \leq |\alpha| \leq n-1\}$, indexed by multi-indices α of degree less than n. We group these vectors by degree to form matrices of size $d_s \times r$ that we call $v_{[s]}$, i.e.,

$$v_{[s]} = [v_\alpha]_{|\alpha|=s} = \begin{bmatrix} v_{\alpha_1, 1} & \cdots & v_{\alpha_1, r} \\ \vdots & \ddots & \vdots \\ v_{\alpha_{d_s}, 1} & \cdots & v_{\alpha_{d_s}, r} \end{bmatrix}.$$

To these we associate the matrix of polynomials

$$y_{[s]}(x) = \left[\sum_{t=0}^{s} \sum_{|\beta|=t} (-1)^{s-t} \binom{\alpha}{\beta} x^{\alpha-\beta} v_\beta \right]_{|\alpha|=s} = \sum_{t=0}^{s} Q_{[s,t]}(x) v_{[t]}.$$

Note that

$$v_{[t]} f(x+k) = [v_\alpha]_{|\alpha|=s} f(x+k) = [v_\alpha f(x+k)]_{|\alpha|=s}.$$

Since $v_\alpha f(x+k) = v_{\alpha,1} f_1(x+k) + \cdots + v_{\alpha,r} f_r(x+k)$, we see that $v_{[t]} f(x+k)$ is a vector whose entries are finite linear combinations of translates of $f_1, \ldots, f_r$. For simplicity, we shall simply say that it is a vector of finite linear combinations of translates of f.

Next, we place the evaluations of the matrix of polynomials $y_{[s]}$ at lattice points into an "infinite row vector" that we call $Y_{[s]}$, i.e.,

$$Y_{[s]} = (y_{[s]}(k))_{k \in \mathbf{Z}^d}.$$

Although $Y_{[s]}$ is indexed by $\mathbf{Z}^d$, we treat it as a row vector in calculations. For example, the product of the row vector $Y_{[s]}$ with the column vector $F(x)$ is computed like the usual dot product of a row vector with a column vector:

$$\begin{aligned} Y_{[s]} F(x) &= (y_{[s]}(k))_{k \in \mathbf{Z}^d} \, [f(x+k)]_{k \in \mathbf{Z}} \\ &= \sum_{k \in \mathbf{Z}^d} y_{[s]}(k) f(x+k) \\ &= \sum_{k \in \mathbf{Z}^d} \sum_{t=0}^{s} Q_{[s,t]}(x) v_{[t]} f(x+k). \end{aligned}$$

6. Polynomial Reproduction   151

This is a vector whose entries are infinite linear combinations of translates of f. Therefore, if we knew, say, that $Y_{[s]} F(x) = X_{[s]}(x)$, then we would know that every monomial x^α of degree $|\alpha| = s$ could be exactly reproduced from lattice translates of f. Hence, if we could achieve this statement for $s = 0, \ldots, n-1$, then we could conclude that f has accuracy n.

4.1   Results for arbitrary functions

In this section we state some results that are valid for all functions $f: \mathbf{R}^d \to \mathbf{C}^r$, without the need to assume refinability.

The following result generalizes Proposition 2.

Theorem 6 ([CHM96]) *Assume that $f: \mathbf{R}^d \to \mathbf{C}^r$ has compact support and that translates of f along $\mathbf{Z}^d$ are independent. If f has accuracy n, then there exist row vectors $\{v_\alpha \in \mathbf{C}^{1\times r} : 0 \le |\alpha| \le n-1\}$ such that $v_0 \ne 0$ and*

$$X_{[s]}(x) = \sum_{k \in \mathbf{Z}^d} y_{[s]}(k)\, f(x+k) = Y_{[s]}\, F(x), \qquad 0 \le s \le n-1.$$

In particular, if p is any polynomial with $\deg(p) < n$, then there exists a unique row vector of polynomials $u_p: \mathbf{R}^d \to \mathbf{C}^{1\times r}$, with $\deg(u_p) = \deg(p)$, such that

$$p(x) = \sum_{k \in \mathbf{Z}^d} u_p(k)\, f(x+k).$$

Note that since translates of f are assumed to be independent, the coefficients $y_{[s]}(k)$ in statement ii) of Theorem 6 which reproduce the monomials of degree s are unique.

Remark 4 Suppose that $f: \mathbf{R}^d \to \mathbf{C}^r$ is compactly supported with independent translates, and that f has accuracy p. Let $\Pi_{n,r}$ be the space of all row vectors of polynomials $p: \mathbf{R}^d \to \mathbf{C}^r$ with $\deg(p) < n$. Then Theorem 6 implies that the linear mapping $\mathbf{u}: \Pi_{n,1} \to \Pi_{n,r}$ defined by $\mathbf{u}(p) = u_p$ is injective and preserves degree. The dimensions of $\Pi_{n,1}$ and $\Pi_{n,r} = \Pi_{n,1} \times \cdots \times \Pi_{n,1}$ are equal only when $r = 1$. Therefore $\mathbf{u}$ is surjective if and only if $r = 1$. As a consequence, if $r = 1$ then for each polynomial $u \in \Pi_{n,1}$ we have that the function $q(x) = \sum_{k \in \mathbf{Z}^d} u(k)\, f(x+k)$ is itself a multivariate polynomial with $\deg(u) = \deg(q)$. However, $\mathbf{u}$ cannot be surjective when $r > 1$. As a consequence, if $r > 1$ then there must exist polynomials $u \in \Pi_{n,r}$ such that $q(x) = \sum_{k \in \mathbf{Z}^d} u(k)\, f(x+k)$ is not a polynomial (one example is given in [CHM96]).

The next result extends Proposition 3.

Theorem 7 ([CHM96]) *Assume that $f: \mathbf{R}^d \to \mathbf{C}^r$ has compact support, and let α be any multi-index. If $u: \mathbf{R}^d \to \mathbf{C}^{1 \times r}$ is a row vector of polynomials such that*

$$x^\alpha = \sum_{k \in \mathbf{Z}^d} u(k) f(x+k),$$

then for each $0 \le \beta \le \alpha$ we have

$$x^\beta = C_\beta \sum_{k \in \mathbf{Z}^d} (D^{\alpha-\beta} u)(k) f(x+k),$$

where

$$D^\gamma u = \left(\frac{\partial^{|\gamma|}}{\partial x^\gamma} u_1, \ldots, \frac{\partial^{|\gamma|}}{\partial x^\gamma} u_r \right)$$

and

$$C_\gamma = (-1)^{|\alpha - \gamma|} \frac{\gamma!}{\alpha!} = (-1)^{|\alpha-\gamma|} \frac{\gamma_1!}{\alpha_1!} \cdots \frac{\gamma_d!}{\alpha_d!}.$$

4.2 Strang–Fix Conditions

We briefly discuss the Strang–Fix conditions in higher dimensions. For a single function, i.e., when $r = 1$, if $f: \mathbf{R}^d \to \mathbf{C}$ is a compactly supported function in $\mathcal{L}^2(\mathbf{R}^d)$, then f satisfies the Strang–Fix conditions of order n if

$$\hat{f}(0) \ne 0 \quad \text{and} \quad \hat{f}^{(s)}(2k\pi) = 0, \; \forall k \in \mathbf{Z}^d - \{0\}, \; 0 \le s \le n-1. \quad (44)$$

When f is continuously differentiable, these conditions are equivalent to accuracy [SF73].

In the general case, if $f = (f_1, \ldots, f_r)^{\mathrm{t}}: \mathbf{R}^d \to \mathbf{C}^r$ is a vector of compactly supported functions, we say that f satisfies the Strang–Fix conditions of order n if there exists a function g which is a finite linear combination of the translates of $f_1, \ldots, f_r$, i.e.,

$$g(x) = \sum_{i=1}^{r} \sum_{k=N_1}^{N_2} a_{k,i} f_i(x+k),$$

and which satisfies the Strang–Fix conditions (44). The following result states that the Strang–Fix conditions are still equivalent to accuracy.

Theorem 8 (see [Jia95]) *Assume that $f_1, \ldots, f_r$ are compactly supported functions in $\mathcal{L}^2(\mathbf{R}^d)$, and that translates of $f = (f_1, \ldots, f_r)^{\mathrm{t}}$ are independent. Then the following statements are equivalent:*

i) f has accuracy n.

ii) f satisfies the Strang–Fix conditions of order n.

4.3 Strang–Fix Conditions and Refinable Functions

The relationship between the Strang–Fix conditions and accuracy for *refinable* functions depends on whether one function or multiple functions are being considered.

If $f: \mathbf{R}^d \to \mathbf{C}$ is refinable, i.e., we are considering only *one* function, then the Strang–Fix conditions relate to accuracy as in the one-dimensional case. The proof of this result can be found in [Jia97] and [CGV97]

Theorem 9 *Let $m = |\det(A)|$ and let $d_1 = 0, d_2, \ldots, d_m \in \mathbf{Z}^d$ be a full set of digits. Define $B = (A^{-1})^{\mathrm{t}}$. If $f: \mathbf{R}^d \to \mathbf{C}$ satisfies the refinement equation (38), then the following statements are equivalent:*

a) f satisfies the Strang–Fix conditions of order n.

b) $m_0^{(\alpha)}(2\pi B d_i) = 0$ for $0 \leq |\alpha| \leq n-1$ and $i = 1, \ldots, m$.

For the case of multiple functions in *one* dimension, Plonka [Plo97] proved the following characterization of accuracy. Note that the symbol $m_0(\omega) = \sum c_k e^{-i\omega k}$ is now matrix-valued, since the c_k are $r \times r$ matrices.

Theorem 10 *Assume that $f = (f_1, \ldots, f_r)^{\mathrm{t}}: \mathbf{R} \to \mathbf{C}^r$ is continuous and compactly supported, and that translates of f are independent. Then the following statements are equivalent:*

a) f has accuracy n.

b) There exist row vectors $Y^0, \ldots, Y^{n-1} \in \mathbf{C}^{1 \times r}$ such that for $s = 0, \ldots, n-1$,

$$\sum_{t=0}^{s} \binom{s}{t} Y^t (2i)^{t-s} m_0^{(s-t)}(0) = \frac{1}{2^s} Y^t,$$

$$\sum_{t=0}^{s} \binom{s}{t} Y^t (2i)^{t-s} m_0^{(s-t)}(\pi) = 0.$$

Additionally, Plonka found that accuracy implies a fundamental factorization of the matrix-valued symbol m_0. This generalizes the one-dimensional, single-function case, and has led to important advances in the construction of multiwavelets [PS95], [MS97].

4.4 A Time-Domain Characterization of Accuracy

An equivalent time-domain version of Theorem 10 of the previous section appears in [HSS96] and was the starting point for the following result, which generalizes Theorem 3 both to higher dimensions and to multiple functions. We use the notation introduced at the beginning of Section 4.

Additionally, we must generalize the bi-infinite matrix L defined in Section 2.5. This follows the same type of generalizations as used before: the index set becomes $\mathbf{Z}^d$ instead of $\mathbf{Z}$, and the dilation 2 is replaced by the dilation matrix A. Hence L becomes the "$\mathbf{Z}^d$ by $\mathbf{Z}^d$ matrix"

$$L = [c_{Ai-j}]_{i,j \in \mathbf{Z}^d}.$$

We compute the product of this "matrix" L with an "infinite column vector" such as $F(x)$ by adapting the rules of ordinary matrix-vector multiplication. For example,

$$LF(x) = [c_{Ai-j}]_{i,j \in \mathbf{Z}^d} \, [f(x+j)]_{j \in \mathbf{Z}^d} = \left[\sum_{j \in \mathbf{Z}^d} c_{Ai-j} \, f(x+j) \right]_{i \in \mathbf{Z}^d}.$$

With this notation, the refinement equation can be recast, analogously to the one-dimensional case, in the form

$$LF(Ax) = F(x), \qquad x \in \mathbf{R}^d.$$

Together with our earlier machinery, this allows the techniques used in the proof of Theorem 3 to be extended to the general setting, and yields the following result.

Theorem 11 ([CHM96]) *Let $m = |\det(A)|$, and let $d_1, \ldots, d_m \in \mathbf{Z}^d$ be a full set of digits. Assume that $f = (f_1, \ldots, f_r)^t \colon \mathbf{R}^d \to \mathbf{C}^r$ satisfies the refinement equation (42), that f is integrable and compactly supported, and that translates of $f_1, \ldots, f_r$ along $\mathbf{Z}^d$ are independent. Define $\Gamma_i = A(\mathbf{Z}^d) - d_i$. Then the following statements are equivalent.*

a) f has accuracy n.

b) There exist row vectors $\{v_\alpha \in \mathbf{C}^{1 \times r} : 0 \leq |\alpha| \leq n-1\}$ such that $v_0 \neq 0$ and

$$Y_{[n-1]} = A_{[n-1]} \, Y_{[n-1]} \, L.$$

c) There exist row vectors $\{v_\alpha \in \mathbf{C}^{1 \times r} : 0 \leq |\alpha| \leq n-1\}$ such that $v_0 \neq 0$ and

$$Y_{[s]} = A_{[s]} \, Y_{[s]} \, L, \qquad 0 \leq s \leq n-1.$$

d) There exist row vectors $\{v_\alpha \in \mathbf{C}^{1 \times r} : 0 \leq |\alpha| \leq n-1\}$ such that $v_0 \neq 0$ and

$$v_{[s]} = \sum_{k \in \Gamma_i} \sum_{t=0}^{s} Q_{[s,t]}(k) \, A_{[t]} \, v_{[t]} \, c_k, \qquad 0 \leq s \leq n-1, \ i = 1, \ldots, m.$$

If $r = 1$, then the statements above are further equivalent to the following statement:

e) $\sum_{k \in \mathbf{Z}^d} c_k = m$ and $\sum_{k \in \Gamma_1} k^\alpha c_k = \cdots = \sum_{k \in \Gamma_m} k^\alpha c_k$, $0 \le |\alpha| \le n-1$.

Since only finitely many matrices c_k are nonzero, the summations in statement d) of Theorem 11 are all finite.

For the case $s = 0$, statement d) in Theorem 11 reduces to the requirement that

$$v_0 = v_0 \sum_{k \in \Gamma_i} c_k, \qquad i = 1, \ldots, m.$$

Since $\mathbf{Z}^d$ is the disjoint union of the cosets Γ_i, this implies that $v_0 = v_0 \Delta$, where $\Delta = \sum_{k \in \Lambda} c_k = m_0(0)$. Hence v_0 is a left 1-eigenvector of this matrix Δ.

An important implication of statement d) in Theorem 11 is that *the vectors v_α are determined directly by the matrices c_k and can be computed without explicit knowledge of f*. These vectors determine the coefficients $y_{[s]}(k)$ needed to reproduce the vector of monomials $X_{[s]}(x)$ from translates of f. Hence these coefficients can be derived directly from the matrices c_k. Further, the system of equations in statement d) has a block triangular structure, i.e., the equation for $v_{[s]}$ involves only $v_{[0]}, \ldots, v_{[s]}$. Hence the system can be checked recursively: $v_{[s+1]}$ is solved for after $v_{[0]}, \ldots, v_{[s]}$ have been found. Each step that can be solved implies one more degree of accuracy.

Finally, note that the condition $Y_{[s]} = A_{[s]} Y_{[s]} L$ in statement c) of Theorem 11 is no longer an eigenvector equation, as it is in one dimension, because $A_{[s]}$ is now a matrix and not a scalar. However, by changing to a basis in which $A_{[s]}$ is in Jordan form, it is possible to derive necessary conditions on the eigenvalues of the matrix L similar to those that hold in one dimension.

Proposition 5 ([CHM97]) *Let $\lambda = (\lambda_1, \ldots, \lambda_d)^t$ be the vector containing all eigenvalues of A. If there exist row vectors $Y_{[s]} \in ((\mathbf{C}^{1 \times r})^{d_s \times 1})^{1 \times \mathbf{Z}^d}$ such that $Y_{[s]} = A_{[s]} Y_{[s]} L$ for $0 \le s \le n-1$, then $\lambda^{-\alpha}$ is a left eigenvalue for L for each multi-index α with $0 \le |\alpha| \le n-1$.*

Considering Theorem 11 and Proposition 5 together, we see that if f has accuracy n, then $\lambda^{-\alpha}$ must be a left eigenvalue for L for each $0 \le |\alpha| \le n-1$. An example from [JRZ96] shows that even in the case $d = 1$, $r = 1$, the existence of such eigenvalues alone is not sufficient to imply accuracy for f; the corresponding left eigenvectors must have the polynomial structure specified in Theorem 11.

Since L is an infinite matrix, it is conceivable that the determination of its eigenvalues could be a difficult task. In fact, the eigenvalues and eigenvectors of L are completely determined by a particular finite submatrix L_0 of L [JRZ96], [CHM97]. Therefore, we have the following alternative test for accuracy: Once an upper bound for n has been computed by checking the eigenvalues of L_0, the left eigenvectors for L_0 lead to the vectors $Y_{[n-1]}$ such that $Y_{[n-1]} = A_{[n-1]} Y_{[n-1]} L$. If these vectors have a polynomial structure, then the accuracy is n. If they do not have a polynomial structure, then the test must be repeated replacing n by $n-1$. This test does require the computation of the eigenvalues of a finite matrix, which cannot be done using only systems of linear equations.

5 Implications of Accuracy

5.1 Accuracy and Order of Approximation

The concept of accuracy has been studied in the context of approximation theory and is closely related to properties of approximation of shift-invariant spaces. In this section we will discuss the connection between accuracy and *order of approximation*. Excellent reviews on this topic and on other related concepts are the papers [Jia95] and [dB90]. These also contain extensive and useful bibliographies.

Let $f_1, \ldots, f_r \in \mathcal{L}^q(\mathbf{R}^d)$ be a fixed set of functions, and define $f = (f_1, \ldots, f_r)^t$ as usual. Let $\mathcal{S}(f)$ be the shift-invariant space defined by (43). Define $\mathcal{S} = \mathcal{S}(f) \cap \mathcal{L}^q(\mathbf{R}^d)$, and set $\mathcal{S}^h = \{g(x/h) : g \in \mathcal{S}\}$. Let $W_n^q(\mathbf{R}^d)$ denote the Sobolev space consisting of all functions whose weak derivatives up to order n all lie in $\mathcal{L}^q(\mathbf{R}^d)$. Then we say that $\mathcal{S}(f)$ provides $\mathcal{L}^q$-*approximation order* n if for each $g \in W_n^q(\mathbf{R}^d)$ there exists a constant c_g such that

$$\forall h > 0, \quad \inf_{k \in \mathcal{S}^h} \|g - k\|_q \leq c_g h^n.$$

The following result states that if the functions $f_1, \ldots, f_r$ are compactly supported and have linearly independent translates, then order of approximation and accuracy are equivalent concepts.

Theorem 12 *Assume that $f_1, \ldots, f_r \in \mathcal{L}^q(\mathbf{R}^d)$ are compactly supported, and that translates of $f = (f_1, \ldots, f_r)^t$ are independent. Then the following statements are equivalent:*

i) f has accuracy n.

ii) $\mathcal{S}(f)$ provides $\mathcal{L}^q$-approximation order n for each $1 \leq q \leq +\infty$.

de Boor and Höllig [dBH83] showed that the assumption of linear independence is necessary. In particular, they gave an example of two functions defined on $\mathbf{R}^2$ which together have accuracy 4 but order of approximation 3. If we drop the assumption on linear independence, the result remains true (with appropriate minor hypothesis) when either $d = 1$ or $r = 1$ [Jia95].

5.2 Accuracy and Smoothness

One motivation from wavelet theory for studying accuracy is that the accuracy of the scaling function is related to the smoothness of the corresponding wavelet. In particular, the scaling function and wavelet have the same amount of smoothness, and accuracy is a necessary condition for the scaling function to be smooth. Hence, in order to construct smooth wavelets, we need scaling functions which have sufficiently high accuracy.

The following result which for simplicity, we present here only for the one-dimensional, single-function case, is due to Meyer [Mey92].

Theorem 13 *Let ϕ be a compactly supported scaling function with orthonormal integer translates, and let ψ be the corresponding wavelet. If ψ is k-times continuously differentiable, then $\int x^s \psi_i(x)\,dx = 0$ for $s = 0, \ldots, k$.*

In light of Theorem 4, the zero-moment condition of this theorem implies that the associated scaling function φ has accuracy $k+1$.

We conclude by presenting some results which show that smoothness of a refinable function implies accuracy, for the particular case of the uniform dilation $A = 2I$.

Theorem 14 ([CDM91]) *Let $f \in \mathbf{C}^k(\mathbf{R}^d)$ be a compactly supported refinable function such that $\hat{f}(0) \neq 0$. Then f has accuracy $k+1$.*

Theorem 15 ([Jia96a]) *If $f \in W_1^k(\mathbf{R}^d)$ is a compactly supported refinable function such that $\hat{f}(0) \neq 0$, then f has accuracy $k+1$.*

This result is extended to higher dimensions with isotropic dilation matrices A in [Jia97]. The case of multiple functions in higher dimensions, again with the uniform dilation $A = 2I$, is discussed in [Ron97].

6 Acknowledgments

C. Cabrelli and U. Molter would like to thank Ka-Sing Lau for the possibility of participating at the excellent workshop in Hong Kong. We also want to thank R.-Q. Jia for pointing out references relevant to Section 5.2.

7 References

[Baj57]   M. Bajraktarevic. Sur une équation fonctionnelle. *Glasnik Mat.-Fiz. I Astr.*, 12(3):201–205, 1957.

[Bar86]   M. F. Barnsley. Fractal functions and interpolation. *Constructive Approximation*, 2:303–329, 1986.

[BW97]   E. Belogay and Y. Wang. Arbitrarily smooth orthogonal nonseparable wavelets in $\mathcal{R}^2$. Preprint, 1997.

[CD93] A. Cohen and I. Daubechies. Non-separable bidimensional wavelet bases. *Rev. Mat. Iberoamericana*, 9:51–137, 1993.

[CDM91] A. S. Cavaretta, W. Dahmen, and C. Micchelli. Stationary subdivision. *Memoirs Amer. Math. Soc.*, 93:1–186, 1991.

[CGV97] A. Cohen, K. Gröchenig, and L. F. Villemoes. Regularity of multivariate refinable functions. Preprint, 1997.

[CH94] D. Colella and C. Heil. Characterizations of scaling functions: Continuous solutions. *SIAM J. Matrix Anal. Appl.*, 15:496–518, 1994.

[CHM96] C. Cabrelli, C. Heil, and U. Molter. Accuracy of lattice translates of several multidimensional refinable functions. *J. Approx. Th.*, 1996. To appear.

[CHM97] C. Cabrelli, C. Heil, and U. Molter. Accuracy of several multidimensional refinable distributions. Impresiones Previas 101, Deto. de Matemática, F.C.E.y N., University of Buenos Aires, (1428) Buenos Aires, ARGENTINA, 1997.

[CHM98] C. Cabrelli, C. Heil, and U. Molter. Self-similarity and multiwavelets in higher dimensions. Preprint 1998.

[CM98] C. A. Cabrelli and U. M. Molter. Generalized self-similarity *Journal of Math. Analysis and Applications.* To appear.

[Dau88] I. Daubechies. Orthonormal bases of compactly supported wavelets. *Comm. Pure Appl. Math.*, 41:909–996, 1988.

[Dau92] I. Daubechies. *Ten Lectures on Wavelets*. SIAM, Philadelphia, 1992.

[dB90] C. de Boor. Quasiinterpolants and approximation power of multivariate splines. In M. Gasca and C. A. Micchelli, editors, *Computation of Curves and Surfaces*, pages 313–345. Kluwer Academic Publishers, The Netherlands, 1990.

[dBH83] C. de Boor and K. Höllig. Approximation order from bivariate C^1- cubics: a counterexample. *Proc. Amer. Math. Soc.*, 87:649–655, 1983.

[dBR92] C. de Boor and A. Ron. The exponentials in the span of the integer translates of a compactly supported function. *J. London Math. Soc.*, 45:519–535, 1992.

[dBVR94a] C. de Boor, R. De Vore, and A. Ron. Approximation from shift-invariant subspaces of $L_2(\mathbf{R}^d)$. *Trans. Amer. Math. Soc.*, 341:787–806, 1994.

[dBVR94b] C. de Boor, R. De Vore, and A. Ron. The structure of finitely generated shift-invariant subspaces of $L_2(\mathbf{R}^d)$. *J. Funct. Anal.*, 119:37–78, 1994.

[DGHM96] G. Donovan, J. S. Geronimo, D. P. Hardin, and P. R. Massopust. Construction of orthogonal wavelets using fractal interpolation functions. *SIAM J. Math. Anal.*, 47:1158–1192, 1996.

[DGL91] N. Dyn, J. A. Gregory, and D. Levin. Analysis of uniform binary subdivision schemes for curve design. *Constr. Approx.*, 7:127–147, 1991.

[DL91] I. Daubechies and J. C. Lagarias. Two-scale difference equations: I. Existence and global regularity of solutions. *SIAM J. Math. Anal.*, 22:1388–1410, 1991.

[DL92] I. Daubechies and J. C. Lagarias. Two-scale difference equations: II. Local regularity, infinite products and fractals. *SIAM J. Math. Anal.*, 23:1031–1079, 1992.

[Dub85] S. Dubuc. Functional equations connected with peculiar curves. In *Iteration Theory and its Functional Equations*, volume 1163 of *Lecture Notes in Mathematics*, pages 33–44. Springer–Verlag, 1985.

[Dub86] S. Dubuc. Interpolation through an iterative scheme. *J. Math. Anal. Appl.*, 114:185–204, 1986.

[Eir92] T. Eirola. Sobolev characterization of solutions of dilation equations. *SIAM J. Math. Anal.*, 23:1015–1030, 1992.

[GM92] K. Gröchenig and W. R. Madych. Multiresolution analysis, haar bases, and self-similar tilings of $\mathbf{R}^n$. *IEEE Trans. Inform. Theory*, 38:556–568, 1992.

[HJ96] B. Han and R.-Q. Jia. Multivariate refinement equations and subdivision schemes. preprint, 1996.

[HSS96] C. Heil, G. Strang, and V. Strela. Approximation by translates of refinable functions. *Numer. Math.*, 73:75–94, 1996.

[Jia95] R.-Q. Jia. Refinable shift-invariant spaces: from splines to wavelets. In C. K. Chui and L. L. Schumaker, editors, *Approximation Theory VIII, Vol. 2*, pages 179–208. World Scientific, Singapore, 1995.

[Jia96a] R.-Q. Jia. The subdivision and transition operators associated with a refinement equation. In F. Fontanella, K. Jetter, and P.-J. Laurent, editors, *Advanced Topics in Multivariate Approximation*. World Scientific, Singapore, 1996. To appear.

[Jia96b] Q. Jiang. Multivariate matrix refinable functions with arbitrary matrix dilation. preprint, 1996.

[Jia97] R.-Q. Jia. Approximation properties of multivariate wavelets. *Math. Comp.*, 1997. To appear.

[JRZ96] R.-Q. Jia, S. D. Riemenschneider, and D. X. Zhou. Approximation by multiple refinable functions. Preprint, 1996.

[JW93] R.-Q. Jia and J. Wang. Orthogonality and stability associated with wavelet decompositions. *Proc. Amer. Math. Soc.*, 177:1115–1124, 1993.

[KV92] J. Kovačević and M. Vetterli. Nonseparable multidimensional perfect reconstruction filter banks and wavelet bases for $\mathbf{R}^n$. *IEEE Trans. Inform. Theory*, 38:533–555, 1992.

[LW95] K.-S. Lau and J. Wang. Characterizations of $\mathcal{L}^{\surd}$-solutions for the two scale dilation equations. *SIAM J. Math. Anal.*, 26:1018–1046, 1995.

[Mey92] Y. Meyer. *Wavelets and Operators*. Cambridge University Press, Cambridge, 1992.

[MP89] C. A. Micchelli and H. Prautzsch. Uniform refinement of curves. *Linear Algebra and Applications*, 114/115:841–870, 1989.

[MS97] C. A. Micchelli and T. Sauer. Regularity of multiwavelets. *Adv. Comput. Math.*, 7:455–545, 1997.

[Plo97] G. Plonka. Approximation order provided by refinable function vectors. *Constr. Approx.*, 13:221–244, 1997.

[PS95] G. Plonka and V. Strela. Construction of multi-scaling functions with approximation and symmetry. *SIAM J. Math. Anal.*, 1995. To appear.

[Rio92] O. Rioul. Simple regularity criteria for subdivision schemes. *SIAM J. Math. Anal.*, 23:1544–1576, 1992.

[Ron97] A. Ron. Smooth refinable functions provide good approximation orders. *SIAM J. Math. Anal.*, 28:731–748, 1997.

[Sch46] I. J. Schoenberg. Contributions to the problem of approximation of equidistant data by analytic functions. *Quart. Appl. Math.*, 4:45–99, 1946.

[SF73]   G. Strang and G. Fix. A Fourier analysis of the finite-element variational method. In G. Geymonat, editor, *Constructive Aspects of Functional Analysis*, pages 793–840. C.I.M.E., 1973.

[SS94]   G. Strang and V. Strela. Orthogonal multiwavelets with vanishing moments. *J. Optical Eng.*, 33:2104–2107, 1994.

[Vil92]   L. F. Villemoes. Energy moments in time and frequency for two-scale difference equations. *SIAM J. Math. Anal.*, 23:1519–1543, 1992.

[Vil94]   L. F. Villemoes. Continuity of nonseparable quincunx wavelets. *Appl. Comput. Harmon. Anal.*, 1:180–187, 1994.

[Wal82]   P. Walters. *An Introduction to Ergodic Theory*. Springer-Verlag, New York, 1982.

[Wan95]   Y. Wang. On two-scale dilation equations. *Random Comput. Dynam.*, 3:289–307, 1995.

7
From Cardinal Hermite Splines to Multiwavelets

Say Song Goh, S. L. Lee and W. S. Tang[1]

ABSTRACT This paper traces the origin of multiwavelets. It surveys the evolution of multiwavelets from cardinal Hermite splines and reviews the recent developments.

1 Cardinal splines and wavelets: a historical perspective

1.1 Forward B-splines and wavelets

As the theory of wavelets and shift-invariant subspaces unfolds, it becomes increasingly transparent that they belong to the extended family of cardinal spline functions. Spline functions were introduced by Schoenberg in 1946 [60] and further developed by him in a series of papers in the seventies (see [62] and the references therein). A popular approach in the construction of wavelets is via the multiresolution analysis of Meyer and Mallat [53]. Particularly relevant to our discussion are the B-splines. The forward B-spline, $Q_n(x)$ of degree n, is refinable and it is the solution of the refinement equation

$$Q_n(x) = \sum_{k=0}^{n+1} \frac{1}{2^n} \binom{n+1}{k} Q_n(2x-k), \quad x \in \mathbb{R}. \tag{1.1}$$

The polynomial $\Pi_n(z) := \sum_{j=0}^{n-1} Q_n(j+1) z^j$ is called the *Euler-Frobenius polynomial* of degree $n-1$. There is evidence to suggest that the use of piecewise polynomials for numerical approximation was known to Euler (see [62]). The coefficients of $\Pi_n(z)$ are known as the *Eulerian numbers* (see [5]).

For a fixed integer n, Q_n is a refinable function with impulse response $H_n(z) = \left(\frac{1+z}{2}\right)^{n+1}$. The first encounter with splines as wavelets was made

---

[1]Department of Mathematics, National University of Singapore, 10 Kent Ridge Crescent, Singapore 119260, Republic of Singapore

by Lemarié [51] who constructed an infinite linear combination of B-splines with orthonormal integer shifts. Recognizing that the integer shifts of Q_n form a Riesz basis of their closed linear span in $L^2(\mathbb{R})$, and that it is practically more advantageous to work with B-splines than with the Lemaré wavelets, Chui and Wang [9] showed that Q_n generates a multiresolution of $L^2(\mathbb{R})$, and they went on to construct the corresponding spline wavelets and their duals which are now commonly known as Chui and Wang's wavelets. Slightly earlier, without recourse to theory of spline functions, Daubechies [14] constructed a class of orthonormal compactly supported wavelets. Such a wavelet is constructed from a scaling function which is a convolution of a uniform B-spline with a distribution that is chosen such that the resulting scaling function has orthonormal integer shifts.

1.2 Cardinal Hermite splines and multiwavelets

The forward B-splines Q_n are associated with Lagrange interpolation at an infinite number of points by piecewise polynomials with simple integer knots (see [61] and [62] and the references therein). Historically, Lagrange (1736–1813) performed interpolation using polynomials which take the values of a given function at a finite number of points. About one hundred years later, Hermite (1822–1901) extended Lagrange's idea by including the values of the function and its consecutive derivatives. This evolution of the interpolation process prompted Schoenberg (1974) to introduce the cardinal Hermite interpolation problem and the cardinal Hermite B-splines [63], a brief description of which follows.

Take a fixed positive integer n. For simplicity, we assume that n is odd. For a positive integer r satisfying $r \leq (n+1)/2$, let $\mathcal{S}_{n,r}$ be the space of spline functions of degree n defined on $\mathbb{R}$ with integer knots of multiplicity r, i.e. $f \in \mathcal{S}_{n,r}$ if and only if $f \in C^{n-r}(\mathbb{R})$ and is a polynomial of degree n on each of the intervals $[\nu, \nu+1]$, $\nu \in \mathbb{Z}$. The space $\mathcal{S}_{n,r}$ has a spline basis comprising functions N_ℓ, $\ell = 1, \ldots, r$, with compact supports. These functions are uniquely determined by the requirement that they vanish outside the interval $[0, n-2r+3]$ and that they satisfy the Hermite interpolating conditions

$$N_\ell^{(k-1)}(\nu) = c(\nu)\delta_\ell(k), \quad \nu = 1, \ldots, n-2r+2, \tag{1.2}$$

for $k, \ell = 1, \ldots, r$, where $c(\nu)$, $\nu = 1, \ldots, n-2r+2$, are the coefficients of the generalized Euler-Frobenius polynomial

$$\Pi_{n,r}(z) = \sum_{\nu=0}^{n-2r+1} c(\nu+1)z^\nu, \quad z \in \mathbb{C} \setminus \{0\}$$

The generalized Euler-Frobenius polynomial $\Pi_{n,r}(z)$ is defined as the minor of order $(n-r+1) \times (n-r+1)$ obtained by deleting the first r rows and the last r columns of the matrix

$$P_n(z) := \left(\binom{k}{\ell} - z\delta_\ell(k) \right)_{k,\ell=0}^n .$$

The functions N_ℓ, $\ell = 1, \ldots, r$, are called *cardinal Hermite B-splines*. They are nice symmetric/antisymmetric functions. These B-splines were introduced by Schoenberg and Sharma [63], and it was shown in [49] that their integer shifts $N_\ell(\cdot - k)$, $\ell = 1, \ldots, r$, $k \in \mathbb{Z}$, indeed form a spline basis of $\mathcal{S}_{n,r}$, i.e. every $f \in \mathcal{S}_{n,r}$ is uniquely expressible in the form $f(x) = \sum_{\ell=1}^r \sum_{k \in \mathbb{Z}} c_\ell(k) N_\ell(x-k)$.

Example 1: With $n = 3$, $r = 2$, there are two B-splines N_1, N_2 supported on $[0,2]$. They are given explicitly by

$$N_1(x) = \begin{cases} 3x^2 - 2x^3, & 0 \le x \le 1, \\ -4 + 12x - 9x^2 + 2x^3, & 1 \le x \le 2, \end{cases}$$

$$N_2(x) = \begin{cases} -x^2 + x^3, & 0 \le x \le 1, \\ -4 + 8x - 5x^2 + x^3, & 1 \le x \le 2. \end{cases}$$

They are C^1 functions and satisfy the interpolating conditions

$$N_1(\nu+1) = \delta_0(\nu), \qquad N_1'(\nu) = 0, \quad \nu \in \mathbb{Z},$$

$$N_2'(\nu+1) = \delta_0(\nu), \qquad N_2(\nu) = 0, \quad \nu \in \mathbb{Z}.$$

Example 2: For $n = 5$, $r = 2$, the generalized Euler-Frobenius polynomial is given by $\Pi_{5,2}(z) = -1 + 6z - z^2$. There are two B-splines N_1, N_2 which are supported on $[0,4]$; N_1 is symmetric while N_2 is antisymmetric about 2. To express N_1 and N_2 explicitly, note that $N_j(\cdot + 2)$, $j = 1, 2$, are supported on $[-2,2]$, so that $N_1(\cdot + 2)$ is an even function while $N_2(\cdot + 2)$ is odd. Therefore, they are completely described by

$$N_1(x+2) = 8(1-x)_+^5 - 50(1-x)_+^4 + 4(2-x)_+^5 - 5(2-x)_+^4, \quad x \ge 0,$$

$$N_2(x+2) = 10(1-x)_+^5 - 26(1-x)_+^4 + (2-x)_+^5 - (2-x)_+^4, \quad x \ge 0.$$

They are C^3 functions satisfying the interpolating conditions:

$$N_1(1) = -1, N_1(2) = 6, N_1(3) = -1, \qquad N_1'(\nu) = 0, \quad \nu \in \mathbb{Z},$$

$$N_2'(1) = -1, N_2'(2) = 6, N_2'(3) = -1, \qquad N_2(\nu) = 0, \quad \nu \in \mathbb{Z}.$$

When $r = 1$, there is only one such B-spline, N_1, which is equal to the forward B-spline, Q_n, up to a constant multiple. Since wavelets are closely related to the forward B-splines which are associated with cardinal Lagrange interpolation, it is natural to follow the course of history to search for the corresponding 'wavelets' related to the cardinal Hermite B-splines which are associated with cardinal Hermite interpolation. This has led us ([27], [25]) to introduce the idea of multiresolution of multiplicity r and to construct spline wavelets of multiplicity r, the precursor to multiwavelets.

Let $V_0 \subset L^2(\mathbb{R})$ be the shift invariant subspace generated by the integer shifts of N_ℓ, $\ell = 1, \ldots, r$, and let $V_m := \{f(2^m \cdot) : f \in V_0\}$, $m \in \mathbb{Z}$. It was shown in [27] that the sequence of subspaces V_m satisfies
(MR1) $V_m \subset V_{m+1}$, $m \in \mathbb{Z}$,
(MR2) $\cap_{m \in \mathbb{Z}} V_m = \{0\}$,
(MR3) $\cup_{m \in \mathbb{Z}} V_m$ is dense in $L^2(\mathbb{R})$,
(MR4) $f(\cdot) \in V_m \iff f(2\cdot) \in V_{m+1}$,
(MR5) the integer shifts of N_ℓ, $\ell = 1, \ldots, r$, form a Riesz basis of V_0.
We say that N_ℓ, $\ell = 1, \ldots, r$, generate a *multiresolution of $L^2(\mathbb{R})$ of multiplicity r*.

Let W_0 be the orthogonal complement of V_0 in V_1. If $\psi_1, \ldots, \psi_r \in W_0$, and their integer shifts form a Riesz basis of W_0, then $\psi_1, \ldots, \psi_r$ will be called *multiwavelets*. If their integer shifts form an orthonormal basis of W_0, then they will be called *orthonormal multiwavelets*. For convenience we shall also call the vector function $\Psi = (\psi_1, \ldots, \psi_r)^T$ multiwavelets. Spline multiwavelets with minimal support analogous to Chui and Wang's wavelets were constructed in [25].

We now form from the B-splines N_ℓ, $\ell = 1, \ldots, r$, the vector function $N := (N_1, \ldots, N_r)^T$. Since $\{V_m\}_{m \in \mathbb{Z}}$ forms a multiresolution of $L^2(\mathbb{R})$, the condition (MR1) implies that the vector function N satisfies a *matrix refinement equation*

$$N(x) = \sum_{k \in \mathbb{Z}} 2h(k) N(2x - k), \quad x \in \mathbb{R}, \qquad (1.3)$$

where $h(k)$, $k \in \mathbb{Z}$, are $r \times r$ matrices such that the entries $h_{i,j}(k)$, $k \in \mathbb{Z}$, are sequences in $\ell^2(\mathbb{Z})$. The vector function N is called a *scaling vector*. The sequence $(h(k))_{k \in \mathbb{Z}}$ is called the *mask* for N. In general, the mask h is an infinite sequence, even though N has compact support. However, it can be shown that $h(k) \to 0$ exponentially as $k \to \infty$.

Further, since $\psi_j \in V_1$, $j = 1, \ldots, r$, there is a matrix sequence $(g(k))_{k \in \mathbb{Z}}$ with entries $g_{i,j} \in \ell^2(\mathbb{Z})$ such that

$$\Psi(x) = \sum_{k \in \mathbb{Z}} 2g(k) N(2x - k), \quad x \in \mathbb{R}. \qquad (1.4)$$

The sequence $(g(k))_{k \in \mathbb{Z}}$ is called a *high pass multifilter*. In this context, $(h(k))_{k \in \mathbb{Z}}$ is also called a *low pass multifilter*.

2 Multiwavelets in wandering subspaces

It turns out that the above process of finding multiwavelets from a multiresolution of $L^2(\mathbb{R})$ (generated by N_ℓ, $\ell = 1, \ldots, r$) can be generalized to a Hilbert space setting. There is a close connection between orthonormal multiwavelets and wandering subspaces for unitary operators.

Let H be a complex Hilbert space, and let $U = (U_1, \ldots, U_d)$ be an ordered d-tuple of distinct unitary operators on H such that $U_k U_j = U_j U_k$, $k, j = 1, \ldots, d$. We shall use the multi-index notation $U^n = U_1^{n_1} \cdots U_d^{n_d}$ for $n = (n_1, \ldots, n_d) \in \mathbb{Z}^d$, with the convention that U_j^0 is the identity operator on H, $j = 1, \ldots, d$. We also assume that U^n is the identity operator only if $n = 0$. For a subset S of H, let $\langle S \rangle$ denote the closed linear span of S in H, and write $U^{\mathbb{Z}^d}(S) := \{U^n s : n \in \mathbb{Z}^d, s \in S\}$. We say that a closed linear subspace V of H is a *wandering subspace* for U if $U^p(V) \perp U^n(V)$ for all $p, n \in \mathbb{Z}^d$, $p \neq n$. Further, if $W = \sum_{n \in \mathbb{Z}^d} U^n(V)$, then we say that V is a complete wandering subspace of W for U.

The following theorem was proved in [28, Theorem 3.1], by first extending the results of Robertson [58] on wandering subspaces for unitary operators. (The case $d = 1$ was previously considered in [27, 50].)

Theorem 2.1 *Let* $X := \{x_1, \ldots, x_r\}$ *be a finite subset of H, and $V_0 = \langle U^{\mathbb{Z}^d}(X) \rangle$. Suppose that there is a unitary operator D on H such that*

$$V_0 \subset DV_0 =: V_1 \tag{2.1}$$

and

$$U^n D = D U^{Mn}, \quad n \in \mathbb{Z}^d, \tag{2.2}$$

where M is a $d \times d$ matrix with integer entries and $m := |\det(M)| \geq 2$. Let W_0 be the orthogonal complement of V_0 in V_1.

(a) If $U^{\mathbb{Z}^d}(X)$ is a Riesz basis of V_0, then there exists a finite subset Γ of W_0, with $r(m-1)$ elements, such that $U^{\mathbb{Z}^d}(\Gamma)$ is a Riesz basis of W_0, and $U^{\mathbb{Z}^d}(X \cup \Gamma)$ is a Riesz basis of V_1.

(b) If $U^{\mathbb{Z}^d}(X)$ is an orthonormal basis of V_0, then there exists a finite subset Γ of W_0, with $r(m-1)$ elements, such that $U^{\mathbb{Z}^d}(\Gamma)$ is an orthonormal basis of W_0, and $U^{\mathbb{Z}^d}(X \cup \Gamma)$ is an orthonormal basis of V_1.

Part (b) of Theorem 2.1 can be rephrased as follows: If $\langle X \rangle$ is a complete wandering subspace of V_0 for U, then there exists a finite subset Γ of W_0, with $r(m-1)$ elements, such that $\langle \Gamma \rangle$ is a complete wandering subspace of W_0 for U, and $\langle X \cup \Gamma \rangle$ is a complete wandering subspace of V_1 for U. This connection between orthonormal multiwavelets and wandering subspaces was first observed in [27], where part (b) of Theorem 2.1 for the case $d = 1$ was derived as a consequence of [58, Theorem 2].

Let us apply Theorem 2.1 to the concrete setting of $L^2(\mathbb{R}^d)$. If $H = L^2(\mathbb{R}^d)$,
$$(U_k f)(x) = f(x - e_k),$$
where $e_k = (\delta_k(j))_{j=1}^d$, $k = 1, \ldots, d$, and
$$(Df)(x) = |\det(M)|^{\frac{1}{2}} f(Mx),$$
for x in $\mathbb{R}^d$ and f in $L^2(\mathbb{R}^d)$, then
$$D^j U^n f = |\det(M)|^{\frac{j}{2}} f(M^j \cdot -n), \quad j \in \mathbb{Z}, \, n \in \mathbb{Z}^d,$$
and (2.2) is satisfied. If $\{\phi_\ell(\cdot - n) : n \in \mathbb{Z}^d, \ell = 1, \ldots, r\}$ is a Riesz basis of V_0 and $V_0 \subset V_1 := \{f(M\cdot) : f \in V_0\}$, then the above theorem gives the existence of $r(m-1)$ multiwavelets in W_0. In particular, if $M = 2I_d$, where I_d is the $d \times d$ identity matrix, then we have $r(2^d - 1)$ multiwavelets. Our previous discussion in Subsection 1.2 involves the special case $d = 1$.

Recently, Dai and Larson [13] have obtained generalizations of part (b) of Theorem 2.1 to other systems of unitary operators. Part (a) of Theorem 2.1 has also been extended to the biorthogonal setting in [70] and [75].

3 Matrix refinement equations with finite mask

Multifilters are the most important ingredients in the applications of multiwavelets. Therefore, the design of 'good multifilters' is an important aspect in the study of multiwavelets. Good scaling vectors and multiwavelets are associated with good multifilters. Therefore, the problem of multifilter design reduces to the problem of constructing good scaling vectors and hence multiwavelets. Thus to design good multifilters we need to know the properties for good scaling vectors in terms of the mask. In practice, we require the multifilters to be finite.

The most general matrix refinement equation (MRE) with finite mask that is being actively studied at present is of the form
$$\Phi(x) = \sum_{k \in \mathbb{Z}^d} \det(M) h(k) \Phi(Mx - k), \quad x \in \mathbb{R}^d, \tag{3.1}$$
where the mask $(h(k))_{k \in \mathbb{Z}^d}$ is a finitely supported sequence of $r \times r$ matrices, $\Phi = (\phi_1, \ldots, \phi_r)^T$ is a vector of tempered distributions on $\mathbb{R}^d$, and M is a $d \times d$ dilation matrix with integer entries and all its eigenvalues are of modulus greater than 1.

In the Fourier domain, (3.1) becomes
$$\widehat{\Phi}(u) = H((M^{-1})^T u) \widehat{\Phi}((M^{-1})^T u), \quad u \in \mathbb{R}^d, \tag{3.2}$$

where $\widehat{\Phi}(u) := (\hat{\phi}_1(u), \ldots, \hat{\phi}_r(u))^T$ and $H(u) := \sum_k h(k) e^{-ik^T u}$. A solution of the MRE (3.1), or equivalently (3.2), is called an (M, h)-*scaling vector*.

The problems addressed in connection with the matrix refinement equation with finite mask include (1) existence of solutions of matrix refinement equations, (2) convergence of the cascade algorithm, (3) stability, orthonormality, and biorthogonality of (M, h)-refinable vectors, (4) approximation order of the shift invariant subspaces generated by (M, h)-refinable vectors, (5) regularity of (M, h)-refinable vectors, (6) construction of (M, h)-refinable vectors, (7) multiwavelet construction, (8) applications.

Problems 1, 2, 3, 4 and 5 were recently addressed in [30], [31], [12], [56], [64], [43], [54], [39], [38], [36], [37]. The theory of multiwavelets mushroomed in the last 5 years, but few examples of multiwavelets are known. In one dimension, symmetric/antisymmetric spline multiwavelets or spline wavelets of multiplicity r were constructed in [25]. The method of construction in [25] exploits the properties of spline functions, and it does not seem to work for other multiwavelets. A general construction of orthonormal multiwavelets from a given orthonormal scaling vector in one dimension can be found in [46]. The results in [46] have been extended in [23] to the biorthogonal setting. Not-a-spline scaling functions with $r = 2$ were first constructed by Geronimo, Hardin and Massopust [21]. Subsequently the corresponding wavelets were constructed in [18], [68] and [46]. They are commonly called GHM multiwavelets. They are orthonormal symmetric pair with supports in $[0, 2]$. This example is simple, but it has provided much inspiration for more constructions ([8], [18]). Recently, Jiang has constructed a large class of scaling vectors and multiwavelets with optimum time-frequency localization in a series of papers [40], [41], [42]. Multiwavelets with 'good filter characteristics' have also been constructed recently in [71]. They are symmetric/antisymmetric pairs of orthonormal multiwavelets. The corresponding multifilters are proven to have high energy concentration, a feature useful for image and video compression [71]. In higher dimensions, very few examples of multiwavelets are known. An example of linear spline multiwavelets ($r = 2$) in two dimensions can be found in [28], and pairs of refinable splines in $\mathbb{R}^2$ were constructed in [24]. Recently, another example in two dimensions has been constructed in [19] These are the only examples known in more than one dimension. Much more has to be done.

In the rest of this section, we shall review some recent developments of multiwavelets in more detail. For simplicity, we shall restrict to the case $d = 1$ and $M = (2)$. The general case is more elaborate but the results are similar.

3.1 Existence of solution of MRE

For $d = 1$ and $M = (2)$, the MRE becomes

$$\Phi(x) = \sum_{k=0}^{N} 2h(k)\Phi(2x - k), \quad x \in \mathbb{R}, \tag{3.3}$$

where $h(k)$, $k = 0, \ldots, N$, is a finite sequence of $r \times r$ matrices, and $\Phi = (\phi_1, \ldots, \phi_r)^T$ is a vector of tempered distributions. In the Fourier domain, (3.3) is equivalent to

$$\widehat{\Phi}(u) = H(u/2)\widehat{\Phi}(u/2), \quad u \in \mathbb{R}. \tag{3.4}$$

Iterating (3.4) leads to

$$\widehat{\Phi}(u) = \prod_{j=1}^{n} H(u/2^j)\widehat{\Phi}(u/2^n), \tag{3.5}$$

for any positive integer n.

The existence of $\widehat{\Phi}$, and hence the distributional solution of the MRE (3.3), depends on the convergence of the infinite product $\prod_{j=1}^{\infty} H(u/2^j)$. This problem was first considered by Heil and Colella [30] in one dimension and substantially extended by Jiang and Shen [43] to higher dimensions. We first introduce some terminologies. A square matrix A is said to satisfy *Condition E^{**}* if it has unit spectral radius and all its eigenvalues on the unit circle are nondegenerate. If in addition, 1 is the only eigenvalue on the unit circle, then A is said to satisfy *Condition E^**. We say that A satisfies *Condition E* if it satisfies Condition E^* and 1 is a simple eigenvalue.

The following result gives a necessary and sufficient condition for the convergence of infinite product of matrices of continuous functions. It can be deduced from Theorem 2.1 in [22].

Proposition 3.1 *Let $A(u)$ be a square matrix of continuous functions on $\mathbb{R}$, satisfying*

$$\sum_{j=1}^{\infty} \|A(u/2^j) - A(0)\| < \infty \quad \text{for all } u \in \mathbb{R}. \tag{3.6}$$

Then $\lim_{n \to \infty} \prod_{j=1}^{n} A(u/2^j)$ exists, and the convergence is uniform on compact sets if and only if $A(0)$ satisfies Condition E^*.

For a MRE with finite mask, the frequency response $H(u)$ satisfies (3.6). It follows that the infinite product $\prod_{j=1}^{\infty} H(u/2^j)$ converges locally uniformly if and only if $H(0)$ satisfies Condition E^*.

Example: Consider the sequence of matrices $h(k)$ with 3 nonzero terms

$$h(0) = \frac{1}{16}\begin{pmatrix} 4 & 6 \\ -1 & -1 \end{pmatrix}, \quad h(1) = \frac{1}{4}\begin{pmatrix} 2 & 0 \\ 0 & 1 \end{pmatrix}, \quad h(2) = \frac{1}{16}\begin{pmatrix} 4 & -6 \\ 1 & -1 \end{pmatrix}.$$

7. From Cardinal Hermite Splines to Multiwavelets

Then $H(u) = h(0) + h(1)^{-iu} + h(2)e^{-i2u}$, and

$$H(0) = h(0) + h(1) + h(2) = \begin{pmatrix} 1 & 0 \\ 0 & 1/8 \end{pmatrix},$$

which satisfies Condition E^*. In fact $H(0)$ satisfies Condition E.

Suppose that $H(0)$ satisfies Condition E^* and its eigenvalue 1 has multiplicity s with linearly independent eigenvectors $v_1, \ldots, v_s$. Then $P(u) := \prod_{j=1}^{\infty} H(u/2^j)$, $u \in \mathbb{R}$, defines an $r \times r$ matrix of continuous functions. By (3.5),

$$\widehat{\Phi}(u) = P(u)\widehat{\Phi}(0). \tag{3.7}$$

The matrix $P(u)$ has polynomial growth as $|u| \to \infty$. The proof is standard and can be found in [15] for the scalar case and [30] for the vector case. Consider $P(u)v_k = \prod_{j=1}^{\infty} H(u/2^j)v_k$, $k = 1, \ldots, s$, which is an r-vector whose components are continuous functions with polynomial growth at infinity. There is a vector $\Phi_k = (\phi_{k,1}, \ldots, \phi_{k,r})^T$ of (tempered) distributions whose Fourier transform satisfies $\widehat{\Phi}_k(u) = P(u)v_k$, $u \in \mathbb{R}$. This follows from the fact that the Fourier transform is an isomorphism on the space of tempered distributions $\mathcal{S}'$, and that continuous functions with polynomial growth belong to $\mathcal{S}'$.

For each $k = 1, \ldots, s$, Φ_k is a distributional solution of the MRE (3.3) since $\widehat{\Phi}_k(u) = H(u/2)\widehat{\Phi}_k(u/2)$. Now, take any v in $\mathbb{C}^r$. By the same argument as above, there is a vector Φ_v of tempered distributions whose Fourier transform satisfies $\widehat{\Phi}_v(u) = P(u)v$, $u \in \mathbb{R}$, which gives the relation $\widehat{\Phi}_v(u) = H(u/2)\widehat{\Phi}_v(u/2)$. This means that Φ_v is also a distributional solution of the MRE (3.3). It is easy to see that $\widehat{\Phi}_v$ belongs to the linear span of $\{\widehat{\Phi}_1, \ldots, \widehat{\Phi}_s\}$.

If $H(0)$ satisfies Condition E^*, then the solutions of the MRE are, in general, vectors of tempered distributions. It is therefore natural to search for simple conditions for the solutions to belong to $L^2(\mathbb{R})^r$. As far as we know, necessary and sufficient conditions in terms of the spectrum of transition operators, are not available, but a simple sufficient condition may be deduced from the characterization of weak convergence of the cascade algorithm.

3.2 Cascade algorithms and transition operators

The cascade algorithm for the MRE (3.3) is a means to establish the existence of its solutions and to compute them. It is very much like the Picard iteration for the solution of differential equations. To define the cascade algorithm, we choose a starting vector function $\Phi_0 = (\phi_{0,1}, \ldots, \phi_{0,r})^T$ in

$L^2(\mathbb{R})^r$ with support in $[0, N]$, and define the cascade sequence Φ_n by

$$\Phi_n(x) = \sum_{k=0}^{N} 2h(k)\Phi_{n-1}(2x-k), \quad x \in \mathbb{R}, \ n = 1, 2, \ldots. \tag{3.8}$$

Each Φ_n belongs to $L^2(\mathbb{R})^r$, and has support in $[0, N]$. Note that if $\Phi_n \to \Phi$, then Φ is a solution of the MRE (3.3). Consider the $r \times r$ matrix

$$f_n(t) = \int_{\mathbb{R}} \Phi_n(x)\Phi_n(x-t)^* dx. \tag{3.9}$$

Equations (3.8) and (3.9) lead to

$$f_n(t) = \sum_{k=0}^{N} \sum_{\ell=k-N}^{k} 2h(k) f_{n-1}(2t-\ell) h(k-\ell)^*. \tag{3.10}$$

Setting $t = \nu$ in (3.10) gives

$$f_n(\nu) = \sum_{k=0}^{N} \sum_{\ell=k-N}^{k} 2h(k) f_{n-1}(2\nu-\ell) h(k-\ell)^*. \tag{3.11}$$

In order to simplify (3.11), we define the *transition operator* T_h for all sequences $(b(n))_{n \in [-N+1, N-1]}$ of $r \times r$ matrices by

$$(T_h b)(\nu) = \sum_{k=0}^{N} \sum_{\ell=k-N}^{k} 2h(k) b(2\nu-\ell) h(k-\ell)^*. \tag{3.12}$$

Then (3.11) simplifies to

$$f_n = T_h f_{n-1}, \quad n = 1, 2, \ldots. \tag{3.13}$$

To describe the transition operator T_h in the Fourier domain which is easier to deal with at times, we introduce the space $\mathcal{V}_N$ of all $r \times r$ matrices with trigonometric polynomial entries whose Fourier coefficients are real and supported in $[-N+1, N-1]$. Let

$$T_H B(u) := \sum_{k=-N+1}^{N-1} T_h b(k) e^{-iku},$$

where $B(u) := \sum_{k=-N+1}^{N-1} b(k) e^{-iku}$. Then T_H is a linear operator on $\mathcal{V}_N$ given by

$$T_H B(u) = H(\tfrac{u}{2}) B(\tfrac{u}{2}) H^*(\tfrac{u}{2}) + H(\tfrac{u}{2}+\pi) B(\tfrac{u}{2}+\pi) H^*(\tfrac{u}{2}+\pi), \quad B \in \mathcal{V}_N. \tag{3.14}$$

Equation (3.13) becomes

$$F_n = T_H F_{n-1}, \quad n = 1, 2 \ldots, \tag{3.15}$$

where F_n is the Fourier series with Fourier coefficients f_n.

We shall say that T_h satisfies Condition E^{**} (Condition E^* or Condition E) if its representation matrix satisfies Condition E^{**} (Condition E^* or Condition E respectively). Theorems 3.1 and 3.2 below are special cases of convergence of nonstationary vector cascade algorithms considered in [26].

Theorem 3.1 *The cascade sequence Φ_n converges weakly in $L^2(\mathbb{R})^r$ for any starting vector Φ_0 if and only if T_h satisfies Condition E^{**}.*

The following corollary gives a sufficient condition for the solutions of the MRE to belong to $L^2(\mathbb{R})^r$.

Corollary 3.1 *If T_h satisfies Condition E^{**}, then the solutions of the MRE (3.3) belong to $L^2(\mathbb{R})^r$.*

Stronger conditions are required for strong convergence in $L^2(\mathbb{R})^r$. We say that h is *fundamental* with respect to a nonzero column vector $v \in \mathbb{R}^r$ if

$$v^T \sum_{j \in \mathbb{Z}} h(2j) = v^T \sum_{j \in \mathbb{Z}} h(2j+1) = \frac{v^T}{2}.$$

Theorem 3.2 *Suppose that*
(a) T_h satisfies Condition E and h is fundamental with respect to v. Then
(b) for any initial vector Φ_0 satisfying

$$v^T \sum_{j \in \mathbb{Z}} \widehat{\Phi}_0(\,\cdot\, - j) = 1 \text{ almost everywhere}, \tag{3.16}$$

the cascade sequence Φ_n converges strongly in $L^2(\mathbb{R})^r$ to a solution of the MRE (3.3).
Conversely, if $H(0)$ satisfies Condition E, then (b) implies (a).

Remark 1 In [26], Theorems 3.1 and 3.2 were proved in its full generality for multidimensional nonstationary cascade algorithms with a general dilation matrix M.

Remark 2 If $H(0)$ satisfies Condition E, then the MRE (3.3) has a unique solution. Under this assumption the statements (a) and (b) in Theorem 3.2 are equivalent, a result obtained in [64]. Shen's result in [64] generalizes corresponding results in [66] for the scalar case in one dimension and in [48] for the scalar case in higher dimensions.

3.3 Stability and orthonormality of scaling vectors

Let $G_\Phi(u)$ denote the Gram matrix of the $(2,h)$-refinable vector $\Phi \in L^2(\mathbb{R})^r$ defined by

$$G_\Phi(u) := \sum_{k \in \mathbb{Z}} \widehat{\Phi}(u + 2\pi k)\widehat{\Phi}^*(u + 2\pi k).$$

Then Φ is stable, i.e. the integer shifts of ϕ_j, $j = 1, \ldots, r$, form a Riesz basis of their closed linear span in $L^2(\mathbb{R})$, if and only if $G_\Phi(u)$ is positive definite for all $u \in \mathbb{T}$; and Φ is orthonormal, i.e. the integer shifts of ϕ_j, $j = 1, \ldots, r$, form an orthonormal set, if and only if $G_\Phi(u) = I_r$ for all $u \in \mathbb{T}$, where $\mathbb{T}$ is the unit circle and I_r denotes the $r \times r$ identity matrix (see [28]). If Φ is orthonormal, then h satisfies

$$h(u)h^*(u) + h(u+\pi)h^*(u+\pi) = I_r, \quad u \in \mathbb{T}. \tag{3.17}$$

A filter sequence h that satisfies (3.17) is called a matrix *conjugate quadrature filter* (CQF).

If the compactly supported $(2,h)$-refinable vector Φ is *stable*, then it is well known that Φ generates a multiresolution of $L^2(\mathbb{R})$ (see [2]). It is also known that $H(0)$ satisfies Condition E and its left eigenvector e corresponding to the eigenvalue 1 satisfies $e\,H(\pi) = 0$ (see, for instance, [43]). In this case, the cascade sequence Φ_n converges strongly in $L^2(\mathbb{R})^r$, and $\lim_{n \to \infty} \Phi_n = \Phi$.

As in the scalar case, it is useful to characterize the stability and orthonormality of Φ in terms of h.

Theorem 3.3 *Suppose that $H(0)$ satisfies Condition E and its left eigenvector e corresponding to the eigenvalue 1 satisfies $eH(\pi) = 0$. Then*
(a) Φ is stable if and only if the transition operator T_H satisfies Condition E and the eigenvector $V(u)$ corresponding to the eigenvalue 1 is nonsingular for all $u \in \mathbb{R}$,
(b) Φ is orthonormal if and only if h is a CQF and T_H satisfies Condition E.

Remark 3 *The multidimensional vector case of Theorem 3.3 with $M = 2I$ was due to Shen [64]. Solutions of matrix refinement equations and their properties with dilation $M = 2I$ was also considered in [52]. The corresponding result with a general dilation matrix was established by Jiang [39]. These results generalize the corresponding multivariate results in [47] and [34] for the scalar case of $r = 1$. The search for a characterization of orthonormality of a scaling function in terms of its mask was started by Lawton [44]. The result in part (b) of Theorem 3.3 for the scalar case in one dimension was due to him, and the condition in (b) for the orthonormality of a scaling function is commonly known as* Lawton's condition.

Since the eigenvalues of a finite matrix can be computed easily, it is useful in practice to represent the operator T_H as a finite matrix. Such a representation can be found in [38], [29] and [57]. We shall describe it here. For any $r \times r$ matrix A, let $A(j)$ be its jth column, i.e. $A = (A(1), \ldots, A(r))$, and define the $r^2 \times 1$ vector $\text{vec}(A)$ by

$$\text{vec}(A) := (A(1)^T, \ldots, A(r)^T)^T. \tag{3.18}$$

For $B(u) = \sum_{k=-N+1}^{N-1} b(k) e^{-iku} \in \mathcal{V}_N$, let $\text{vec}(B)$ be the $((2N-1)r^2) \times 1$ vector defined by

$$\text{vec}(B) := \left(\text{vec}(b(-N+1))^T, \ldots, \text{vec}(b(N-1))^T\right)^T.$$

For any two matrices $C = (c_{ij})$ and $D = (d_{ij})$, let $C \otimes D := (c_{ij}D)$ denote the Kronecker product of C and D. Then for any compatible matrices C, D, E, we have

$$\text{vec}(CDE^T) = (E \otimes C)\text{vec}(D). \tag{3.19}$$

It can be easily shown (see [38], [29] and [57]) that the matrix $\mathcal{T}_h$ representing the operator T_H is given by

$$\mathcal{T}_h := (2a(2i-j))_{i,j=-N+1}^{N-1}, \tag{3.20}$$

and that

$$\text{vec}(T_H B) = \mathcal{T}_h \text{vec}(B), \quad B \in \mathcal{V}_N,$$

where $a(j)$ is the $r^2 \times r^2$ matrix defined by $a(j) := \sum_{\ell=0}^{N} h(\ell-j) \otimes h(\ell)$.

3.4 Approximation order and vanishing moments

Let $\Phi = (\phi_1, \ldots, \phi_r)^T$ be a $(2, h)$-refinable vector and V_0 be the closed linear span of the integer shifts of ϕ_j, $j = 1, \ldots, r$. For the definition of approximation order for shift-invariant subspaces generated by Φ, see [3], [4], [32] and [33].

The multifilter h is said to have *vanishing moments of order m* if there exist real $1 \times r$ row vectors ℓ_0^k, $k = 0, \ldots, m-1$, with $\ell_0^0 \neq 0$, such that

$$\begin{cases} \sum_{j=0}^{k} \binom{k}{j} (2i)^{j-k} \ell_0^j D^{k-j} H(0) = 2^{-k} \ell_0^k, \\ \sum_{j=0}^{k} \binom{k}{j} (2i)^{j-k} \ell_0^j D^{k-j} H(\pi) = 0, \end{cases} \tag{3.21}$$

where $D^{k-j}H(u)$ denotes the matrix formed by the $(k-j)$th derivatives of the entries of $H(u)$.

The relationship between approximation order and vanishing moments has been studied by many researchers ([31], [56], [36]). The following results are due to Heil, Strang and Strela [31] and Plonka [56].

Theorem 3.4 *Suppose that the compactly supported $(2,h)$-refinable vector $\Phi = (\phi_1, \ldots, \phi_r)^T$ is stable. Then the following are equivalent:*
(a) *Φ has approximation of order m,*
(b) *polynomials of degree less than m lie in the closed linear span of the integer shifts of ϕ_j, $j = 1, \ldots, r$,*
(c) *h has vanishing moments of order m.*

3.5 Regularity

In the one-dimensional scalar case, there is a simple characterization of regularity of a scaling function that provides a formula for computing the Sobolev exponent readily (see [15], [20], [73]). Interest in regularity therefore spilled over to refinable vectors (see [12], [64], [54], [38], [39], [37], [59]). This is a natural historical development. Various regularity estimates of a refinable vector Φ were given in [12], [64], [54], [38], [39], [37], [59]. The standard method motivated by the scalar case in one dimension is to estimate the decay of the Fourier transform $\widehat{\Phi}(u) = \prod_{j=1}^{\infty} H(u/2^j)\widehat{\Phi}(0)$. This requires a factorization of the mask $H(u)$ corresponding to the B-spline factor in the scalar case. Such a factorization for the vector case was obtained by Plonka [12], who was obviously also motivated by the cardinal Hermite splines [55]. In [55], she used the nonuniform B-splines with multiple knots of same multiplicity at the integers. Unlike the cardinal Hermite B-splines described here, these B-splines are refinable with finite masks which factorize completely into simple factors. Shen [64] avoided the use of factorization of the masks to obtain regularity estimates for scaling vectors. The results in [64] were subsequently extended by Jiang ([38], [39]) to cover general dilation matrices. Sharper regularity estimates were obtained in [38] and [39] in terms of the spectral radius of transition operators restricted to invariant subspaces that are smaller than those in [64]. For the scalar case $r = 1$, Jia [35] gave a characterization of regularity of (M, h)-refinable functions in $\mathbb{R}^d$ with isotropic dilation matrix M, i.e. M is similar to a diagonal matrix $\mathrm{diag}(\lambda_1, \ldots, \lambda_d)$, and $|\lambda_1| = \cdots = |\lambda_d|$.

Recently, Jia, Riemenschneider and Zhou [37] as well as Ron and Shen [59] gave a complete characterization of regularity of refinable vectors. The results in [59] also cover the case of infinite mask and provide an estimate for each component of the scaling vector. In what follows we shall describe the regularity estimates in [39] in more detail. This choice is just our personal preference, because we find the results in [39] simpler and amenable to computation.

For $s \geq 0$, let $W^s(\mathbb{R})$ be the Sobolev space of functions f with $(1 + |u|^2)^{\frac{s}{2}} \widehat{f}(u) \in L^2(\mathbb{R})$. Let $\gamma = n + \lambda$ with $n \in \mathbb{Z}_+$ and $0 \leq \lambda < 1$. We define $f \in C^{\gamma}(\mathbb{R})$ if and only if $f \in C^{(n)}(\mathbb{R})$ and $f^{(n)}$ is uniformly Hölder continuous with exponent λ, i.e.

$$|D^n f(x+y) - D^n f(x)| \leq C|y|^{\lambda},$$

7. From Cardinal Hermite Splines to Multiwavelets   177

for some constant C independent of $x, y \in \mathbb{R}$, where D denotes derivative. Then we have the well-known inclusion

$$W^s(\mathbb{R}) \subset C^\gamma(\mathbb{R}), \quad \text{for } s > \gamma + \frac{1}{2}.$$

Suppose that the finite mask h has vanishing moments of order m, i.e. its corresponding matrix H satisfies (3.21) for some $1 \times r$ vectors ℓ_0^k, $k = 0, \ldots, m-1$, with $\ell_0^0 \neq 0$. Let $m_0 \leq m$ be the largest nonnegative integer such that there exist $1 \times r$ vectors ℓ_0^k, $k = m, \ldots, m+m_0-1$, satisfying

$$\sum_{j=0}^{k} \binom{k}{j} (2i)^{j-k} \ell_0^j D^{k-j} H(0) = 2^{-k} \ell_0^k. \quad (3.22)$$

Note that if all the numbers 2^{-k}, $k = m, \ldots, m+m_0-1$, are not eigenvalues of $H(0)$ for some $m_0 \in \mathbb{Z}_+$, then the vectors ℓ_0^k, $k = m, \ldots, m+m_0-1$, can be chosen iteratively by

$$\ell_0^k \left(2^{-k} I_r - H(0) \right) = \sum_{j=0}^{k-1} \binom{k}{j} (2i)^{j-k} \ell_0^j D^{k-j} H(0).$$

For $\nu \in \mathbb{Z}$, define $1 \times r$ row vectors ℓ_ν^k by

$$\ell_\nu^k := \sum_{j=0}^{k} \binom{k}{j} (-\nu)^{k-j} \ell_0^k, \quad k = 0, \ldots, m+m_0-1. \quad (3.23)$$

Writing

$$\ell^k(\nu) := \sum_{j=0}^{k} (-1)^j \binom{k}{j} \ell_{-\nu}^j \otimes \ell_0^{k-j}, \quad \nu \in \mathbb{Z},$$

let L_N^k be the $1 \times ((2N-1)r^2)$ vectors given by

$$L_N^k := (\ell^k(-N+1), \ldots, \ell^k(N-1)). \quad (3.24)$$

Then as shown in [39], $L_N^k \mathcal{T}_h = 2^{-k} L_N^k$. Therefore if $L_N^k \neq 0$, then L_N^k is a left eigenvector of $\mathcal{T}_h$ corresponding to the eigenvalue 2^{-k}.

Let $e_j := (\delta_j(k))_{k=1}^r$, $j = 1, \ldots, r$, be the standard unit row vectors in $\mathbb{R}^r$, and for $k < m$, let

$$p_j^k(\nu) := e_j \otimes \ell_\nu^k, \quad q_j^k(\nu) := \ell_{-\nu}^k \otimes e_j, \quad \nu \in \mathbb{Z}.$$

Then define the $1 \times ((2N-1)r^2)$ vectors $P_{j,N}^k$, $Q_{j,N}^k$, $k < m$, respectively by

$$P_{j,N}^k := (p_j^k(-N+1), \ldots, p_j^k(N-1)),$$
$$Q_{j,N}^k := (q_j^k(-N+1), \ldots, q_j^k(N-1)).$$

Let $\mathcal{L}_N$ be the $r^2(2N-1) \times (m+m_0)$ matrix defined by

$$\mathcal{L}_N := ((L_N^0)^T, \ldots, (L_N^{m+m_0-1})^T),$$

where L_N^k are the vectors in (3.24). For $j = 1, \ldots, r$, let $\mathcal{P}_{j,N}$ and $\mathcal{Q}_{j,N}$ be the $r^2(2N-1) \times m$ matrices defined respectively by

$$\mathcal{P}_{j,N} := ((P_{j,N}^0)^T, \ldots, (P_{j,N}^{m-1})^T), \quad \mathcal{Q}_{j,N} := ((Q_{j,N}^0)^T, \ldots, (Q_{j,N}^{m-1})^T).$$

Finally, we define the $r^2(2N-1) \times ((2r+1)m + m_0)$ matrix

$$\mathcal{M}_N := (\mathcal{L}_N, \mathcal{P}_{1,N}, \ldots, \mathcal{P}_{r,N}, \mathcal{Q}_{1,N}, \ldots, \mathcal{Q}_{r,N}).$$

Let $\mathcal{V}_N^0$ denote the subspace of $\mathcal{V}_N$ which comprises $B \in \mathcal{V}_N$ satisfying $(\mathcal{M}_N)^T \text{vec}(B) = 0$. Then $\mathcal{V}_N^0$ is invariant under T_H. Let $T_H|_{\mathcal{V}_N^0}$ denote the restriction of T_H to $\mathcal{V}_N^0$, and let

$$s_0 := -\log_4(\rho(T_H|_{\mathcal{V}_N^0})), \qquad (3.25)$$

where $\rho(T_H|_{\mathcal{V}_N^0})$ is the spectral radius of $T_H|_{\mathcal{V}_N^0}$. Restricting Jiang's results in [39] to one dimension with dilation 2, we have

Theorem 3.5 *The components ϕ_j, $j = 1, \ldots, r$, of the $(2, h)$-refinable vector Φ belong to $W^{s_0 - \epsilon}(\mathbb{R})$ for any $\epsilon > 0$.*

Corollary 3.2 *The components ϕ_j, $j = 1, \ldots, r$, of the $(2, h)$-refinable vector Φ belong to $C^{s_0 - \frac{1}{2} - \epsilon}(\mathbb{R})$ for any $\epsilon > 0$.*

Note that the results in [39] cover multidimensional scaling vectors and multiwavelets with an arbitrary dilation matrix M whose eigenvalues are all nondegenerate.

4 Periodic multiwavelets

In many practical applications, one often deals with periodic functions and periodic filters. Thus the study of periodic multiwavelets of the space of 2π-periodic functions, $L^2([0, 2\pi)^d)$, is of interest. Analogous to the construction of multiwavelets of $L^2(\mathbb{R}^d)$, periodic multiwavelets can be constructed via a periodic multiresolution $\{V_m\}_{m \geq 0}$ of $L^2([0, 2\pi)^d)$ of multiplicity r with dilation matrix M generated by periodic scaling functions. In [22], a general theory of periodic multiresolutions and multiwavelets was derived. Here, we shall describe some of the results. For simplicity, we shall concentrate on the case $d = 1$ and $M = (2)$.

For $m \geq 0$, the multiresolution subspace V_m is generated by the $\frac{2\pi}{2^m}$-shifts of r scaling functions $\phi_1^m, \ldots, \phi_r^m$, i.e.

$$V_m = \langle \{\phi_j^m(\cdot - 2\pi k/2^m) : k = 0, \ldots, 2^m - 1, j = 1, \ldots, r\} \rangle.$$

For different levels m of the multiresolution, the scaling functions for the corresponding subspaces V_m need not be related by dilation. Thus the periodic multiresolution $\{V_m\}_{m\geq 0}$ is nonstationary. The sequence of scaling functions $\phi_1^m, \ldots, \phi_r^m$, $m \geq 0$, satisfies the *periodic matrix refinement equation*

$$\Phi^m(x) = \sum_{k=0}^{2^{m+1}-1} h_{m+1}(k)\Phi^{m+1}(x - 2\pi k/2^{m+1}), \quad x \in [0, 2\pi), \tag{4.1}$$

for every $m \geq 0$, where $h_{m+1}(k)$, $k = 0, \ldots, 2^{m+1}-1$, is a periodic sequence of $r \times r$ matrices of period 2^{m+1}, and $\Phi^m := (\phi_1^m, \ldots, \phi_r^m)^T$. By considering Fourier coefficients, we see that (4.1) is equivalent to

$$\widehat{\Phi}^m(n) = H_{m+1}(n)\widehat{\Phi}^{m+1}(n), \quad n \in \mathbb{Z}, \tag{4.2}$$

where $\widehat{\Phi}^m(n) := (\hat{\phi}_1^m(n), \ldots, \hat{\phi}_r^m(n))^T$, $n \in \mathbb{Z}$, and H_{m+1} denotes the finite Fourier transform (FFT) of h_{m+1}. The equations (4.1) and (4.2) are periodic analogues of (3.3) and (3.4) respectively.

The analysis of a periodic multiresolution is enriched by a class of linearly independent functions called *polyphase splines*, defined by

$$v_j^{m,k}(x) := \sum_{p \in \mathbb{Z}} \hat{\phi}_j^m(k + 2^m p)e^{i(k+2^m p)x}, \quad x \in [0, 2\pi), \tag{4.3}$$

for $m \geq 0$, $k = 0, \ldots, 2^m-1$, $j = 1, \ldots, r$. It was shown in [22] that $\{v_j^{m,k} : k = 0, \ldots, 2^m-1, j = 1, \ldots, r\}$ forms an alternative basis of the multiresolution subspace V_m. The polyphase spline vector $v^{m,k} := (v_1^{m,k}, \ldots, v_r^{m,k})^T$ largely facilitates the construction of periodic multiwavelets.

For $m \geq 0$, let W_m be the wavelet subspace defined by the orthogonal complement of V_m in V_{m+1}. We seek multiwavelets $\psi_1^m, \ldots, \psi_r^m$ such that their $\frac{2\pi}{2^m}$-shifts form a basis of W_m.

Theorem 4.1 *Let $\{V_m\}_{m\geq 0}$ be a periodic multiresolution of $L^2[0, 2\pi)$ of multiplicity r with dilation $M = 2$. Then there exists a sequence G_{m+1}, $m \geq 0$, of $r \times r$ periodic matrices of period 2^{m+1} such that for every $m \geq 0$, the $\frac{2\pi}{2^m}$-shifts of the functions $\psi_1^m, \ldots, \psi_r^m$, defined by*

$$\Psi^m := \sum_{k=0}^{2^m-1} u^{m,k}, \tag{4.4}$$

where $\Psi^m := (\psi_1^m, \ldots, \psi_r^m)^T$ and

$$u^{m,k} := G_{m+1}(k)v^{m+1,k} + G_{m+1}(k + 2^m)v^{m+1,k+2^m} \tag{4.5}$$

for $k = 0, \ldots, 2^m - 1$, form a basis of W_m.

The above theorem was proved in [22] under the most general multidimensional setting with an arbitrary dilation matrix M. Its proof is constructive which leads to an algorithmic approach of constructing multiwavelets once a periodic multiresolution is known. The polyphase spline basis $\{v_j^{m,k} : k = 0, \ldots, 2^m - 1, j = 1, \ldots, r\}$ of V_m, $m \geq 0$ enables one to reduce the multiwavelet construction problem to a tractable matrix extension problem, which gives the desired matrices G_{m+1}, $m \geq 0$.

5 References

[1] Bartle, G., *Cardinal spline interpolation and the block spin construction of wavelets*, in Wavelets: A Tutorial in Theory and Applications, C. K. Chui (ed.), Academic Press, 1992, 73–90.

[2] de Boor, C., R. DeVore and A. Ron, *On the construction of multivariate (pre)wavelets*, Constr. Approx. 9(1993), 123–166.

[3] de Boor, C., R. DeVore and A. Ron, *The structure of finitely generated shift-invariant spaces in $L_2(\mathbb{R})$*, J. Funct. Anal. 119(1994), 37–78.

[4] de Boor, C., R. DeVore and A. Ron, *Approximation orders of FSI spaces in $L_2(\mathbb{R})$*, Preprint, 1996.

[5] Carlitz, L., *Eulerian numbers and polynomials*, Math. Mag. 32(1959), 247–260.

[6] Cavaretta, A. S., W. Dahmen and C. A. Micchelli, *Stationery subdivision*, Mem. Amer. Math. Soc. 93(1991), 1–186.

[7] Chui, C. K., *An Introduction to Wavelets*, Academic Press, Boston, 1992.

[8] Chui, C. K. and J. Lian, *A study on orthonormal multi-wavelets*, J. Appl. Numer. Math. 20(1996), 272–298.

[9] Chui, C. K. and J. Wang, *On compactly supported spline wavelets and a duality principle*, Trans. Amer. Math. Soc. 330(1992), 903–915.

[10] Cohen, A. and I. Daubechies, *A stability criterion for biorthogonal wavelet bases and their subband coding scheme*, Duke Math. J. 68(1992), 313–335.

[11] Cohen, A., I. Daubechies and J. C. Feauveau, *Biorthogonal basis of compactly supported wavelets*, Comm. Pure and Appl. Math. 45(1992) 485–560.

[12] Cohen, A., I. Daubechies and G. Plonka, *Regularity of refinable functions*, J. Fourier Anal. and Appl. (1996).

[13] Dai, X. and D. R. Larson, *Wandering vectors for unitary systems and orthogonal wavelets*, Mem. Amer. Math. Soc. (1996).

[14] Daubechies, I., *Orthonormal bases of compactly supported wavelets*, Comm. Pure and Appl. Math. 41(1988), 909–996.

[15] Daubechies, I., *Ten lectures on Wavelets*, CBMS-NSF Series in Appl. Math. #61, SIAM Publ., Philadelphia, 1992.

[16] Deslauriers, G. and S. Dubuc, *Symmetric iterative interpolation processes*, Constr. Approx. 5(1989), 49–68.

[17] Donovan, G., J. S. Geronimo and D. H. Hardin, *Intertwining multiresolution analysis and the construction of piecewise polynomial wavelets*, SIAM J. Math. Anal. 27(1996), 1791–1815.

[18] Donovan, G., J. S. Geronimo, D. H. Hardin and P. R. Massopust, *Construction of orthogonal wavelets using fractal interpolation functions*, SIAM J. Math. Anal. 27(1996), 1158–1192.

[19] Donovan, G., J. S. Geronimo, D. H. Hardin, *A class of orthogonal multiresolution analyses in 2D*, Preprint, 1997.

[20] Eirola, T., *Sobolev characterization of solutions of dilation equations*, SIAM J. Math. Anal. 23(1992), 1015–1030.

[21] Geronimo, J. S., D. H. Hardin and P. R. Massopust, *Fractal functions and wavelet expansions based on several scaling functions*, J. Approx. Theory 78(1994), 373–401.

[22] Goh, S. S., S. L. Lee and K. M. Teo, *Multidimensional periodic multiwavelets*, Preprint, National University of Singapore, 1997.

[23] Goh, S. S. and V. B. Yap, *Matrix extension and biorthogonal multiwavelet construction*, Linear Algebra and its Applic. (to appear).

[24] Goodman, T. N. T., *Characterizing pairs of bivariate refinable splines*, Preprint, University of Dundee, 1997.

[25] Goodman, T. N. T. and S. L. Lee, *Wavelets of multiplicity r*, Trans. Amer. Math. Soc. 342(1994), 307–324.

[26] Goodman, T. N. T. and S. L. Lee, *Convergence of nonstationary vector cascade algorithms*, Preprint, National University of Singapore, 1997.

[27] Goodman, T. N. T., S. L. Lee and W. S. Tang, *Wavelets in wandering subspaces,* Trans. Amer. Math. Soc. 338(1993), 639–654.

[28] Goodman, T. N. T., S. L. Lee and W. S. Tang, *Wavelet bases for a set of commuting unitary operators,* Adv. in Comput. Math. 1(1993), 109–126.

[29] Goodman, T. N. T., R. Q. Jia and C. A. Micchelli, *On the spectral radius of a bi-infinite periodic and slanted matrix,* South East Asian Math. Bulletin (to appear).

[30] Heil, C. and D. Colella, *Matrix refinement equations: existence and uniqueness,* J. Fourier Anal. and Appl. 2(1996), 363–377.

[31] Heil, C., G. Strang and V. Stella, *Approximation by translates of refinable functions,* Numer. Math. 73(1996), 75–94.

[32] Jia, R. Q. and J. J. Lei, *Approximation by multi-integer translates of functions having global support,* J. Approx. Theory 72(1993), 2–23.

[33] Jia, R. Q., *Refinable shift-invariant spaces: from splines to multiwavelets,* Approximation Theory VIII, Vol. 2 (C. Chui and L. Schumaker eds.) 1995, pp. 179–208.

[34] Jia, R. Q. and Bin Han, *Multivariate refinement equations and subdivision schemes,* SIAM J. Math. Analysis (to appear).

[35] Jia, R. Q., *Characterization of smoothness of multivariate refinable functions in Sobolev space,* Trans. Amer. Math. Soc. (to appear).

[36] Jia, R. Q., S. Riemenschneider and D. X. Zhou, *Approximation by multiple refinable functions,* Canadian J. Math. (to appear).

[37] Jia, R. Q., S. Riemenschneider and D. X. Zhou, *Smoothness of multiple refinable functions and multiwavelets,* Preprint, University of Alberta, 1996.

[38] Jiang, Q., *On the regularity of matrix refinable functions,* SIAM J. Math. Anal. (to appear).

[39] Jiang, Q., *Multivariate matrix refinable functions with arbitrary matrix dilation,* Trans. Amer. Math. Soc. (to appear).

[40] Jiang, Q., *Orthogonal multiwavelets with optimum time-frequency resolution,* IEEE Trans. Signal Proc. (to appear).

[41] Jiang, Q., *On the design of multifilter banks and orthogonal multiwavelet bases,* Preprint, National University of Singapore, 1997.

[42] Jiang, Q., *On the construction of biorthogonal multiwavelet bases*, Preprint, National University of Singapore, 1997.

[43] Jiang, Q. and Z. Shen, *On the existence and weak stability of matrix refinable functions*, Constr. Approx. (to appear).

[44] Lawton, W., *Necessary and sufficient conditions for constructing orthogonal wavelets*, J. Math. Phys. 32(1991), 52–61.

[45] Lawton, W., *Multilevel properties of the wavelet-Galerkin operator*, J. Math. Phys. 32(1991), 1440–1443.

[46] Lawton, W., S. L. Lee and Z. Shen, *An algorithm for matrix extension and wavelet construction*, Math. of Comput. 65(1996), 723–737.

[47] Lawton, W., S. L. Lee and Z. Shen, *Stability and orthonormality of multivariate refinable functions*, SIAM J. Math. Anal. 28(1997), 999–1014.

[48] Lawton, W., S. L. Lee and Z. Shen, *Convergence of multidimensional cascade algorithm*, Numerische Math. (to appear).

[49] Lee, S. L., *B-spline for cardinal Hermite interpolation*, Linear Algebra and its Applic. 12(1975), 269–280.

[50] Lee, S. L., H. H. Tan and W. S. Tang, *Wavelet bases for a unitary operator*, Proc. Edinburgh Math. Soc. 38(1995), 233–260.

[51] Lemarié, P., *Ondelettes á localization exponentielle*, J. Math. Pures et Appl. 67(1988), 227–236.

[52] R. Long, W. Chen and S. Yuan, *Wavelets generated by vector multiresolution analysis*, Appl. Comput. Harmon. Anal. 4(1997), 317–350.

[53] Mallat, S., *Multiresolution approximation and wavelets*, Trans. Amer. Math. Soc. 315(1989), 628–666.

[54] Micchelli, C. A. and T. Saur, *Regularity of multiwavelets*, Advances in Comp. Math. 7(1997), 455–545.

[55] Plonka, G., *Two-scale symbol and autocorrelation symbol for B-splines with multiple knots*, Adv. in Comput. Math. 3(1995), 1–22.

[56] Plonka, G., *Approximation order provided by refinable function vectors*, Constr. Approx. 13(1997), 221–244.

[57] Plonka, G., *On stability of scaling vectors*, in Surface Fitting and Multiresolution Methods, A. Le Méhauté, C. Rabut and L. L. Schumaker (eds.), Vanderbilt University Press, 1997, 293–300.

[58] Robertson, J. B., *On wandering subspaces for unitary operators*, Proc. Amer. Math. Soc. 16(1965), 233–236.

[59] Ron, A. and Z. W. Shen, *The Sobolev regularity of refinable functions*, Preprint, National University of Singapore, 1997.

[60] Schoenberg, I. J., *Contributions to the problem of approximation of equidistant data by analytic functions*, Quart. Appl. Math. 4(1946), 45–99, 112–141.

[61] Schoenberg, I. J., *Cardinal interpolation and spline functions*, J. Approx. Theory 2(1969), 167–206.

[62] Schoenberg, I. J., *Cardinal Spline Interpolation*, CBMS-NSF Series in Appl. Math. #12, SIAM Publ., Philadelphia, 1973.

[63] Schoenberg, I. J. and A. Sharma, *Cardinal interpolation and spline functions V. The B-splines for cardinal Hermite interpolation*, Linear Algebra and its Applic. 7(1973), 1–42.

[64] Shen, Z., *Refinable function vectors*, SIAM J. Math. Anal. (to appear).

[65] Strang, G., *The optimal coefficients in Daubechies wavelets*, Physica D. 60(1992), 239–244.

[66] Strang, G., *Eigenvalues of $(\downarrow 2)H$ and convergence of the cascade algorithm*, IEEE Trans. Signal Proc. 44(1996).

[67] Strang, G. and T. Nguyen, *Wavelets and Filter Banks*, Wellesley-Cambridge Press, 1996.

[68] Strang, G. and V. Strela, *Short wavelets and matrix dilation equations*, IEEE Trans. Signal Proc. 43(1995), 108–115.

[69] Strela, V., P. Heller, G. Strang, P. Topiwala and C. Heil, *The application of multiwavelet filter banks to image processing*, IEEE Trans. Image Proc. (to appear).

[70] Tang, W. S., *Oblique projections, biorthogonal Riesz bases and multiwavelets in Hilbert spaces*, Preprint, National University of Singapore, 1997.

[71] Tham, J. Y., L. Shen, S. L. Lee and H. H. Tan, *On the design and applications of multiwavelets in image compression*, Preprint, National University of Singapore, 1997.

[72] Villemoes, L., *Energy moments in time and frequency for two-scale difference equation solutions and wavelets*, SIAM J. Math. Anal. 23(1992), 1519–1543.

[73] Villemoes, L., *Wavelet analysis of refinement equations,* SIAM J. Math. Anal. 25(1994), 1433–1460.

[74] Xia, X., D. Hardin, J. Geronimo and B. Suter, *Design of prefilters for discrete multiwavelet transforms,* IEEE Trans. Signal Proc. 4(1996), 25–35.

[75] Yap, V. B., *Biorthogonal Multiwavelets and Matrix Extensions,* M.Sc. Thesis, National University of Singapore, 1997.

8
Orthogonal Multiwavelet Constructions: 101 Things To Do With a Hat Function

G.C. Donovan [1]   J.S. Geronimo [2]
and D.P. Hardin [3]

ABSTRACT We describe a method for constructing compactly supported orthogonal shift invariant bases and apply this construction to find orthogonal multiwavelets.

1 Introduction

In this paper we describe a method developed in [2] for constructing orthogonal multiwavelets. Our examples will all start with the piecewise linear "hat" function $h(x) := (1 - |x|)^+$ and hence the title of this paper.

First we introduce some notation. We call a finite length vector $\Phi = (\phi_1, \ldots, \phi_r)^T$ with components in $L^2(\mathbf{R})$ a *generator*. For a generator Φ, let
$$B(\Phi) := \{\phi_i(\cdot - n) | n \in \mathbf{Z}, i = 1, \ldots, r\}$$
and
$$S(\Phi) := \operatorname{clos}_{L^2(\mathbf{R})} \operatorname{span}(B(\Phi)).$$

A space V is called a *finitely generated shift-invariant space* (FSI) if $V = S(\Phi)$ for some generator Φ. If $B(\Phi)$ is an orthogonal system, we say that Φ is an *orthogonal generator*.

---

[1] Department of Mathematics, Princeton University, Princeton, New Jersey 08544-1000 (e-mail:donovan@math.princeton.edu)
The research was partially supported by an NSF postdoctoral fellowship
[2] School of Mathematics, Georgia Institute of Technology, Atlanta, Georgia 30332 (e-mail:geronimo@math.gatech.edu)
The research was partially supported by a grant from the NSF
[3] Department of Mathematics, Vanderbilt University, Nashville, Tennessee 37240-0001 (e-mail:hardin@math.vanderbilt.edu)
The research was partially supported by a grant from the NSF

Let $Df = f(\cdot/2)$. An FSI space V is said to be *refinable* if

$$D(V) \subset V.$$

A generator $\Phi = (\phi_1, \ldots, \phi_r)^T$ is said to be *refinable* if Φ satisfies a refinement equation of the form

$$\Phi(x/2) = \sum_k c(k)\Phi(x-k), \qquad x \in \mathbf{R},$$

for some finitely supported sequence $c : \mathbf{Z} \to \mathbf{R}^{r \times r}$. The matrices $c(k)$ are called the scaling coefficients for Φ. If Φ is refinable then $V = S(\Phi)$ is refinable as well. If Φ is a refinable generator then we also call Φ a *scaling vector*.

Let Φ be a compactly supported refinable generator, $V = S(\Phi)$, and $V_j = D^{-j}V$ for $j \in \mathbf{Z}$. Then the sequence of nested spaces V_j is called a *multiresolution analysis of $L^2(\mathbf{R})$*. (The compact support of Φ shows that the usual conditions $\bigcap V_j = 0$ and clos $\bigcup V_j = L^2(\mathbf{R})$ are automatically satisfied, see [6] and references therein.) Define $W_j := V_{j+1} \cap V_j^\perp$, for $j \in \mathbf{Z}$. If Ψ is a generator such that $B(\Psi)$ is an orthogonal basis for W_0 then Ψ is called an orthogonal multiwavelet associated with Φ. If Ψ is an orthogonal multiwavelet then there must exist a sequence of matrices $d : \mathbf{Z} \to \mathbf{R}^{r \times r}$ so that

$$\Psi(x/2) = \sum_k d(k)\Phi(x-k), \qquad x \in \mathbf{R}.$$

The matrices $d(k)$ are called the wavelet coefficients for Ψ.

The focus of this paper is a method for constructing orthogonal scaling vectors which was first developed in [2] for the purpose of constructing piecewise polynomial orthogonal scaling vectors. As we show here, this method can also be used to give a simple construction of the family of orthogonal scaling vectors constructed in [4] using fractal interpolation functions. There are two parts to our construction:

- In Section 2 we present a construction of orthogonal generators that is based on adding appropriate components to a given generator.

- In Section 3 we discuss two ways to preserve refinability: (a) fractal interpolation functions, and (b) intertwining multiresolution analyses.

Finally, in Section 4 we present a simple method for finding the wavelet coefficients for an orthogonal multiwavelet given the scaling coefficients of an orthogonal scaling vector of the type considered in this paper.

2 Orthogonal generators

In this section we describe the procedure from [2] for creating orthogonal generators. The idea is to start with a generator supported on $[-1, 1]$ and add components that are supported on $[0,1]$ in such a way that the resulting FSI space has an orthogonal, compactly-supported generator.

We illustrate the method by first considering the special case of starting with the hat function h and adding a function $w \in L^2(\mathbf{R})$ that is supported on $[0,1]$. Let $V = S((h,w)^T)$. We construct a new generator $\Phi = \{\phi_1, \phi_2\}$ by replacing h with $(I - P_{S(w)})h$:

$$\phi_1 = (I - P_{S(w)})h \tag{1}$$

$$= h - \frac{\langle h, w \rangle}{\langle w, w \rangle} w - \frac{\langle h, w(\cdot + 1) \rangle}{\langle w, w \rangle} w(\cdot + 1) \tag{2}$$

$$\phi_2 = w \tag{3}$$

Since $h = \phi_1 + P_{S(\phi_2)}h \in S(\Phi)$, it follows that $V = S(\Phi)$. Also, ϕ_1 is orthogonal to the shifts of ϕ_2 by construction. Hence, $\Phi = \{\phi_1, \phi_2\}$ will be an orthogonal generator iff

$$\langle \phi_1, \phi_1(\cdot - 1) \rangle = 0$$

which, substituting for ϕ_1 using (1), is equivalent to

$$\langle h, h(\cdot - 1) \rangle = \frac{\langle h, w \rangle \langle w, h(\cdot - 1) \rangle}{\langle w, w \rangle}. \tag{4}$$

To reiterate, we have the following lemma:

Lemma 1 *Suppose $w \in L^2(\mathbf{R})$ is supported on $[0,1]$ and satisfies (4) then the generator Φ given by (1) is an orthogonal generator for $V = S((h,w)^T)$.*

2.1 Example A

Here we find a specific example of a w satisfying the hypotheses of Lemma 1. Note that the condition (4) is a condition on the span of w. For this example we add the additional constraints that w be piecewise polynomial and symmetric about $x = 1/2$.

Let $q : \mathbf{R} \to \mathbf{R}$ be the piecewise quadratic polynomial given by $q(x) = x(1-x)\chi_{[0,1]}(x)$. In order for w to be symmetric we choose $w \in \text{span}\{q, q^2\}$. Then

$$w = c_1 q + c_2 q^2$$

for some constants c_1 and c_2 and we directly calculate

$$\langle h, w \rangle = \langle w, h(\cdot - 1) \rangle$$
$$= \frac{1}{60}(5c_1 + c_2)$$
$$\langle w, w \rangle = \frac{1}{630}(21c_1^2 + 9c_1 c_2 + c_2^2).$$

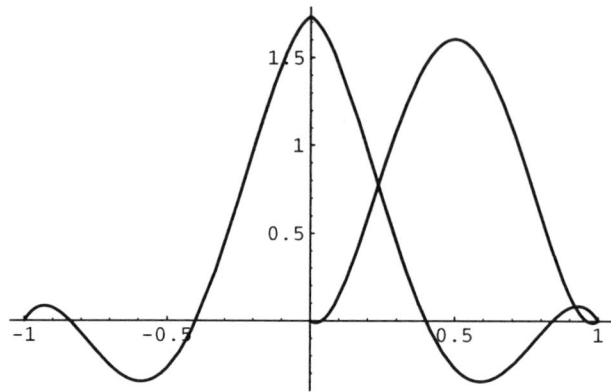

FIGURE 1. Orthogonal (not refinable) generator from Example A

Substituting the above into Equation (4) gives the following quadratic equation in the variable $\alpha := c_2/c_1$:

$$\alpha^2 + 30\alpha + 105 = 0.$$

Hence, $\alpha = -15 \pm 2\sqrt{30}$. The resulting orthogonal generator for the choice $\alpha = -15 - 2\sqrt{30}$ is shown in Figure 1.

2.2 General construction

More generally, suppose $G = (g_1, \ldots, g_s)^T$ is a generator supported on $[-1, 1]$ such that no component of G is supported on $[-1,0]$ or $[0,1]$, and further suppose $U = (u_1, \ldots, u_t)^T$ is a generator supported on $[0, 1]$. Let $\widetilde{\Phi} = (G, U)$. We wish to know when there exists some orthogonal generator Φ such that $S(\Phi) = S(\widetilde{\Phi})$. If

$$\langle (I - P_{S(U)})g_i, (I - P_{S(U)})g_j \rangle = 0, \qquad 1 \leq i \neq j \leq s \tag{5}$$

then an orthogonal generator Φ for $S(\widetilde{\Phi})$ can be found as follows:
Let $(\phi_1, \ldots, \phi_s)$ be an orthogonal basis for the span of

$$\{(I - P_{S(U)})g_1, \ldots, (I - P_{S(U)})g_s\}$$

and let $(\phi_{s+1}, \ldots, \phi_r)$, $r = s + t$, be an orthogonal basis for the span of the elements of W. Then $\Phi = (\phi_1, \ldots, \phi_r)^T$ is an orthogonal generator and since $H \in S(\Phi)$ it follows that $S(H) \subset S(\Phi)$ and thus $V = S(\Phi)$. Under the additional assumption of local linear independence of the generators, it is shown in [2] that the condition (5) is both necessary and sufficient for the existence of an orthogonal Φ.

3 Refinable generators

Suppose H is a refinable generator supported on $[-1, 1]$ with

$$H(x/2) = \sum_k q(k) H(x - k).$$

As in the previous section we add a generator W supported on $[0, 1]$ in order to get orthogonality. However, now we also want the space (H, W) to be refinable. Note that the orthogonal generator constructed in Example A is not refinable. We find two strategies for choosing W so that (H, W) is refinable. First, the most general approach is based on the fact that if (H, W) is refinable then

$$\begin{pmatrix} H \\ W \end{pmatrix}(x/2) = \sum_{k=0,1} \begin{pmatrix} q(k) & 0 \\ r(k) & s(k) \end{pmatrix} \begin{pmatrix} H \\ W \end{pmatrix}(x - k)$$

which implies that W satisfies an inhomogeneous refinement equation of the form:

$$W(x/2) = \sum_{k=0,1} r(k) H(x-k) + \sum_{k=0,1} s(k) W(x-k). \tag{6}$$

In the case H and W are scalar functions, the solutions to such equations fall into a class of functions known as fractal interpolation functions that were introduced by Barnsley in [1].

The second approach is to choose the components of W from V_1 (where V_p is the multiresolution analysis generated by H). This leads to notion of intertwining multiresolution analyses.

3.1 Fractal Interpolation Functions

To illustrate the way fractal interpolation functions arise in our construction, we consider the case $H = (h)$ and $W = (w)$. We further require that $w \in C_0[0, 1]$ where $C_0[0, 1]$ denotes the space of continuous functions on $\mathbf{R}$ whose support lies in $[0, 1]$. Note that h is a scaling function with refinement equation:

$$h(x/2) = (1/2) h(x + 1) + h(x) + (1/2) h(x - 1). \tag{7}$$

If $(h, w)^T$ is to be refinable then $w(\cdot/2)$ must be a linear combination of $h(\cdot - 1)$, w, and $w(\cdot - 1)$. Therefore w satisfies an inhomogeneous refinement equation of the form

$$w(x/2) = h(x - 1) + \sum_{i=0,1} s_i w(x - i) \tag{8}$$

for some constants s_0 and s_1. (We have chosen the normalization $w(1/2) = 1$. If $w(1/2) = 0$, then w is zero at all dyadic rational points and so, by continuity, must be identically zero.) Using $w(1/2^{n+1}) = 1/2^n + s_0 w(1/2^n)$ and the continuity of w at 0 we get $|s_0| < 1$. Similarly, the continuity of w at 1 shows $|s_1| < 1$.

Let $\rho = \max(|s_0|, |s_1|) < 1$. Consider the operator $\Gamma : C_0[0,1] \to C_0[0,1]$ given by

$$\Gamma(f) = h(2 \cdot -1) + \sum_{i=0,1} s_i f(2 \cdot -i). \tag{9}$$

Then Γ is a contraction on $C_0[0,1]$ (in the sup norm) with contractivity ρ. By the contraction mapping principle, Γ has a unique fixed point w, i.e., w is the unique solution of the inhomogeneous dilation equation (8). One can similarly show that (8) has a solution $w \in L^2(\mathbf{R})$ iff $s_0^2 + s_1^2 < 2$.

Equation (8) may be used to calculate exact expressions for various inner products involving w. For example:

$$\begin{aligned} \langle w, 1 \rangle &= \langle h(2 \cdot -1), 1 \rangle + \sum_{i=0,1} s_i \langle w(2 \cdot -i), 1 \rangle \\ &= \frac{1}{2} + \frac{1}{2} \langle w, 1 \rangle \sum_{i=0,1} s_i. \end{aligned}$$

Let $\bar{s} = (s_0 + s_1)/2$. Then solving the above for $\langle w, 1 \rangle$ gives

$$\langle w, 1 \rangle = \frac{1}{2} \frac{1}{1 - \bar{s}}. \tag{10}$$

Similarly, we find

$$\begin{aligned} \langle w, h \rangle &= (2 - s_1)/4(1 - \bar{s})(2 - \bar{s}) \\ \langle w, h(\cdot - 1) \rangle &= (2 - s_0)/4(1 - \bar{s})(2 - \bar{s}) \\ \langle w, w \rangle &= \frac{4 - s_0^2 + s_0 s_1 - s_1^2}{3(1 - \bar{s})(2 - \bar{s})(2 - s_0^2 - s_1^2)}. \end{aligned} \tag{11}$$

Substituting the above into (4) we get:

$$\begin{aligned} 8 + 12(s_0 + s_1) - 28(s_0^2 + s_1^2) - 14(s_0 s_1) + 6(s_0^3 + s_1^3) + \\ 18(s_0^2 s_1 + s_0 s_1^2) + 2(s_0^4 + s_1^4) - 7(s_0^3 s_1 + s_0 s_1^3) = 0 \end{aligned} \tag{12}$$

This gives the one parameter family of orthogonal refinable generators constructed in [4]:

Lemma 2 *Suppose $s_0, s_1 \in (-1, 1)$ satisfy (12). Let w be the unique solution to (8) and let $\Phi = (\phi_1, \phi_2)^T$ be as in (1). Then Φ is an orthogonal refinable generator.*

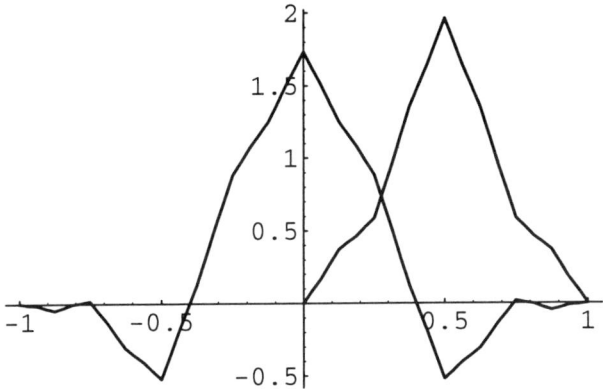

FIGURE 2. Orthogonal scaling vector from Example B

3.2 Example B

For symmetric w we choose $s_0 = s_1 = s$. Substituting into (12) yields $s = -1/5$. The resulting Φ (this was first constructed in [5]) is shown in Figure 2.

The scaling coefficients for the (normalized) Φ are:

$$c(-2) = \tfrac{1}{20}\begin{pmatrix} 0 & -1 \\ 0 & 0 \end{pmatrix} \quad c(-1) = \tfrac{1}{20}\begin{pmatrix} -3\sqrt{2} & 9 \\ 0 & 0 \end{pmatrix}$$

$$c(0) = \tfrac{1}{20}\begin{pmatrix} 10\sqrt{2} & 9 \\ 0 & 6\sqrt{2} \end{pmatrix} \quad c(1) = \tfrac{1}{20}\begin{pmatrix} -3\sqrt{2} & -1 \\ 16 & 6\sqrt{2} \end{pmatrix}$$

3.3 Intertwining Multiresolution Analyses

Suppose (V_p) is a multiresolution analysis with a compactly supported scaling vector. As observed in [2], any multiresolution analysis generated by a compactly supported scaling vector can be reindexed so that it is generated by a scaling vector supported on $[-1, 1]$. Suppose H is a scaling vector supported on $[-1, 1]$. Let $V = S(H)$. Suppose the components of W are in $V_1 = D^{-1}V$. Let $\tilde{V} = S((H, W))$. Then $V_0 \subset \tilde{V}_0 \subset V_1$. Applying D^{-1} yields $V_1 \subset \tilde{V}_1 \subset V_2$ which shows that $\tilde{V}$ is refinable. The idea is then to choose $W \in V_1$ so that (5) holds and therefore obtain an orthogonal scaling vector.

We say a multiresolution analysis is *orthogonal* if it has some orthogonal compactly supported generator. The following theorem shows that an orthogonal multiresolution can always be constructed from any multiresolution analysis that is generated by a compactly supported generator.

Theorem 1 *[2] If (V_p) is a multiresolution analysis generated by a compactly supported scaling vector then there is some pair of integers (q, n) and* some orthogonal multiresolution analysis $(\widetilde{V}_p)$ such that

$$V_q \subset \widetilde{V}_0 \subset V_{q+n}.$$

We say that (V_p) and $(\widetilde{V}_p)$ are *intertwining multiresolution analyses*. Note that (V_0) can be chosen to be any classical spline space with integer knots. Hence, Theorem 1 shows that compactly supported orthogonal piecewise polynomial scaling functions and wavelets of arbitrary regularity and approximation order exist. In [3] we explicitly carry through the intertwining construction starting with the spline spaces $V_0 = S_d^r$, $r = 0, 1$ and $d > r+1$ (here S_d^r denotes the piecewise polynomial functions in $C^r(\mathbf{R}) \cap L^2(\mathbf{R})$ of degree at most d with integer knots). Using the theory of orthogonal polynomials (in particular, the theory of hyperspherical polynomials) we are able to give explicit formulas for the resulting orthogonal generators. Hence, these families provide families of C_0 and C_1 piecewise polynomial orthogonal generators and orthogonal multiwavelets with arbitrary approximation order.

3.4 Example C

Let $H = (h, q)^T$ where, as before, h is the hat function and q is the piecewise quadratic given by $q(x) = x(1-x)\chi_{[0,1]}(x)$. Then $S(H)$ is the space of continuous square-integrable piecewise quadratic functions with integer knots and clearly H is refinable. We choose $w \in V_1$ with supp $w \subset [0, 1]$. Hence w is in the $\mathcal{A} = \text{span } \{h(2 \cdot -1), q(2\cdot), q(2 \cdot -1)\}$. We will eventually need an orthogonal basis for the span of q and w so we also require $w \perp q$. It is easy to verify directly that $v_1 = q(2\cdot) - q(2 \cdot -1)$ and $v_2 = q(2\cdot) + q(2 \cdot -1) - (7/25)h(2 \cdot -1)$ form an orthogonal basis for $\mathcal{A} \cap q^\perp$. In the notation of Section 2.2, we have $G = h$ and $U = (q, w)^T$ and $\Phi = (\phi_1, \phi_2, \phi_3)^T = ((I - P_{S(H)})h, q, w)^T$. Using $w = c_1 v_1 + c_2 v_2$, the orthogonality condition (5) reduces to the quadratic equation $125 - 256\alpha^2 = 0$ in the variable $\alpha = c_2/c_1$. Choosing $\alpha = 5\sqrt{5}/16$ gives the orthogonal generator shown in Figure 3.

4 Multiwavelets

There is a general procedure for constructing the wavelet coefficients from the scaling coefficients. This procedure is based on completing a rectangular matrix polynomial to a square matrix polynomial using the factorization for "paraunitary" matrices of Vaidyanathan (see [8] and [7]). Here we present an alternate construction of the wavelet coefficients in the case the scaling vector is of the type constructed in this paper.

8. 101 Things to do with a Hat Function

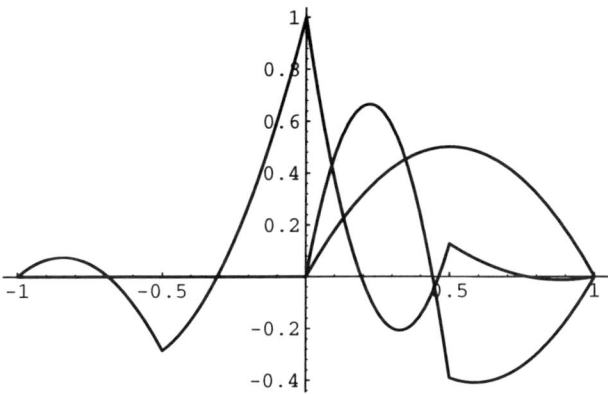

FIGURE 3. Orthogonal scaling vector from Example C

Assume Φ is an orthonormal scaling vector of the type constructed in this paper, that is, we assume

1. Φ is continuous and supported on $[-1, 1]$,
2. $\Phi_{2,r} := (\phi_2, \ldots, \phi_r)^T$ is supported on $[0, 1]$, and
3. $\phi_1(0) \neq 0$.

Then the scaling coefficients for Φ are of the form

$$c(k) = \begin{pmatrix} p(k) & q(k) \\ r(k) & s(k) \end{pmatrix}, \qquad k = -2, \ldots, 1$$

where we have written $c(k)$ in terms of the blocks corresponding to ϕ_1 and $\Phi_{2,r}$ (so $p(k)$ is 1×1, $q(k)$ is $1 \times (r-1)$, etc.). Because $\Phi_{2,r}$ is supported on $[0, 1]$ we have $r(k), s(k) = 0$ for $k = -2, -1$. Because ϕ_1 is continuous and nonzero at 0, we have $p(0) = 1/\sqrt{2}$. Let $\alpha_L = (q(-2)\, p(-1)\, q(-1))$, $\alpha_R = (q(0)\, p(1)\, q(1))$ and $\beta = (s(0)\, r(1)\, s(1))$ (here α_L and α_R denote row vectors in $\mathbf{R}^{2r-1}$ and β is a $r-1 \times 2r-1$ matrix). The orthonormality of Φ implies

- α_L, α_R and the $r-1$ rows of β are orthogonal.
- $\|\phi_1\|^2 = \alpha_L \alpha_L^T + 1/2 + \alpha_R \alpha_R^T = 1$
- the rows of β together with $\hat{\alpha}_L = \alpha_L/\|\alpha_L\|$ and $\hat{\alpha}_R = \alpha_R/\|\alpha_R\|$ form an orthonormal set.

We construct the wavelet coefficients in two steps. If $r > 2$ we find $r - 2$ wavelets that are supported on $[0, 1]$. Let

$$\eta = \begin{pmatrix} \hat{\alpha}_L \\ \hat{\alpha}_R \\ \beta \end{pmatrix}.$$

Then the rows of η form an orthonormal set. Let $\gamma \in \mathbf{R}^{(r-2)\times(2r-1)}$ be such that its the rows together with the rows of η form an orthonormal basis of $\mathbf{R}^{2r-1}$. We write γ in the block form $\gamma = (h(0)\, g(1)\, h(1))$ where $h(0)$ and $h(1)$ are, respectively, the first and last $r-1$ columns of η, and $g(1)$ is the rth column of η. We then construct $r-2$ wavelets supported on $[0,1]$

$$(\psi_3,\ldots,\psi_r)^T = (0 h(0))\sqrt{2}\Phi(2\cdot) + (g(1) h(1))\sqrt{2}\Phi(2\cdot -1).$$

(Note that by construction the functions $\psi_3,\ldots,\psi_r$ are orthogonal to Φ and $\Phi(\cdot -1)$ and hence to V_0.)

The remaining two wavelets ψ_1 and ψ_2 are supported on $[-1,1]$. Suppose ψ is such a wavelet. Then it must be of the form

$$\psi = \mu(-1)\sqrt{2}\Phi_{2,r}(2\cdot +1) + c_1\sqrt{2}\phi_1(2\cdot) + \mu(1)\sqrt{2}\Phi_{2,r}(2\cdot -1).$$

Since ψ is orthogonal to $\Phi_{2,r}$, $\phi_1(\cdot -1)$ and $\psi_3,\ldots,\psi_r$ we have $\mu(1)$ is orthogonal to α_L and the rows of γ, η. Hence, $\mu(1)$ must be a multiple of α_R. Similarly, $\mu(-1)$ must be a multiple of α_L and so ψ must be of the form

$$\psi = c_0\alpha_L\sqrt{2}\Phi_{2,r}(2\cdot +1) + c_1\sqrt{2}\phi_1(2\cdot) + c_2\alpha_R\sqrt{2}\Phi_{2,r}(2\cdot -1).$$

Since $\alpha_L \perp \alpha_R$ it follows that any ψ of the above type is orthogonal to the nonzero shifts of any other ψ (that is, for some different choice of c_0, c_1, c_2). The only orthogonality condition that remains is

$$\langle \psi, \phi_1 \rangle = c_0 \|\alpha_L\|^2 + c_1/\sqrt{2} + c_2 \|\alpha_R\|^2 = 0.$$

Let $\theta = \|\alpha_L\|/\|\alpha_R\|$. Using $\|\alpha_L\|^2 + \|\alpha_R\|^2 = 1/2$ it follows that we may choose $(c_0, c_1, c_2) = (1, -1/\sqrt{2}, 1)$ to get ψ_1 and $(c_0, c_1, c_2) = (\sqrt{2}\theta^{-1}, 0, -\sqrt{2}\theta)$ to get ψ_2. Let $\Psi = (\psi_1,\ldots,\psi_r)$. Recalling the definitions

$$\alpha_L = ((q(-2)\, p(-1)\, q(-1))), \qquad \alpha_R = (q(0)\, p(1)\, q(1))$$

we get the following wavelet coefficients for Ψ:

$$\underbrace{\begin{pmatrix} 0 & q(-2) \\ 0 & \sqrt{2}\theta^{-1}q(-2) \\ 0 & 0 \end{pmatrix}}_{d(-2)}, \quad \underbrace{\begin{pmatrix} p(-1) & q(-1) \\ \sqrt{2}\theta^{-1}p(-1) & \sqrt{2}\theta^{-1}q(-1) \\ 0 & 0 \end{pmatrix}}_{d(-1)},$$

$$\underbrace{\begin{pmatrix} -1/\sqrt{2} & q(0) \\ 0 & -\sqrt{2}\theta q(0) \\ 0 & h(0) \end{pmatrix}}_{d(0)}, \quad \underbrace{\begin{pmatrix} p(1) & q(1) \\ -\sqrt{2}\theta p(1) & -\sqrt{2}\theta q(1) \\ g(1) & h(1) \end{pmatrix}}_{d(1)}.$$

Applying this procedure to the scaling coefficients from Example B we get the following wavelet coefficients (note $\theta = 1$):

$$d(-2) = \tfrac{1}{20}\begin{pmatrix} 0 & -1 \\ 0 & -\sqrt{2} \end{pmatrix} \qquad d(-1) = \tfrac{1}{20}\begin{pmatrix} -3\sqrt{2} & 9 \\ -6 & 9\sqrt{2} \end{pmatrix}$$

$$d(0) = \tfrac{1}{20}\begin{pmatrix} -10\sqrt{2} & 9 \\ 0 & -9\sqrt{2} \end{pmatrix} \qquad d(1) = \tfrac{1}{20}\begin{pmatrix} -3\sqrt{2} & -1 \\ 6 & \sqrt{2} \end{pmatrix}$$

5 References

[1] M. F. Barnsley, Fractal functions and interpolations, Constr. Approx. **2** (1986), 303–329.

[2] G. C. Donovan, J. S. Geronimo, and D. P. Hardin, Intertwining multiresolution analyses and the construction of piecewise polynomial wavelets, *SIAM Journ. Math. Analysis* **27**:6, (1996) 1791–1815.

[3] G. C. Donovan, J. S. Geronimo, and D. P. Hardin, "Orthogonal Polynomials and the Construction of Piecewise Polynomial Smooth Wavelets", preprint (1997).

[4] G. C. Donovan, J. S. Geronimo, D. P. Hardin, and P. R. Massopust, "Construction of Orthogonal Wavelets Using Fractal Interpolation Functions", *SIAM Journ. Math. Analysis* **27**, 1158–1192 (1996)

[5] J. S. Geronimo, D. P. Hardin, and P. R. Massopust, "Fractal functions and wavelet expansions based on several scaling functions", *J. Approx. Theory* **78**, pp. 373–401 (1994)

[6] R.-Q. Jia, "Refinable Shift-invariant Spaces: From Splines to Wavelets", *Approximation Theory VIII, Vol. 2: Wavelets and Multilevel Approximation*, pp. 179-208 (1995)

[7] G. Strang and V. Strela, Short wavelets and matrix dilation equations, IEEE Trans. SP, **43**, pp. 108–115 (1995)

[8] P. P. Vaidyanathan, *Multirate Systems and Filter Banks*, Simon and Schuster, (1993)

9
Convergence of Vector Subdivision Schemes and Construction of Biorthogonal Multiple Wavelets

Rong-Qing Jia[1]

ABSTRACT Let $\phi = (\phi_1, \ldots, \phi_r)^T$ be a refinable vector of compactly supported functions in $L_2(\mathbb{R})$. It is shown in this paper that there exists a refinable vector $\tilde{\phi}$ of compactly supported functions in $L_2(\mathbb{R})$ such that $\tilde{\phi}$ is dual to ϕ if and only if the shifts of $\phi_1, \ldots, \phi_r$ are linearly independent. This result is established on the basis of a complete characterization of the convergence of vector subdivision schemes associated with exponentially decaying masks. As an application of the general theory, two interesting examples of biorthogonal double wavelets are constructed.

1 Introduction

We are interested in multiple refinable functions and multiple wavelets. Suppose $\phi_1, \ldots, \phi_r$ are complex-valued functions on $\mathbb{R}$. Denote by ϕ the vector $(\phi_1, \ldots, \phi_r)^T$, the transpose of $(\phi_1, \ldots, \phi_r)$. We say that ϕ is refinable if it satisfies the following refinement equation:

$$\phi = \sum_{\alpha \in \mathbb{Z}} a(\alpha)\phi(2\cdot - \alpha), \qquad (1.1)$$

where each $a(\alpha)$ is an $r \times r$ matrix of complex numbers.

Let $L_2(\mathbb{R})$ denote the linear space of all square integrable complex-valued functions on $\mathbb{R}$. It is well-known that $L_2(\mathbb{R})$ is a Hilbert space with the inner product given by

$$\langle f, g \rangle := \int_{\mathbb{R}} f(x)\overline{g(x)}\,dx,$$

---

[1]Department of Mathematical Sciences, University of Alberta, Edmonton, Canada T6G 2G1. Research was supported in part by NSERC Canada under Grant OGP 121336.

where $\overline{g(x)}$ denotes the complex conjugate of $g(x)$. The norm of $f \in L_2(\mathbb{R})$ is given by $\|f\|_2 := \langle f, f \rangle^{1/2}$. More generally, for $1 \le p < \infty$, we define

$$\|f\|_p := \left(\int_{\mathbb{R}} |f(x)|^p \, dx \right)^{1/p}.$$

For $p = \infty$, define $\|f\|_\infty$ to be the essential supremum of $|f|$ on $\mathbb{R}$. Let $L_p(\mathbb{R})$ denote the linear space of all functions f for which $\|f\|_p < \infty$. Equipped with the norm $\|\cdot\|_p$, $L_p(\mathbb{R})$ is a Banach space. By $(L_p(\mathbb{R}))^r$ we denote the linear space of all vectors $f = (f_1, \ldots, f_r)^T$ such that $f_1, \ldots, f_r \in L_p(\mathbb{R})$. The norm on $(L_p(\mathbb{R}))^r$ is defined by

$$\|f\|_p := \left(\sum_{j=1}^r \|f_j\|_p^p \right)^{1/p}, \qquad f = (f_1, \ldots, f_r)^T \in (L_p(\mathbb{R}))^r.$$

Suppose $\phi = (\phi_1, \ldots, \phi_r)^T$ and $\tilde\phi = (\tilde\phi_1, \ldots, \tilde\phi_r)^T$ belong to $(L_2(\mathbb{R}))^r$. We say that the shifts of $\phi_1, \ldots, \phi_r$ and the shifts of $\tilde\phi_1, \ldots, \tilde\phi_r$ are biorthogonal, if

$$\langle \phi_j(\cdot - \alpha), \tilde\phi_k(\cdot - \beta) \rangle = \delta_{jk} \delta_{\alpha\beta} \qquad \forall\, j,k = 1, \ldots, r,\; \alpha, \beta \in \mathbb{Z}, \qquad (1.2)$$

where δ_{jk} and $\delta_{\alpha\beta}$ stand for the Kronecker sign. If this is the case, then $\tilde\phi$ is said to be a dual to ϕ. If, in addition, ϕ and $\tilde\phi$ are refinable, then ϕ and $\tilde\phi$ are a pair of biorthogonal vectors of multiple refinable functions. Biorthogonal multiple wavelets are generated from biorthogonal multiple refinable functions. In the scalar case ($r = 1$), a basic theory of biorthogonal wavelets was established by Cohen, Daubechies, and Feauveau [6].

Suppose $\phi = (\phi_1, \ldots, \phi_r)^T$ is a refinable vector of compactly supported functions in $L_2(\mathbb{R})$. Under what conditions does there exist a refinable dual vector of compactly supported functions? The main purpose of this paper is to address this fundamental question.

Let $\phi = (\phi_1, \ldots, \phi_r)^T$ and $\tilde\phi = (\tilde\phi_1, \ldots, \tilde\phi_r)^T$ be dual vectors of compactly supported functions in $L_2(\mathbb{R})$. Suppose $c_1, \ldots, c_r$ are sequences on $\mathbb{Z}$ such that

$$\sum_{j=1}^r \sum_{\alpha \in \mathbb{Z}} c_j(\alpha) \phi_j(\cdot - \alpha) = 0.$$

Taking inner product of both sides of the above equation with $\tilde\phi_k(\cdot - \beta)$ and employing the dual relation (1.2), we obtain $c_k(\beta) = 0$ for all $\beta \in \mathbb{Z}$ and all $k = 1, \ldots, r$. In other words, the shifts of $\phi_1, \ldots, \phi_r$ are linearly independent. Thus, linear independence is a necessary condition for the existence of a dual vector of compactly supported functions. This fact was observed by Dahmen and Micchelli in [9, Proposition 4].

Let $\phi = (\phi_1, \ldots, \phi_r)^T$ be a refinable vector of compactly supported functions in $L_2(\mathbb{R})$ such that the shifts of $\phi_1, \ldots, \phi_r$ are linearly independent. In Section 3 we shall show that there exists a refinable vector

$\tilde{\phi} = (\tilde{\phi}_1, \ldots, \tilde{\phi}_r)^T$ of compactly supported functions in $L_2(\mathbb{R})$ such that the dual relation (1.2) holds true. In other words, linear independence is also a sufficient condition for the existence of a dual refinable vector of compactly supported functions. Thus, we give a complete answer to the aforementioned fundamental question.

For the scalar case ($r = 1$), Lemarié-Rieusset [30] proved that for any minimally supported refinable function, there exists a compactly supported dual refinable function. See Chui and Wang [3] for a discussion on minimally supported refinable functions, and Jia and Wang [26] for a characterization of the linear independence of the shifts of a refinable function in terms of its mask. However, being minimally supported is not an appropriate condition for either the multivariate setting or the multiple setting.

Suppose $\phi = (\phi_1, \ldots, \phi_r)^T$ is a vector of compactly supported functions in $L_2(\mathbb{R})$ such that the shifts of $\phi_1, \ldots, \phi_r$ are linearly independent. If ϕ satisfies the refinement equation (1.1), then the matrix $M := \sum_{\alpha \in \mathbb{Z}} a(\alpha)/2$ must have a simple eigenvalue 1 and its other eigenvalues are less than 1 in modulus (see [9]). Throughout this paper we assume that this condition is satisfied.

The key to our investigation of multiple refinable functions will be a study of vector subdivision schemes, which are of independent interest. Subdivision schemes have been studied mainly for the case in which the mask a is finitely supported. In the scalar case ($r = 1$), the uniform convergence of stationary subdivision schemes was investigated by Cavaretta, Dahmen, and Micchelli [1]. In [18] Jia gave a characterization for the L_p-convergence of a subdivision scheme ($1 \le p \le \infty$). In particular, the L_2-convergence of a subdivision scheme was characterized in terms of the spectral radius of a certain finite matrix associated to the mask. His results were extended by Han and Jia [15] to the multivariate setting. For the vector case ($r > 1$), Cohen, Daubechies, and Plonka [7] obtained some sufficient conditions for L_∞-convergence and L_2-convergence of cascade algorithms, and Shen [35] gave a characterization for the L_2-convergence of cascade algorithms. In [23], Jia, Riemenschneider, and Zhou provided a characterization for the L_p-convergence of subdivision schemes ($1 \le p \le \infty$).

For the reason which will become clear later, we need to consider the case where the mask a is not finitely supported but decays exponentially fast. Let a be such a mask. Let Q_a be the bounded linear operator on $(L_2(\mathbb{R}))^r$ given by

$$Q_a \phi := \sum_{\alpha \in \mathbb{Z}} a(\alpha) \phi(2 \cdot - \alpha), \qquad \phi = (\phi_1, \ldots, \phi_r)^T \in (L_2(\mathbb{R}))^r. \quad (1.3)$$

Let $y = (y_1, \ldots, y_r)$ be a left eigenvector of M corresponding to the eigenvalue 1, that is, $yM = y$ and $y \ne 0$. In the scalar case ($r = 1$), $M = (1)$, so y is chosen to be 1. The vector y will be fixed throughout this paper. By $L_{2,c}(\mathbb{R})$ we denote the linear space of all compactly supported functions in $L_2(\mathbb{R})$. A vector $\phi = (\phi_1, \ldots, \phi_r)^T \in (L_{2,c}(\mathbb{R}))^r$ is said to satisfy the moment conditions of order 1 if

$$y \sum_{\alpha \in \mathbb{Z}} \phi(\cdot - \alpha) = 1.$$

We say that the (vector) subdivision scheme associated with a converges in the L_2-norm, if there exists a vector $\phi \in (L_2(\mathbb{R}))^r$ such that for any $\phi_0 \in (L_{2,c}(\mathbb{R}))^r$ satisfying the moment conditions of order 1, the sequence $Q_a^n \phi_0$ converges to ϕ in the L_2-norm. If this is the case, then ϕ is a solution of the refinement equation (1.1).

The Kronecker product of two matrices is a useful tool in our study of vector refinement equations. Let us recall some basic properties of the Kronecker product from [28]. Suppose

$$A = (a_{ij})_{1 \le i \le m, 1 \le j \le n} \quad \text{and} \quad B = (b_{ij})_{1 \le i \le r, 1 \le j \le s}$$

are two matrices. The (right) Kronecker product of A and B, written $A \otimes B$, is defined to be the block matrix

$$A \otimes B := \begin{bmatrix} a_{11}B & a_{12}B & \cdots & a_{1n}B \\ a_{21}B & a_{22}B & \cdots & a_{2n}B \\ \vdots & \vdots & \ddots & \vdots \\ a_{m1}B & a_{m2}B & \cdots & a_{mn}B \end{bmatrix}.$$

For three matrices A, B, and C of the same type, we have

$$\begin{aligned}(A+B) \otimes C &= (A \otimes C) + (B \otimes C); \\ A \otimes (B+C) &= (A \otimes B) + (A \otimes C).\end{aligned}$$

If A, B, C, D are four matrices such that the products AC and BD are well-defined, then

$$(A \otimes B)(C \otimes D) = (AC) \otimes (BD).$$

If $\lambda_1, \ldots, \lambda_r$ are the eigenvalues of an $r \times r$ matrix A and $\mu_1, \ldots, \mu_r$ are the eigenvalues of an $r \times r$ matrix B, then the eigenvalues of $A \otimes B$ are $\lambda_j \mu_k$, $j, k = 1, \ldots, r$. See [28, Chap. 12] for a proof of these results.

The Kronecker product was used by Goodman, Jia, and Micchelli [13] in their study of the spectral radius of a bi-infinite periodic and slanted matrix. It was also employed by Jiang [27] in his work on the regularity of matrix refinable functions, and by Zhou [36] in his investigation of the joint spectral radius of a finite collection of matrices.

For $\mu > 0$, let E_μ denote the linear space of all sequences u on $\mathbb{Z}$ for which

$$\|u\|_{E_\mu} := \sum_{\alpha \in \mathbb{Z}} |u(\alpha)| e^{\mu|\alpha|} < \infty.$$

Equipped with the norm $\|\cdot\|_{E_\mu}$, E_μ becomes a Banach space. Note that a similar space was used by Cohen and Daubechies [5]. Let E_μ^r denote the linear space of all mappings u from $\mathbb{Z}$ to $\mathbb{C}^r$ for which there exist $u_1, \ldots, u_r \in E_\mu$ such that $u(\alpha) = (u_1(\alpha), \ldots, u_r(\alpha))^T$ for all $\alpha \in \mathbb{Z}$. The norm on E_μ^r is defined by

$$\|u\|_{E_\mu^r} := \max_{1 \leq j \leq r} \|u_j\|_{E_\mu}.$$

By $E_\mu^{r \times r}$ we denote the linear space of all mappings g from $\mathbb{Z}$ to $\mathbb{C}^{r \times r}$ for which there exist $g_{jk} \in E_\mu$, $j, k = 1, \ldots, r$, such that $g(\alpha) = (g_{jk}(\alpha))_{1 \leq j,k \leq r}$ for all $\alpha \in \mathbb{Z}$. The norm on $E_\mu^{r \times r}$ is defined by

$$\|g\|_{E_\mu^{r \times r}} := \max_{1 \leq j,k \leq r} \|g_{jk}\|_{E_\mu}.$$

Suppose the mask a belongs to $E_\mu^{r \times r}$ for some $\mu > 0$. Let b be defined by

$$b(\alpha) := \sum_{\beta \in \mathbb{Z}} \overline{a(\beta)} \otimes a(\alpha + \beta)/2, \qquad \alpha \in \mathbb{Z}. \tag{1.4}$$

Then b lies in $E_\mu^{r^2 \times r^2}$. Let T_b be the transition operator on $E_\mu^{r^2}$ defined by

$$T_b u(\alpha) := \sum_{\beta \in \mathbb{Z}} b(2\alpha - \beta) u(\beta), \qquad \alpha \in \mathbb{Z}, \tag{1.5}$$

where $u \in E_\mu^{r^2}$. It is easily seen that T_b is a bounded operator. The transition operator plays an important role in our study of refinement equations. When $r = 1$ and b is finitely supported, transition matrices were introduced by Deslauriers and Dubuc [10] in their study of interpolatory subdivision schemes. In [14], Goodman, Micchelli, and Ward connected transition operators with subdivision operators in their work on spectral radius formulas.

The transition operator T_b defined by (1.5) is a compact operator. Indeed, if b is finitely supported, then T_b is the limit of a sequence of finite-rank operators, so T_b is a compact operator. In general, we can find a sequence $b^{(N)}$ ($N = 1, 2, \ldots$) of elements of $E_\mu^{r^2 \times r^2}$ with finite support such that $\|b^{(N)} - b\|_{E_\mu^{r^2 \times r^2}} \to 0$ as $N \to \infty$. It follows that $\|T_{b^{(N)}} - T_b\| \to 0$ as $N \to \infty$. As the limit of a sequence of compact operators, T_b itself is a compact operator. The reader is referred to [34, Chap. 4] for a basic theory of the spectral properties of compact operators. In particular, if we denote by $\rho(T_b)$ the spectral radius of T_b, then $\rho(T_b) = |\sigma|$ for some eigenvalue σ of T_b.

Recall that M is the matrix $\sum_{\alpha \in \mathbb{Z}} a(\alpha)/2$. By (1.4) we have

$$\sum_{\alpha \in \mathbb{Z}} b(\alpha)/2 = \left(\sum_{\beta \in \mathbb{Z}} \overline{a(\beta)}/2\right) \otimes \left(\sum_{\alpha \in \mathbb{Z}} a(\alpha + \beta)/2\right) = \overline{M} \otimes M.$$

Thus, the matrix $\sum_{\alpha \in \mathbb{Z}} b(\alpha)/2$ has a simple eigenvalue 1 and its other eigenvalues are less than 1 in modulus. Also, recall that y is a left eigenvector of M corresponding to the eigenvalue 1. Hence, we have

$$(\overline{y} \otimes y)(\overline{M} \otimes M) = (\overline{y}\overline{M}) \otimes (yM) = \overline{M} \otimes M.$$

In other words, $\bar{y} \otimes y$ is a left eigenvector of $\overline{M} \otimes M$ corresponding to the eigenvalue 1.

In Section 2 we will establish a characterization for the L_2-convergence of a vector subdivision scheme. Suppose $a \in E_\mu^{r \times r}$. Let $b \in E_\mu^{r^2 \times r^2}$ be defined by (1.4), and let T_b be the transition operator on $E_\mu^{r^2}$ given by (1.5). Consider the subspace V of $E_\mu^{r^2}$ defined by

$$V := \left\{ v \in E_\mu^{r^2} : (\bar{y} \otimes y) \sum_{\alpha \in \mathbb{Z}} v(\alpha) = 0 \right\}. \tag{1.6}$$

We will show that the subdivision scheme associated with a converges in the L_2-norm if and only if V is invariant under T_b and $\rho(T_b|_V) < 1$.

Finally, in Section 4 we shall apply the general theory to construction of biorthogonal multiple wavelets. Two examples are given. In the first example, the wavelets are piecewise linear functions with short support. In the second example, the wavelets are almost in C^2, and the dual wavelets are in C^1. All the wavelets and dual wavelets are either symmetric or antisymmetric about the origin. The approximation and smoothness properties of these wavelets will be analyzed.

2 Vector Subdivision Schemes

This section is devoted to a study of vector subdivision schemes. We shall establish a characterization for the L_2-convergence of a vector subdivision scheme in terms of the corresponding transition operator.

Let $\ell(\mathbb{Z})$ denote the linear space of all complex-valued sequences on $\mathbb{Z}$, and let $\ell_0(\mathbb{Z})$ denote the linear space of all finitely supported sequences on $\mathbb{Z}$. The difference operators ∇ and Δ on $\ell(\mathbb{Z})$ are defined by

$$\nabla u := u - u(\cdot - 1) \quad \text{and} \quad \Delta u := -u(\cdot + 1) + 2u - u(\cdot - 1), \qquad u \in \ell(\mathbb{Z}).$$

For $\beta \in \mathbb{Z}$, we denote by δ_β the sequence on $\mathbb{Z}$ given by

$$\delta_\beta(\alpha) = \begin{cases} 1 & \text{for } \alpha = \beta, \\ 0 & \text{for } \alpha \in \mathbb{Z} \setminus \{\beta\}. \end{cases}$$

In particular, we write δ for δ_0.

Let $u \in \ell(\mathbb{Z})$. For $1 \le p < \infty$, we define

$$\|u\|_p := \left(\sum_{\alpha \in \mathbb{Z}} |u(\alpha)|^p \right)^{1/p}.$$

For $p = \infty$, define $\|u\|_\infty$ to be the supremum of $|u|$ on $\mathbb{Z}$. Let $\ell_p(\mathbb{Z})$ denote the linear space of all sequences u for which $\|u\|_p < \infty$. Equipped with the norm $\|\cdot\|_p$, $\ell_p(\mathbb{Z})$ becomes a Banach space. By $\ell_p(\mathbb{Z} \to \mathbb{C}^r)$ we denote the linear space of all sequences u such that $u(\alpha) = (u_1(\alpha), \ldots, u_r(\alpha))^T$ for some $u_1, \ldots, u_r \in \ell_p(\mathbb{Z})$ and for all $\alpha \in \mathbb{Z}$. Obviously, $u \mapsto (u_1, \ldots, u_r)^T$ is a canonical isomorphism between $\ell_p(\mathbb{Z} \to \mathbb{C}^r)$ and $(\ell_p(\mathbb{Z}))^r$. Thus, we may identify $\ell_p(\mathbb{Z} \to \mathbb{C}^r)$ with $(\ell_p(\mathbb{Z}))^r$. The norm of $u = (u_1, \ldots, u_r)^T$ is given by

$$\|u\|_p := \left(\sum_{j=1}^r \|u_j\|_p^p\right)^{1/p}.$$

Equipped with this norm, $(\ell_p(\mathbb{Z}))^r$ becomes a Banach space. We also identify $\ell_p(\mathbb{Z} \to \mathbb{C}^{r \times r})$ with $(\ell_p(\mathbb{Z}))^{r \times r}$. The spaces $(\ell(\mathbb{Z}))^r$, $(\ell_0(\mathbb{Z}))^r$, $(\ell(\mathbb{Z}))^{r \times r}$, and $(\ell_0(\mathbb{Z}))^{r \times r}$ are defined analogously. The difference operators ∇ and Δ can be naturally extended to $(\ell(\mathbb{Z}))^r$ and $(\ell(\mathbb{Z}))^{r \times r}$.

For two functions f, g in $L_2(\mathbb{R})$, $f \odot g$ is defined as follows:

$$f \odot g(x) := \int_{\mathbb{R}} f(x+y)\overline{g(y)}\,dy, \qquad x \in \mathbb{R}.$$

In other words, $f \odot g$ is the convolution of f with the function $y \mapsto \overline{g(-y)}$, $y \in \mathbb{R}$. It is easily seen that $f \odot g$ lies in $C_0(\mathbb{R})$, the space of continuous functions on $\mathbb{R}$ which vanish at ∞ (see [12, p. 232]). In particular, $f \odot g$ is uniformly continuous. Clearly,

$$\|f \odot g\|_\infty \le \|f\|_2 \|g\|_2.$$

Moreover, $\|f\|_2^2 = (f \odot f)(0)$.

For a matrix $A = (a_{ij})_{1 \le i,j \le r}$, the vector

$$(a_{11}, \ldots, a_{r1}, a_{12}, \ldots, a_{r2}, \ldots, a_{1r}, \ldots, a_{rr})^T$$

is said to be the vec-function of A and written as $\operatorname{vec} A$. Suppose A, X, and B are three $r \times r$ matrices. Then we have (see [28, p. 410])

$$\operatorname{vec}(AXB) = (B^T \otimes A)\operatorname{vec} X. \tag{2.1}$$

For $\phi, \psi \in (L_2(\mathbb{R}))^r$, let $\phi \odot \psi^T$ be defined as follows:

$$\phi \odot \psi^T := \begin{bmatrix} \phi_1 \odot \psi_1 & \phi_1 \odot \psi_2 & \cdots & \phi_1 \odot \psi_r \\ \phi_2 \odot \psi_1 & \phi_2 \odot \psi_2 & \cdots & \phi_2 \odot \psi_r \\ \vdots & \vdots & \ddots & \vdots \\ \phi_r \odot \psi_1 & \phi_r \odot \psi_2 & \cdots & \phi_r \odot \psi_r \end{bmatrix}.$$

Suppose $\phi \in (L_2(\mathbb{R}))^r$ is a solution of the refinement equation (1.1), where the mask a is assumed to be in $(\ell_1(\mathbb{Z}))^{r \times r}$ for the time being. It follows from (1.1) that

$$\phi \odot \phi^T = \sum_{\alpha \in \mathbb{Z}} \sum_{\beta \in \mathbb{Z}} a(\alpha) \phi(2 \cdot - \alpha) \odot \phi^T(2 \cdot - \beta) \overline{a(\beta)}^T.$$

Note that
$$\phi(2\cdot-\alpha)\odot\phi^T(2\cdot-\beta)=\frac{1}{2}\phi\odot\phi^T(2\cdot-\alpha+\beta).$$

In light of (2.1) we obtain
$$\operatorname{vec}\left(a(\alpha)\phi(2\cdot-\alpha)\odot\phi^T(2\cdot-\beta)\overline{a(\beta)}^T\right)$$
$$=\tfrac{1}{2}\overline{a(\beta)}\otimes a(\alpha)\operatorname{vec}(\phi\odot\phi^T)(2\cdot-\alpha+\beta).$$

Therefore,
$$\operatorname{vec}(\phi\odot\phi^T)=\sum_{\alpha\in\mathbb{Z}}\sum_{\beta\in\mathbb{Z}}\frac{1}{2}\overline{a(\beta)}\otimes a(\alpha)\operatorname{vec}(\phi\odot\phi^T)(2\cdot-\alpha+\beta). \qquad (2.2)$$

Let $f:=\operatorname{vec}(\phi\odot\phi^T)$. Then f lies in $(C_0(\mathbb{R}))^{r^2}$, the linear space of $r^2\times 1$ vectors of functions in $C_0(\mathbb{R})$. It follows from (2.2) that f satisfies the following refinement equation:
$$f=\sum_{\alpha\in\mathbb{Z}}b(\alpha)f(2\cdot-\alpha),$$
where b is given by (1.4). For $c,d\in(\ell_1(\mathbb{Z}))^{r\times r}$, let $c\diamond d$ be defined by
$$(c\diamond d)(\alpha):=\sum_{\beta\in\mathbb{Z}}\overline{d(\beta)}\otimes c(\alpha+\beta), \qquad \alpha\in\mathbb{Z}.$$

Then $b=a\diamond a/2$.

Iterating (1.3) n times yields
$$Q_a^n\phi=\sum_{\alpha\in\mathbb{Z}}a_n(\alpha)\phi(2^n\cdot-\alpha), \qquad n=1,2,\ldots, \qquad (2.3)$$
where each a_n is independent of the choice of ϕ. In particular, $a_1=a$. Consequently, for $n>1$ we have
$$\begin{aligned}Q_a^n\phi &= Q_a^{n-1}(Q_a\phi)=\sum_{\beta\in\mathbb{Z}}a_{n-1}(\beta)(Q_a\phi)(2^{n-1}\cdot-\beta)\\ &=\sum_{\beta\in\mathbb{Z}}\sum_{\alpha\in\mathbb{Z}}a_{n-1}(\beta)a(\alpha)\phi(2^n\cdot-2\beta-\alpha)\\ &=\sum_{\alpha\in\mathbb{Z}}\left[\sum_{\beta\in\mathbb{Z}}a_{n-1}(\beta)a(\alpha-2\beta)\right]\phi(2^n\cdot-\alpha).\end{aligned}$$

This establishes the following iteration relation for a_n ($n=1,2,\ldots$):
$$a_1=a \quad\text{and}\quad a_n(\alpha)=\sum_{\beta\in\mathbb{Z}}a_{n-1}(\beta)a(\alpha-2\beta), \quad \alpha\in\mathbb{Z}. \qquad (2.4)$$

Similarly, for $f\in(C_0(\mathbb{R}))^{r^2}$ we have
$$Q_b^n f=\sum_{\alpha\in\mathbb{Z}}b_n(\alpha)f(2^n\cdot-\alpha), \qquad (2.5)$$

where b_n $(n = 1, 2, \ldots)$ are given by the following iteration relation:

$$b_1 = b \quad \text{and} \quad b_n(\alpha) = \sum_{\beta \in \mathbb{Z}} b_{n-1}(\beta) b(\alpha - 2\beta), \quad \alpha \in \mathbb{Z}. \quad (2.6)$$

The sequences a_n and b_n are related by the following equation:

$$b_n(\alpha) = \sum_{\beta \in \mathbb{Z}} \overline{a_n(\beta)} \otimes a_n(\alpha + \beta)/2^n, \quad \alpha \in \mathbb{Z}, n = 1, 2 \ldots. \quad (2.7)$$

This will be proved by induction on n. By the definition of b, (2.7) is true for $n = 1$. Suppose $n > 1$ and (2.7) is valid for $n - 1$. For $\alpha \in \mathbb{Z}$, by (2.4) we have

$$\sum_{\beta \in \mathbb{Z}} \overline{a_n(\beta)} \otimes a_n(\alpha + \beta)$$
$$= \sum_{\beta \in \mathbb{Z}} \sum_{\gamma \in \mathbb{Z}} \sum_{\eta \in \mathbb{Z}} \overline{(a_{n-1}(\gamma) a(\beta - 2\gamma))} \otimes (a_{n-1}(\eta) a(\alpha + \beta - 2\eta))$$
$$= \sum_{\beta \in \mathbb{Z}} \sum_{\gamma \in \mathbb{Z}} \sum_{\eta \in \mathbb{Z}} \overline{(a_{n-1}(\gamma)} \otimes a_{n-1}(\eta + \gamma)) \overline{(a(\beta)} \otimes a(\alpha - 2\eta + \beta))$$
$$= \sum_{\eta \in \mathbb{Z}} 2^n b_{n-1}(\eta) b(\alpha - 2\eta),$$

where the induction hypothesis has been used to derive the last equality. This together with (2.6) establishes (2.7).

Let ϕ_0 and ψ_0 be two elements in $(L_2(\mathbb{R}))^r$. By using the same argument as was done in the proof of (2.2), we obtain

$$\text{vec}\big((Q_a^n \phi_0) \odot (Q_a^n \psi_0)^T\big)$$
$$= \sum_{\alpha \in \mathbb{Z}} \sum_{\beta \in \mathbb{Z}} 2^{-n} \overline{a_n(\beta)} \otimes a_n(\alpha) \, \text{vec}(\phi_0 \odot \psi_0^T)(2^n \cdot - \alpha + \beta).$$

This in connection with (2.7) shows that, for $n = 1, 2, \ldots$,

$$\text{vec}\big((Q_a^n \phi_0) \odot (Q_a^n \psi_0)^T\big) = Q_b^n \big(\text{vec}(\phi_0 \odot \psi_0^T)\big). \quad (2.8)$$

We claim that, for $w \in \mathbb{C}^{r^2}$ and $n = 1, 2, \ldots$,

$$T_b^n(w \delta_\beta)(\alpha) = b_n(2^n \alpha - \beta) w \quad \forall \alpha, \beta \in \mathbb{Z}. \quad (2.9)$$

This will be proved by induction on n. When $n = 1$, (2.9) follows from (1.5). Suppose $n > 1$ and (2.9) is valid for $n - 1$. We have

$$T_b^n(w\delta_\beta) = T_b^{n-1}(T_b(w\delta_\beta)) = T_b^{n-1}\left(\sum_{\gamma \in \mathbb{Z}} b(2\gamma - \beta) w \delta_\gamma\right)$$
$$= \sum_{\alpha \in \mathbb{Z}} \sum_{\gamma \in \mathbb{Z}} b_{n-1}(2^{n-1}\alpha - \gamma) b(2\gamma - \beta) w \delta_\alpha$$
$$= \sum_{\alpha \in \mathbb{Z}} b_n(2^n \alpha - \beta) w \delta_\alpha,$$

where (2.6) has been used to derive the last equality. This completes the induction procedure. It follows from (2.9) that

$$T_b^n(w \nabla \delta_\beta)(\alpha) = \nabla b_n(2^n \alpha - \beta) w, \quad \forall \alpha, \beta \in \mathbb{Z}. \quad (2.10)$$

Recall that y is a left eigenvector of the matrix $M = \sum_{\alpha \in \mathbb{Z}} a(\alpha)/2$ corresponding to the eigenvalue 1. Let $e_1, \ldots, e_r$ be a basis for $\mathbb{C}^r$ such that $ye_1 = 1$ and $ye_j = 0$ for $j = 2, \ldots, r$. Let $e_{jk} := \overline{e_k} \otimes e_j$ for $j, k = 1, \ldots, r$. Then $\{e_{jk} : j, k = 1, \ldots, r\}$ is a basis for $\mathbb{C}^{r^2}$ such that $(\overline{y} \otimes y)e_{11} = 1$ and $(\overline{y} \otimes y)e_{jk} = 0$ for $(j, k) \neq (1, 1)$.

The following theorem gives a necessary condition for the L_2-convergence of a vector subdivision scheme with an exponentially decaying mask.

Theorem 2.1 *Let $a \in E_\mu^{r \times r}$ for some $\mu > 0$ and let b be given by (1.4). If the subdivision scheme associated with a converges in the L_2-norm, then*

$$\lim_{n \to \infty} \|T_b^n v\|_\infty = 0 \quad \forall v \in V, \tag{2.11}$$

where T_b is the transition operator defined in (1.5) and V is the linear space given by (1.6).

Proof. Suppose that the subdivision scheme associated with a converges in the L_2-norm. Let ϕ_0 and ψ_0 be two elements in $(L_{2,c}(\mathbb{R}))^r$ satisfying the moment conditions of order 1. Then both sequences $\phi_n := Q_a^n \phi_0$ and $\psi_n := Q_a^n \psi_0$ converge to the same limit function ϕ in the L_2-norm. For $n = 0, 1, \ldots$, let $f_n := \mathrm{vec}(\phi_n \odot \psi_n^T)$, and let $f := \mathrm{vec}(\phi \odot \phi^T)$. Then

$$\begin{aligned}
\|f_n - f\|_\infty &= \|\mathrm{vec}(\phi_n \odot \psi_n^T - \phi \odot \phi^T)\|_\infty \\
&\leq \|\mathrm{vec}(\phi_n \odot (\psi_n - \phi)^T)\|_\infty + \|\mathrm{vec}((\phi_n - \phi) \odot \phi^T)\|_\infty \\
&\leq \|\phi_n\|_2 \|\psi_n - \phi\|_2 + \|\phi_n - \phi\|_2 \|\phi\|_2.
\end{aligned}$$

This shows that f_n converges to f uniformly.

In particular, choose $\phi_0 = \psi_0 = e_1 \chi$, where χ is the characteristic function of the unit interval $[0, 1)$. Then both ϕ_0 and ψ_0 satisfy the moment conditions of order 1. But $\mathrm{vec}((e_1\chi) \odot (e_1\chi)^T) = e_{11}h$, where h is the hat function given by $h(x) := \max\{1 - |x|, 0\}$, $x \in \mathbb{R}$. With the help of (2.8), we see that $Q_b^n(e_{11}h)$ converges to f uniformly. Since f is uniformly continuous, $\|f - f(\cdot - 2^{-n})\|_\infty \to 0$ as $n \to \infty$. Consequently,

$$\lim_{n \to \infty} \|Q_b^n(e_{11}h) - Q_b^n(e_{11}h)(\cdot - 2^{-n})\|_\infty = 0.$$

But (2.5) implies

$$Q_b^n(e_{11}h) - Q_b^n(e_{11}h)(\cdot - 2^{-n}) = \sum_{\alpha \in \mathbb{Z}} \nabla b_n(\alpha) e_{11} h(2^n \cdot - \alpha).$$

It follows that

$$\lim_{n \to \infty} \|\nabla b_n e_{11}\|_\infty = 0. \tag{2.12}$$

Furthermore, we observe that, for $j = 2, \ldots, r$, $e_1\chi$ and $(e_1 + e_j)\chi$ both satisfy the moment conditions of order 1. Hence, $Q_a^n(e_1\chi)$ and $Q_a^n(e_1 + e_j)\chi$ both converge to the same limit ϕ in the L_2-norm. This shows that, for

$j = 2, \ldots, r$, $\|Q_a^n(e_j\chi)\|_2 \to 0$ as $n \to \infty$. Choosing $\phi_0 = e_j\chi$ and $\psi_0 = e_k\chi$ in (2.8), we obtain

$$\text{vec}\big((Q_a^n(e_j\chi)) \odot (Q_a^n(e_k\chi))^T\big) = Q_b^n(e_{jk}h), \quad j,k = 1,\ldots,r, \ n = 1,2,\ldots.$$

Therefore,

$$\lim_{n\to\infty} \|Q_b^n(e_{jk}h)\|_\infty = 0, \quad \text{for } (j,k) \neq (1,1).$$

But by (2.5) we have

$$Q_b^n(e_{jk}h) = \sum_{\alpha \in \mathbb{Z}} b_n(\alpha) e_{jk}h(2^n \cdot - \alpha).$$

This shows that

$$\lim_{n\to\infty} \|b_n e_{jk}\|_\infty = 0, \quad (j,k) \neq (1,1). \tag{2.13}$$

To summarize, we have shown that (2.12) and (2.13) are necessary conditions for the subdivision scheme associated with a to converge in the L_2-norm.

Let v be an element of V. Then v can be expressed as

$$v = \sum_{j=1}^r \sum_{k=1}^r \sum_{\beta \in \mathbb{Z}} c_{jk}(\beta) e_{jk} \delta_\beta,$$

where $c_{jk} \in E_\mu$, $j, k = 1, \ldots, r$. Since $v \in V$, we have $(\overline{y} \otimes y) \sum_{\alpha \in \mathbb{Z}} v(\alpha) = 0$. But $(\overline{y} \otimes y) e_{11} = 1$ and $(\overline{y} \otimes y) e_{jk} = 0$ for $(j,k) \neq (1,1)$. Hence,

$$0 = (\overline{y} \otimes y) \sum_{\alpha \in \mathbb{Z}} v(\alpha) = (\overline{y} \otimes y) \sum_{j=1}^r \sum_{k=1}^r \sum_{\beta \in \mathbb{Z}} c_{jk}(\beta) e_{jk} = \sum_{\beta \in \mathbb{Z}} c_{11}(\beta).$$

So c_{11} can be written as $\sum_{\beta \in \mathbb{Z}} d(\beta) \nabla \delta_\beta$, where d is given by

$$d(\beta) := \begin{cases} c_{11}(\beta) + c_{11}(\beta - 1) + c_{11}(\beta - 2) + \cdots & \text{for } \beta \leq -1, \\ -c_{11}(\beta + 1) - c_{11}(\beta + 2) - c_{11}(\beta + 3) - \cdots & \text{for } \beta \geq 0. \end{cases}$$

Clearly, d belongs to $\ell_1(\mathbb{Z})$. The relations (2.12) and (2.13) together with (2.9) and (2.10) imply that, for every $\beta \in \mathbb{Z}$ and $(j,k) \neq (1,1)$,

$$\lim_{n\to\infty} \|T_b^n(e_{11} \nabla \delta_\beta)\|_\infty = 0 \quad \text{and} \quad \lim_{n\to\infty} \|T_b^n(e_{jk}\delta_\beta)\|_\infty = 0.$$

In light of the expression of v, (2.11) follows at once. □

Condition (2.11) is also sufficient for the L_2-convergence of the vector subdivision scheme associated with mask a. To see this, we first show that, for any $\psi \in (L_{2,c}(\mathbb{R}))^r$,

$$\|Q_a^n \psi\|_2^2 \leq r \|T_b^n v(0)\|_\infty, \quad n = 1, 2, \ldots, \tag{2.14}$$

where

$$v(\alpha) := \text{vec}\,(\psi \odot \psi^T)(\alpha), \quad \alpha \in \mathbb{Z}.$$

Let $g := \text{vec}(\psi \odot \psi^T)$, $\psi_n := Q_a^n \psi$, and $g_n := \text{vec}(\psi_n \odot \psi_n^T)$ $(n = 1, 2, \ldots)$. Then we have
$$\|\psi_n\|_2^2 \leq r \|g_n(0)\|_\infty. \tag{2.15}$$

But (2.8) tells us that $g_n = Q_b^n g$. Hence,
$$g_n(0) = (Q_b^n g)(0) = \sum_{\alpha \in \mathbb{Z}} b_n(\alpha) g(-\alpha) = \sum_{\beta \in \mathbb{Z}} b_n(-\beta) g(\beta). \tag{2.16}$$

Furthermore, for $n = 1, 2, \ldots$, we have
$$T_b^n v(\alpha) = \sum_{\beta \in \mathbb{Z}} b_n(2^n \alpha - \beta) v(\beta), \qquad \alpha \in \mathbb{Z}. \tag{2.17}$$

This will be proved by induction on n. By the definition of T_b, (2.17) is true for $n = 1$. Suppose (2.17) is valid for $n - 1$. For $\alpha \in \mathbb{Z}$, we have
$$\begin{aligned} T_b^n v(\alpha) &= T_b^{n-1}(T_b v)(\alpha) \\ &= \sum_{\beta \in \mathbb{Z}} b_{n-1}(2^{n-1}\alpha - \beta)(T_b v)(\beta) \\ &= \sum_{\beta \in \mathbb{Z}} \sum_{\gamma \in \mathbb{Z}} b_{n-1}(2^{n-1}\alpha - \beta) b(2\beta - \gamma) v(\gamma) \\ &= \sum_{\gamma \in \mathbb{Z}} \left[\sum_{\beta \in \mathbb{Z}} b_{n-1}(\beta) b(2^n \alpha - \gamma - 2\beta) \right] v(\gamma) \\ &= \sum_{\gamma \in \mathbb{Z}} b_n(2^n \alpha - \gamma) v(\gamma), \end{aligned}$$

where (2.6) has been used to derive the last equality. This completes the induction procedure. It follows from (2.16) and (2.17) that $g_n(0) = T_b^n v(0)$. This togetehr with (2.15) implies (2.14).

If a is finitely supported and the matrix $M := \sum_{\alpha \in \mathbb{Z}} a(\alpha)/2$ has a simple eigenvalue 1 and its other eigenvalues are less than 1 in modulus, then we must have $\rho(T_b) \geq 1$. Indeed, if $\rho(T_b) < 1$, then (2.14) tells us that $Q_a^n(e_1\chi)$ would converge to 0 in the L_2-norm. On the other hand, it was proved in [23] that the limit of $Q_a^n(e_1\chi)$ must be a nonzero vector of functions in $L_2(\mathbb{R})$. This contradiction demonstrates $\rho(T_b) \geq 1$.

When $a \in E_\mu^{r \times r}$, this conclusion remains valid. To see this, we recall the following fact from functional analysis (see [34, Theorem 10.20]). Let $A, A_1, A_2, \ldots$ be bounded linear operators on a Banach space. Suppose $\|A_n - A\| \to 0$ as $n \to \infty$. Then for a given $\varepsilon > 0$ there exists some n_0 such that $\rho(A_n) < \rho(A) + \varepsilon$ for all $n \geq n_0$. In order to apply this result, we find sequences $a^{(N)}$ $(N = 1, 2, \ldots)$ such that each $a^{(N)}$ is supported on $[-N, N]$, $y \sum_{\alpha \in \mathbb{Z}} a^{(N)}(\alpha)/2 = y$, and $\|a^{(N)} - a\|_{E_\mu^{r \times r}} \to 0$ as $N \to \infty$. Let $b^{(N)} := a^{(N)} \diamond a^{(N)}/2$. Then $\|T_{b^{(N)}} - T_b\| \to 0$ as $N \to \infty$. If $\rho(T_b) < 1$, then $\rho(T_{b^{(N)}}) < 1$ for sufficiently large N. But the latter is impossible. Therefore, we must have $\rho(T_b) \geq 1$.

Let $a \in E_\mu^{r \times r}$ for some $\mu > 0$. We say that a satisfies the basic sum rule if
$$y \sum_{\alpha \in \mathbb{Z}} a(2\alpha) = y \sum_{\alpha \in \mathbb{Z}} a(2\alpha - 1) = y.$$

Let b be given by (1.4). If a satisfies the basic sum rule, then b satisfies the basic sum rule stated as follows:

$$(\bar{y} \otimes y) \sum_{\alpha \in \mathbb{Z}} b(2\alpha) = (\bar{y} \otimes y) \sum_{\alpha \in \mathbb{Z}} b(2\alpha - 1) = \bar{y} \otimes y.$$

The converse of this statement is also true. Moreover, if b satisfies the basic sum rule, then the space V given by (1.6) is invariant under T_b.

We are in a position to establish the main result of this section.

Theorem 2.2 *Let $a \in E_\mu^{r \times r}$ for some $\mu > 0$ and let b be given by (1.4). Then the subdivision scheme associated with a converges in the L_2-norm if and only if*
(a) *a satisfies the basic sum rule, and*
(b) *$\rho(T_b|_V) < 1$.*

Proof. Suppose the subdivision scheme associated with a converges in the L_2-norm. Then (2.11) is valid, by Theorem 2.1. If V is not invariant under T_b, then there exists $v \in V$ such that $T_b v \notin V$. Note that the codimension of V in $E_\mu^{r^2}$ is 1. Hence, any $u \in E_\mu^{r^2}$ can be represented as $u = w + c(T_b v)$ for some $w \in V$ and $c \in \mathbb{C}$. It follows from (2.11) that

$$\lim_{n \to \infty} \|T_b^n u\|_\infty = 0 \quad \forall u \in E_\mu^{r^2}.$$

Consequently, $\rho(T_b) < 1$. But we have proved $\rho(T_b) \geq 1$. This contradiction shows that V is invariant under T_b.

Since V is invariant under T_b, we have $T_b(e_{jk} \nabla \delta) \in V$ for $j, k = 1, \ldots, r$. It follows that

$$\sum_{\alpha \in \mathbb{Z}} (\bar{y} \otimes y)[b(2\alpha) - b(2\alpha - 1)] e_{jk} = (\bar{y} \otimes y) \sum_{\alpha \in \mathbb{Z}} T_b(e_{jk} \nabla \delta)(\alpha) = 0.$$

Since the above relation is true for all $j, k = 1, \ldots, r$, we deduce that

$$(\bar{y} \otimes y) \sum_{\alpha \in \mathbb{Z}} [b(2\alpha) - b(2\alpha - 1)] = 0.$$

But

$$(\bar{y} \otimes y) \sum_{\alpha \in \mathbb{Z}} [b(2\alpha) + b(2\alpha - 1)] = 2(\bar{y} \otimes y).$$

Therefore, b satisfies the basic sum rule. Consequently, a also satisfies the basic sum rule.

Since T_b is a compact operator, $\rho(T_b|_V) = |\tau|$ for some eigenvalue τ of $T_b|_V$. Suppose $T_b v = \tau v$ for some $v \in V$ with $v \neq 0$. It follows that $T_b^n v = \tau^n v$ for $n = 1, 2, \ldots$. By (2.11), $\|T_b^n v\|_\infty$ converges to 0 as $n \to \infty$. Therefore, $|\tau|^n \to 0$ as $n \to \infty$. This shows $\rho(T_b|_V) = |\tau| < 1$, as desired.

It remains to prove the sufficiency of conditions (a) and (b). For this purpose, let ϕ_0 be an $r \times 1$ vector of compactly supported functions in

$L_2(\mathbb{R})$ such that ϕ_0 satisfies the moment conditions of order 1. We wish to prove that $Q_a^n \phi_0$ is a Cauchy sequence in $(L_2(\mathbb{R}))^r$. We observe that

$$Q_a^{n+1}\phi_0 - Q_a^n\phi_0 = Q_a^n(Q_a\phi_0 - \phi_0) = Q_a^n\psi, \tag{2.18}$$

where $\psi := Q_a\phi_0 - \phi_0$. Since $y\sum_{\alpha\in\mathbb{Z}}\phi_0(\cdot - \alpha) = 1$ and a satisfies the basic sum rule, we have

$$y\sum_{\alpha\in\mathbb{Z}}(Q_a\phi_0)(\cdot - \alpha) = y\sum_{\alpha\in\mathbb{Z}}\sum_{\beta\in\mathbb{Z}}a(\beta)\phi_0(2\cdot - 2\alpha - \beta)$$
$$= \sum_{\beta\in\mathbb{Z}}y\left[\sum_{\alpha\in\mathbb{Z}}a(\beta - 2\alpha)\right]\phi_0(\cdot - \beta) = \sum_{\beta\in\mathbb{Z}}y\phi_0(\cdot - \beta) = 1.$$

In other words, $Q_a\phi_0$ also satisfies the moment conditions of order 1. Consequently,

$$y\sum_{\alpha\in\mathbb{Z}}\psi(\cdot - \alpha) = 0.$$

Let $v(\alpha) := \text{vec}(\psi \odot \psi^T)(\alpha)$, $\alpha \in \mathbb{Z}$. Then the above relation and (2.1) imply

$$(\overline{y}\otimes y)\sum_{\alpha\in\mathbb{Z}}v(\alpha) = (\overline{y}\otimes y)\sum_{\alpha\in\mathbb{Z}}\int_{\mathbb{R}}\text{vec}\left(\psi(\alpha+x)\overline{\psi(x)}^T\right)dx = 0.$$

Hence v lies in V. Since $\rho(T_b|_V) < 1$, by (2.14) we see that there exist two constants $C > 0$ and $t \in (0,1)$ such that

$$\|Q_a^n\psi\|_2 \leq Ct^n, \quad n = 1, 2, \ldots.$$

Since $0 < t < 1$, this together with (2.18) tells us that $Q_a^n\phi_0$ is a Cauchy sequence in $(L_2(\mathbb{R}))^r$. Thus, $Q_a^n\phi_0$ converges in the L_2-norm for every ϕ_0 in $(L_{2,c}(\mathbb{R}))^r$ satisfying the moment conditions of order 1. If ψ_0 is another such vector, then $y\sum_{\alpha\in\mathbb{Z}}(\phi_0 - \psi_0)(\cdot - \alpha) = 0$. By what has been proved, $Q_a^n(\phi_0 - \psi_0)$ converges to 0 in the L_2-norm. In other words, $Q_a^n\phi_0$ and $Q_a^n\psi_0$ converge to the same limit. We conclude that the subdivision scheme associated with a converges in the L_2-norm. □

3 Biorthogonal Multiple Refinable Functions

Let $\phi = (\phi_1, \ldots, \phi_r)^T$ be a refinable vector of compactly supported functions in $L_2(\mathbb{R})$. In this section we show that there exists a dual refinable vector of compactly supported functions in $L_2(\mathbb{R})$ if and only if the shifts of $\phi_1, \ldots, \phi_r$ are linearly independent.

The linear independence of the shifts of a finite number of compactly supported functions was characterized by Jia and Micchelli [21] in terms of the Fourier transform of these functions. The Fourier-Laplace transform of a compactly supported integrable function f is defined by

$$\hat{f}(\zeta) := \int_{\mathbb{R}} f(x)e^{-ix\zeta}\,dx, \quad \zeta \in \mathbb{C}.$$

Let $\phi_1,\ldots,\phi_r$ be compactly supported integrable functions. It was proved in [21] that the shifts of $\phi_1,\ldots,\phi_r$ are linearly independent if and only if, for any $\zeta \in \mathbb{C}$, the sequences $(\hat{\phi}_j(\zeta + 2\beta\pi))_{\beta\in\mathbb{Z}}$ $(j = 1,\ldots,r)$ are linearly independent. This result is also valid if $\phi_1,\ldots,\phi_r$ are compactly supported distributions.

Another important concept is stability. Let $\phi_1,\ldots,\phi_r$ be a finite number of functions in $L_p(\mathbb{R})$ $(1 \le p \le \infty)$. We say that the shifts of $\phi_1,\ldots,\phi_r$ are L_p-stable if there exist two positive constants C_1 and C_2 such that, for arbitrary finitely supported sequences $\lambda_1,\ldots,\lambda_r$ on $\mathbb{Z}$,

$$C_1 \sum_{j=1}^{r} \|\lambda_j\|_p \le \left\| \sum_{j=1}^{r} \sum_{\alpha\in\mathbb{Z}} \phi_j(\cdot - \alpha)\lambda_j(\alpha) \right\|_p \le C_2 \sum_{j=1}^{r} \|\lambda_j\|_p.$$

Given a function ϕ on $\mathbb{R}$, set $\phi^\circ := \sum_{\alpha\in\mathbb{Z}} |\phi(\cdot - \alpha)|$. By $\mathcal{L}_p(\mathbb{R})$ we denote the linear space of all functions ϕ for which $(\phi^\circ)^p$ is integrable on the interval $[0,1]$. Let $\phi_1,\ldots,\phi_r$ be functions in $\mathcal{L}_p(\mathbb{R})$ $(1 \le p \le \infty)$. It was proved by Jia and Micchelli [21] that the shifts of $\phi_1,\ldots,\phi_r$ are L_p-stable if and only if, for any $\xi \in \mathbb{R}$, the sequences $(\hat{\phi}_j(\xi+2\beta\pi))_{\beta\in\mathbb{Z}}$ $(j = 1,\ldots,r)$ are linearly independent. In particular, when $\phi_1,\ldots,\phi_r$ are compactly supported, linear independence implies stability.

In what follows, by I_r we denote the $r \times r$ identity matrix. The complex conjugate of a matrix M is denoted by M^*. For $f = (f_1,\ldots,f_r)^T$ and $g = (g_1,\ldots,g_r)^T$ in $(L_2(\mathbb{R}))^r$, we define

$$\langle f, g^T \rangle := \begin{bmatrix} \langle f_1,g_1\rangle & \langle f_1,g_2\rangle & \cdots & \langle f_1,g_r\rangle \\ \langle f_2,g_1\rangle & \langle f_2,g_2\rangle & \cdots & \langle f_2,g_r\rangle \\ \vdots & \vdots & \ddots & \vdots \\ \langle f_r,g_1\rangle & \langle f_r,g_2\rangle & \cdots & \langle f_r,g_r\rangle \end{bmatrix}.$$

Thus, f and g are dual to each other if and only if

$$\langle f(\cdot - \gamma), g^T \rangle = \delta_{\gamma,0} I_r \quad \forall \gamma \in \mathbb{Z}.$$

Let a be an element in $(\ell_0(\mathbb{Z}))^{r\times r}$ such that $M := \sum_{\alpha\in\mathbb{Z}} a(\alpha)/2$ has a simple eigenvalue 1 and its other eigenvalues are less than 1 in modulus. Let y be a nonzero $1 \times r$ vector such that $yM = y$. It was proved by Heil and Colella [16] that there exists a unique distributional solution ϕ of the refinement equation (1.1) such that ϕ is compactly supported and $y\hat{\phi}(0) = 1$. Similarly, let $\tilde{a}$ be an element in $(\ell_0(\mathbb{Z}))^{r\times r}$ such that the matrix $\widetilde{M} := \sum_{\alpha\in\mathbb{Z}} \tilde{a}(\alpha)/2$ has a simple eigenvalue 1 and its other eigenvalues are less than 1 in modulus. Let $\tilde{y}$ be a $1 \times r$ vector such that $\tilde{y}\widetilde{M} = \tilde{y}$ and $\tilde{y}y^* = 1$. Then there exists a unique distributional solution $\tilde{\phi}$ of the refinement equation

$$\tilde{\phi} = \sum_{\alpha\in\mathbb{Z}} \tilde{a}(\alpha)\tilde{\phi}(2\cdot - \alpha) \tag{3.1}$$

such that $\tilde{\phi}$ is compactly supported and $\tilde{y}\hat{\tilde{\phi}}(0) = 1$.

Theorem 3.1 *The vectors ϕ and $\tilde{\phi}$ belong to $(L_2(\mathbb{R}))^r$ and are dual to each other if and only if*
(a) $\sum_{\alpha \in \mathbb{Z}} a(\alpha) \tilde{a}(\alpha + 2\gamma)^* = 2\delta_{\gamma,0} I_r$ for all $\gamma \in \mathbb{Z}$, and
(b) the subdivision schemes associated with both a and $\tilde{a}$ converge in the L_2-norm.

Proof. Suppose $\phi \in (L_2(\mathbb{R}))^r$ and $\tilde{\phi} \in (L_2(\mathbb{R}))^r$ are dual to each other. Then we have $\langle \phi(\cdot - \gamma), \tilde{\phi}^T \rangle = \delta_{\gamma,0} I_r$ for all $\gamma \in \mathbb{Z}$. By using the refinement equations (1.1) and (3.1) we see that, for each $\gamma \in \mathbb{Z}$,

$$\langle \phi(\cdot - \gamma), \tilde{\phi}^T \rangle = \sum_{\alpha \in \mathbb{Z}} \sum_{\beta \in \mathbb{Z}} \langle a(\alpha) \phi(2 \cdot - 2\gamma - \alpha), \tilde{\phi}(2 \cdot - \beta)^T \tilde{a}(\beta)^T \rangle$$
$$= \sum_{\alpha \in \mathbb{Z}} a(\alpha) \tilde{a}(\alpha + 2\gamma)^* / 2.$$

Hence, condition (a) is satisfied. Moreover, ϕ and $\tilde{\phi}$ being dual to each other implies that the shifts of $\phi_1, \ldots, \phi_r$ are stable. By [23, Theorem 3.2], the subdivision scheme associated with a is L_2-convergent. The same reason shows that the subdivision scheme associated with $\tilde{a}$ also converges in the L_2-norm.

Now suppose conditions (a) and (b) are satisfied. If $\phi_0 \in (L_2(\mathbb{R}))^r$ and $\tilde{\phi}_0 \in (L_2(\mathbb{R}))^r$ are dual to each other, then condition (a) tells us that $Q_a \phi_0$ and $Q_{\tilde{a}} \tilde{\phi}_0$ are also dual to each other. We choose the initial vectors ϕ_0 and $\tilde{\phi}_0$ as follows. Let $f_1 := \chi_{[0,1)}$ and, for $j = 2, \ldots, r$, let

$$f_j := \sum_{k=0}^{2^j - 1} (-1)^k \chi_{[k/2^j, (k+1)/2^j)},$$

where $\chi_{[s,t)}$ denotes the characteristic function of the interval $[s, t)$. It is easily seen that $f := (f_1, \ldots, f_r)^T$ is dual to itself (see [23, Theorem 8.1] for the construction of f). Since $y \tilde{y}^* = 1$, we can find two $r \times r$ matrices A and $\tilde{A}$ such that the first column of A is $\tilde{y}^*$, the first column of $\tilde{A}$ is y^*, and $\tilde{A}^* A = I_r$. Let $\phi_0 := Af$ and $\tilde{\phi}_0 := \tilde{A}f$. Then ϕ_0 and $\tilde{\phi}_0$ satisfy the moment conditions of order 1 (with respect to y and $\tilde{y}$). Moreover, ϕ_0 and $\tilde{\phi}_0$ are dual to each other. Therefore, for $n = 1, 2, \ldots$, $Q_a^n \phi_0$ and $Q_{\tilde{a}}^n \tilde{\phi}_0$ are dual to each other. In other words,

$$\langle (Q_a^n \phi_0)(\cdot - \gamma), (Q_{\tilde{a}}^n \tilde{\phi}_0)^T \rangle = \delta_{\gamma,0} I_r \qquad \forall \gamma \in \mathbb{Z}. \tag{3.2}$$

But condition (b) tells us that $\|Q_a^n \phi_0 - \phi\|_2 \to 0$ and $\|Q_{\tilde{a}}^n \tilde{\phi}_0 - \tilde{\phi}\|_2 \to 0$ as $n \to \infty$. Letting $n \to \infty$ in (3.2), we obtain

$$\langle \phi(\cdot - \gamma), \tilde{\phi}^T \rangle = \delta_{\gamma,0} I_r \qquad \forall \gamma \in \mathbb{Z}.$$

This proves that ϕ and $\tilde{\phi}$ are dual to each other. □

Taking the Fourier transforms of both sides of (1.1) and (3.1), we obtain

$$\hat{\phi}(\xi) = H(\xi/2)\hat{\phi}(\xi/2) \quad \text{and} \quad \hat{\tilde{\phi}}(\xi) = \tilde{H}(\xi/2)\hat{\tilde{\phi}}(\xi/2), \qquad \xi \in \mathbb{R}, \quad (3.3)$$

where

$$H(\xi) := \sum_{\alpha \in \mathbb{Z}} a(\alpha) e^{-i\alpha\xi}/2 \quad \text{and} \quad \tilde{H}(\xi) := \sum_{\alpha \in \mathbb{Z}} \tilde{a}(\alpha) e^{-i\alpha\xi}/2. \quad (3.4)$$

It is easily seen that condition (a) is equivalent to

$$H(\xi)\tilde{H}(\xi)^* + H(\xi + \pi)\tilde{H}(\xi + \pi)^* = I_r \qquad \forall \xi \in \mathbb{R}. \quad (3.5)$$

Moreover, Theorem 2.2 tells us that condition (b) is equivalent to

$$\rho(T_b|_V) < 1 \quad \text{and} \quad \rho(T_{\tilde{b}}|_{\tilde{V}}) < 1, \quad (3.6)$$

where $\tilde{b} := \tilde{a} \diamond \tilde{a}/2$, V is the space given by (1.6), and

$$\tilde{V} := \left\{ v \in E_\mu^{r^2} : (\overline{\tilde{y}} \otimes \tilde{y}) \sum_{\alpha \in \mathbb{Z}} v(\alpha) = 0 \right\}. \quad (3.7)$$

Thus, Theorem 3.1 can be restated as follows: The vectors ϕ and $\tilde{\phi}$ belong to $(L_2(\mathbb{R}))^r$ and are dual to each other if and only if both (3.5) and (3.6) hold true. Since a is finitely supported, [23, Theorem 7.1] tells us that, in (3.6), V can be chosen to be the minimum invariant subspace of T_b generated by $e_{11}(\Delta\delta), e_{22}\delta, \ldots, e_{rr}\delta$, and $\tilde{V}$ can be chosen in a similar way.

For the scalar case ($r = 1$), Lawton [29] first gave a characterization for orthogonality of the shifts of a refinable function in terms of the spectral radius of the corresponding transition matrix. Cohen and Daubechies [4] established the above form of characterization of biorthogonality of a pair of refinable functions. In [31], Long and Chen extended the results in [4] to the multivariate setting. Note that an essential ingredient of the proof of [4, Theorem 4.3] is the fact that a univariate trigonometric polynomial has only finitely many zeros in the interval $[-\pi, \pi]$. So the extension to the multivariate setting is not trivial. For the vector case ($r > 1$), on the basis of the work of Long, Chen, and Yuan [32], Shen [35] proved that ϕ and $\tilde{\phi}$ are dual to each other is equivalent to conditions (3.5), (3.6), and an additional condition on the eigenvectors of T_b and $T_{\tilde{b}}$ corresponding to the eigenvalue 1. However, as was demonstrated above, the condition on the eigenvectors of T_b and $T_{\tilde{b}}$ is redundant.

Let $\phi = (\phi_1, \ldots, \phi_r)^T$ be a vector of compactly supported functions in $L_2(\mathbb{R})$ such that the shifts of $\phi_1, \ldots, \phi_r$ are linearly independent. Suppose ϕ satisfies the refinement equation (1.1) with a finitely supported mask a. Then there exists $\tilde{a} \in (\ell_0(\mathbb{Z}))^r$ such that the $r \times r$ matrices $H(\xi)$ and $\tilde{H}(\xi)$ given by (3.4) satisfy (3.5). To see this, we first show that the $r \times (2r)$ matrix

$$[H(\xi) \quad H(\xi + \pi)] \quad (3.8)$$

has full rank r for every $\xi \in \mathbb{C}$. Suppose to the contrary that there exists some $\xi \in \mathbb{C}$ such that this matrix has rank less than r. Then there exists a nonzero $1 \times r$ vector $t = (t_1, \ldots, t_r)$ such that $tH(\xi) = 0$ and $tH(\xi+\pi) = 0$. By (3.3) we have

$$t\hat{\phi}(2\xi + 4\beta\pi) = tH(\xi)\hat{\phi}(\xi + 2\beta\pi) = 0 \qquad \forall \beta \in \mathbb{Z}$$

and

$$t\hat{\phi}(2\xi + 2\pi + 4\beta\pi) = tH(\xi + \pi)\hat{\phi}(\xi + \pi + 2\beta\pi) = 0 \qquad \forall \beta \in \mathbb{Z}.$$

It follows that $t\hat{\phi}(2\xi+2\beta\pi) = 0$ for all $\beta \in \mathbb{Z}$. Thus, the shifts of $\phi_1, \ldots, \phi_r$ would be linearly dependent. This contradiction shows that the matrix in (3.8) has full rank r for every $\xi \in \mathbb{C}$. Let

$$P(z) := \sum_{\alpha \in \mathbb{Z}} a(\alpha) z^\alpha / 2, \qquad z \in \mathbb{C} \setminus \{0\}.$$

Then $P(z)$ is an $r \times r$ matrix of Laurent polynomials and $H(\xi) = P(e^{-i\xi})$ for $\xi \in \mathbb{C}$. By what has been proved, the matrix

$$[P(z) \quad P(-z)]$$

has full rank r for every $z \in \mathbb{C} \setminus \{0\}$. Hence, by a well-known result from algebra, there exist two $r \times r$ matrices $U(z)$ and $V(z)$ of Laurent polynomials such that

$$[P(z) \quad P(-z)] \begin{bmatrix} U(z) \\ V(z) \end{bmatrix} = I_r, \qquad z \in \mathbb{C} \setminus \{0\}.$$

Let $Q(z) := (U(z) + V(-z))/2$. Then we have

$$[P(z) \quad P(-z)] \begin{bmatrix} Q(z) \\ Q(-z) \end{bmatrix} = I_r, \qquad z \in \mathbb{C} \setminus \{0\}.$$

Let $K(\xi) := Q(e^{-i\xi})^*$, $\xi \in \mathbb{C}$. Then

$$H(\xi)K(\xi)^* + H(\xi+\pi)K(\xi+\pi)^* = I_r \qquad \forall \xi \in \mathbb{C}. \qquad (3.9)$$

We may express $K(\xi)$ as $\sum_{\alpha \in \mathbb{Z}} c(\alpha) e^{-i\alpha\xi}/2$ for some $c \in (\ell_0(\mathbb{Z}))^{r \times r}$. However, there is no guarantee that the subdivision scheme associated with c will converge in the L_2-norm. To overcome this difficulty, we first construct an exponentially decaying mask $\tilde{a}$ such that the subdivision scheme associated with $\tilde{a}$ converges in the L_2-norm and the $r \times r$ matrices $H(\xi)$ and $\tilde{H}(\xi)$ given by (3.4) satisfy (3.5). In the process, the bracket product of two functions in $\mathcal{L}_2(\mathbb{R})$ introduced by Jia and Micchelli [20, Theorem 3.2] will be used. For $f, g \in \mathcal{L}_2(\mathbb{R})$, their bracket product is defined by

$$[f, g](e^{-i\xi}) := \sum_{\beta \in \mathbb{Z}} \hat{f}(\xi + 2\beta\pi)\overline{\hat{g}(\xi + 2\beta\pi)}, \qquad \xi \in \mathbb{R}.$$

We are in a position to establish the main result of this section.

Theorem 3.2 *Let $\phi = (\phi_1, \ldots, \phi_r)^T$ be an $r \times 1$ vector of compactly supported functions in $L_2(\mathbb{R})$ with linearly independent shifts. Suppose that ϕ satisfies the refinement equation (1.1) with a finitely supported mask a. Then there exists a refinable vector $\tilde{\phi} = (\tilde{\phi}_1, \ldots, \tilde{\phi}_r)^T$ of compactly supported functions in $L_2(\mathbb{R})$ such that $\tilde{\phi}$ is dual to ϕ.*

Proof. Let

$$G(\xi) := \big([\phi_j, \phi_k](e^{-i\xi})\big)_{1 \le j,k \le r}, \qquad \xi \in \mathbb{R}.$$

Then $G(\xi)$ is 2π-periodic. Since the shifts of $\phi_1, \ldots, \phi_r$ are stable, the Gram matrix $G(\xi)$ is positive definite for every $\xi \in \mathbb{R}$ (see [20, Theorem 4.1]). Let $\tilde{\phi} = (\tilde{\phi}_1, \ldots, \tilde{\phi}_r)^T$ be given by

$$\widehat{\tilde{\phi}}(\xi) := G(\xi)^{-1} \hat{\phi}(\xi), \qquad \xi \in \mathbb{R}.$$

Then $\tilde{\phi}_1, \ldots, \tilde{\phi}_r$ decay exponentially fast and have stable shifts. In particular, $\tilde{\phi}_1, \ldots, \tilde{\phi}_r$ belong to $\mathcal{L}_2(\mathbb{R})$. For every $\xi \in \mathbb{R}$ we have

$$\begin{aligned}\sum_{\beta \in \mathbb{Z}} \widehat{\tilde{\phi}}(\xi + 2\beta\pi)\hat{\phi}(\xi + 2\beta\pi)^* &= G(\xi)^{-1} \sum_{\beta \in \mathbb{Z}} \hat{\phi}(\xi + 2\beta\pi)\hat{\phi}(\xi + 2\beta\pi)^* \\ &= G(\xi)^{-1} G(\xi) = I_r.\end{aligned}$$

This shows that $\tilde{\phi}$ is dual to ϕ. Moreover, $\tilde{\phi}$ is refinable. Indeed, we have

$$\begin{aligned}\widehat{\tilde{\phi}}(\xi) &= G(\xi)^{-1} \hat{\phi}(\xi) = G(\xi)^{-1} H(\xi/2) \hat{\phi}(\xi/2) \\ &= G(\xi)^{-1} H(\xi/2) G(\xi/2) \widehat{\tilde{\phi}}(\xi/2).\end{aligned}$$

Consequently,

$$\widehat{\tilde{\phi}}(\xi) = \tilde{H}(\xi/2) \widehat{\tilde{\phi}}(\xi/2),$$

where

$$\tilde{H}(\xi) = G(2\xi)^{-1} H(\xi) G(\xi), \qquad \xi \in \mathbb{R}.$$

Clearly, $\tilde{H}$ is 2π-periodic. Since $\tilde{\phi}$ is dual to ϕ, H and $\tilde{H}$ satisfy the relation (3.5). Suppose $\tilde{H}(\xi) = (\tilde{h}_{jk}(\xi))_{1 \le j,k \le r}$, where

$$\tilde{h}_{jk}(\xi) = \sum_{\alpha \in \mathbb{Z}} \tilde{a}_{jk}(\alpha) e^{-i\alpha\xi}/2, \qquad \xi \in \mathbb{R}.$$

Each sequence $\tilde{a}_{jk}$ decays exponentially fast. That is, there exists some $\mu > 0$ such that $\tilde{a}_{jk} \in E_\mu$ for all $j, k = 1, \ldots, r$. Let $\tilde{a} := (\tilde{a}_{jk})_{1 \le j,k \le r}$. Then $\tilde{a} \in E_\mu^{r \times r}$. Since the shifts of $\tilde{\phi}_1, \ldots, \tilde{\phi}_r$ are stable, from the proof of [23, Theorem 3.1] we see that the subdivision scheme associated with $\tilde{a}$ is L_2-convergent. Let $\tilde{b} := \tilde{a} \diamond \tilde{a}/2$ and let $\tilde{V}$ be the linear space given in (3.7). By Theorem 2.2 we have $\rho(T_{\tilde{b}}|_{\tilde{V}}) < 1$.

Let

$$M := \sum_{\alpha \in \mathbb{Z}} a(\alpha)/2 \quad \text{and} \quad \tilde{M} := \sum_{\alpha \in \mathbb{Z}} \tilde{a}(\alpha)/2.$$

There exists a unique $1 \times r$ vector y such that $yM = y$ and $y\hat{\phi}(0) = 1$. Similarly, there exists a unique $1 \times r$ vector $\tilde{y}$ such that $\tilde{y}\tilde{M} = \tilde{y}$ and $\tilde{y}\hat{\tilde{\phi}}(0) = 1$. The duality of ϕ and $\tilde{\phi}$ implies $y\tilde{y}^* = 1$. Since the subdivision scheme associated with $\tilde{a}$ converges in the L_2-norm, Theorem 2.2 tells us that $\tilde{a}$ satisfies the basic sum rule:

$$\tilde{y}\sum_{\alpha\in\mathbb{Z}}\tilde{a}(2\alpha) = \tilde{y}\sum_{\alpha\in\mathbb{Z}}\tilde{a}(2\alpha-1) = \tilde{y}.$$

For $N = 1, 2, \ldots$, we can find $\tilde{a}^{(N)} \in (\ell_0(\mathbb{Z}))^{r\times r}$ such that each $\tilde{a}^{(N)}$ is finitely supported and $\|\tilde{a}^{(N)} - \tilde{a}\|_{E_\mu^{r\times r}} \to 0$ as $N \to \infty$. For $\xi \in \mathbb{R}$, let

$$\tilde{H}_N(\xi) := \sum_{\alpha\in\mathbb{Z}}\tilde{a}^{(N)}(\alpha)e^{-i\alpha\xi}/2, \qquad \xi \in \mathbb{R},$$

and

$$\varepsilon_N(\xi) := I_r - \left[H(\xi)\tilde{H}_N(\xi)^* + H(\xi+\pi)\tilde{H}_N(\xi+\pi)^*\right], \qquad \xi \in \mathbb{R}. \quad (3.10)$$

Then ε_N is π-periodic: $\varepsilon_N(\xi) = \varepsilon_N(\xi+\pi)$ for all $\xi \in \mathbb{R}$. Let

$$F_N(\xi) := \tilde{H}_N(\xi) + \varepsilon_N(\xi)^* K(\xi), \qquad \xi \in \mathbb{R},$$

where K is an $r \times r$ matrix of trigonometric polynomials satisfying (3.9). Thus, by (3.9) and (3.10) we have

$$H(\xi)F_N(\xi)^* + H(\xi+\pi)F_N(\xi+\pi)^*$$
$$= \left[H(\xi)\tilde{H}_N(\xi)^* + H(\xi+\pi)\tilde{H}_N(\xi+\pi)^*\right]$$
$$+ \left[H(\xi)K(\xi)^* + H(\xi+\pi)K(\xi+\pi)^*\right]\varepsilon_N(\xi)$$
$$= (I_r - \varepsilon_N(\xi)) + \varepsilon_N(\xi) = I_r.$$

Write

$$F_N(\xi) = \sum_{\alpha\in\mathbb{Z}}c^{(N)}(\alpha)e^{-i\alpha\xi}/2, \qquad \xi \in \mathbb{R},$$

where each $c^{(N)} \in (\ell_0(\mathbb{Z}))^{r\times r}$. Since $\|\tilde{a}^{(N)} - \tilde{a}\|_{E_\mu^{r\times r}} \to 0$ as $N \to \infty$, by the construction of F_N we also have $\|c^{(N)} - \tilde{a}\|_{E_\mu^{r\times r}} \to 0$ as $N \to \infty$. Observe that $\tilde{a}$ satisfies the basic sum rule. Hence, we may choose $\tilde{a}^{(N)}$ ($N = 1, 2, \ldots$) in such a way that each $c^{(N)}$ satisfies the basic sum rule (with respect to $\tilde{y}$):

$$\tilde{y}\sum_{\alpha\in\mathbb{Z}}c^{(N)}(2\alpha) = \tilde{y}\sum_{\alpha\in\mathbb{Z}}c^{(N)}(2\alpha-1) = \tilde{y}.$$

Let $\widetilde{M}^{(N)} := \sum_{\alpha\in\mathbb{Z}}c^{(N)}(\alpha)/2$. For sufficiently large N, 1 is a simple eigenvalue of $\widetilde{M}^{(N)}$ and its other eigenvalues are less than 1 in modulus. Let $\tilde{b}^{(N)} := c^{(N)} \diamond c^{(N)}/2$. Then $\tilde{b}^{(N)} \to \tilde{b}$ in the space $E_\mu^{r^2 \times r^2}$ as $N \to \infty$. But $\rho(T_{\tilde{b}}|_{\tilde{V}}) < 1$, where $\tilde{V}$ is the linear space given in (3.7); hence $\rho(T_{\tilde{b}^{(N)}}|_{\tilde{V}}) < 1$ for sufficiently large N. Therefore, by Theorem 2.2, the subdivision scheme associated with $c^{(N)}$ converges in the L_2-norm. By Theorem 3.1, the limit f is an $r \times 1$ vector of compactly supported functions in $L_2(\mathbb{R})$ and f is dual to ϕ. The proof of the theorem is complete. □

4 Biorthogonal Multiple Wavelets

In this section we apply the general theory developed so far to the construction of biorthogonal multiple wavelets.

The first nontrivial example of continuous symmetric orthogonal double wavelets was constructed by Donovan, Geronimo, Hardin, and Massopust in [11] by means of fractal interpolation. In [2], Chui and Lian constructed orthogonal double wavelets with symmetry by using refinement equations. However, they did not prove that the double refinable functions they constructed are functions in $L_2(\mathbb{R})$ with orthogonal shifts. In [23], Jia, Riemenschneider, and Zhou did that and constructed an entire family of orthogonal double wavelets that are continuous and have symmetry.

Biorthogonal wavelets have advantages over orthogonal wavelets in several aspects. In particular, biorthogonal wavelets can be constructed from spline functions, and the coefficients in the corresponding filters can be chosen to be rational numbers. For Hermite cubic splines, Dahmen, Han, Jia, and Kunoth [8] found a refinable dual vector of continuous functions and constructed biorthogonal double wavelets on the interval.

In this section, we will give two examples of biorthogonal double wavelets. In the first example, the wavelets are piecewise linear functions with short support. In the second example, the wavelets are almost in C^2, and the dual wavelets are in C^1. All the wavelets and dual wavelets are either symmetric or anti-symmetric about the origin.

Let us start with multiresolution of $L_2(\mathbb{R})$. Suppose $\phi = (\phi_1, \ldots, \phi_r)^T$ is a refinable vector of compactly supported functions in $L_2(\mathbb{R})$. Let S denote the closed linear subspace of $L_2(\mathbb{R})$ generated by $\phi_1, \ldots, \phi_r$. For $k \in \mathbb{Z}$, let S_k be the 2^k-dilate of S:

$$S_k := \{g(2^k \cdot) : g \in S\}.$$

It was proved by Jia and Shen [25] that $(S_k)_{k\in\mathbb{Z}}$ forms a multiresolution of $L_2(\mathbb{R})$. In other words, $S_k \subset S_{k+1}$ for $k \in \mathbb{Z}$, $\cup_{k\in\mathbb{Z}} S_k$ is dense in $L_2(\mathbb{R})$, and $\cap_{k\in\mathbb{Z}} S_k = \{0\}$. Suppose the shifts of $\phi_1, \ldots, \phi_r$ are linearly independent. By Theorem 3.2, there exists a refinable vector $\tilde{\phi} = (\tilde{\phi}_1, \ldots, \tilde{\phi}_r)^T$ of compactly supported functions in $L_2(\mathbb{R})$ such that $\tilde{\phi}$ and ϕ are dual to each other. Let $\tilde{S}$ denote the closed linear subspace of $L_2(\mathbb{R})$ generated by $\tilde{\phi}_1, \ldots, \tilde{\phi}_r$. For $k \in \mathbb{Z}$, let $\tilde{S}_k$ be the 2^k-dilate of $\tilde{S}$. Then $(\tilde{S}_k)_{k\in\mathbb{Z}}$ also forms a multiresolution of $L_2(\mathbb{R})$. We wish to find $\psi_1, \ldots, \psi_r \in S_1$ and $\tilde{\psi}_1, \ldots, \tilde{\psi}_r \in \tilde{S}_1$ such that

$$\langle \psi_j, \tilde{\phi}_k(\cdot - \gamma) \rangle = 0 \quad \forall\, j, k = 1, \ldots, r \quad \text{and} \quad \gamma \in \mathbb{Z}, \quad (4.1)$$

$$\langle \phi_j, \tilde{\psi}_k(\cdot - \gamma) \rangle = 0 \quad \forall\, j, k = 1, \ldots, r \quad \text{and} \quad \gamma \in \mathbb{Z}, \quad (4.2)$$

and

$$\langle \psi_j, \tilde{\psi}_k(\cdot - \gamma) \rangle = \delta_{\gamma,0} \delta_{jk} \quad \forall\, j, k = 1, \ldots, r \quad \text{and} \quad \gamma \in \mathbb{Z}. \quad (4.3)$$

Let W be the closed linear space of $L_2(\mathbb{R})$ generated by the shifts of $\psi_1, \ldots, \psi_r$, and let $\tilde{W}$ be the closed linear space of $L_2(\mathbb{R})$ generated by the shifts of $\tilde{\psi}_1, \ldots, \tilde{\psi}_r$. If (4.1), (4.2), and (4.3) are true, then S_1 is the direct sum of S_0 and W, and $\tilde{S}_1$ is the direct sum of $\tilde{S}_0$ and $\tilde{W}$. As was done in [4, Theorem 5.1], it can be proved that

$$\{2^{k/2}\psi_j(2^k \cdot - \alpha) : j = 1, \ldots, r, \ k \in \mathbb{Z}, \ \alpha \in \mathbb{Z}\}$$

forms a Riesz basis for $L_2(\mathbb{R})$, and

$$\{2^{k/2}\tilde{\psi}_j(2^k \cdot - \alpha) : j = 1, \ldots, r, \ k \in \mathbb{Z}, \ \alpha \in \mathbb{Z}\}$$

forms the dual basis.

Suppose

$$\psi = \sum_{\alpha \in \mathbb{Z}} c(\alpha)\phi(2 \cdot - \alpha) \quad \text{and} \quad \tilde{\psi} = \sum_{\alpha \in \mathbb{Z}} \tilde{c}(\alpha)\tilde{\phi}(2 \cdot - \alpha). \tag{4.4}$$

Then (4.1), (4.2), and (4.3) are respectively equivalent to the following equations:

$$\sum_{\beta \in \mathbb{Z}} c(\beta)\tilde{a}(2\gamma + \beta)^* = 0 \qquad \forall \gamma \in \mathbb{Z}, \tag{4.5}$$

$$\sum_{\beta \in \mathbb{Z}} \tilde{c}(\beta)a(2\gamma + \beta)^* = 0 \qquad \forall \gamma \in \mathbb{Z}, \tag{4.6}$$

$$\sum_{\beta \in \mathbb{Z}} c(\beta)\tilde{c}(2\gamma + \beta)^* = 2\delta_{\gamma,0}I_r \qquad \forall \gamma \in \mathbb{Z}. \tag{4.7}$$

Before giving two examples of biorthogonal double wavelets, we take a brief review of the approximation and smoothness properties of multiple refinable functions and multiple wavelets.

Let $\phi = (\phi_1, \ldots, \phi_r)^T$ be a vector of compactly supported distributions on $\mathbb{R}$. We say that ϕ has accuracy k if the shift-invariant space generated by $\phi_1, \ldots, \phi_r$ contains all polynomials of degree less than k. If, in addition, $\phi_1, \ldots, \phi_r$ are functions in $L_p(\mathbb{R})$ ($1 \le p \le \infty$), then ϕ has accuracy k if and only if the shift-invariant space generated by $\phi_1, \ldots, \phi_r$ provides approximation order k (see [19]).

Now suppose ϕ satisfies the refinement equation (1.1) with mask a. The optimal accuracy of ϕ was characterized in terms of the mask by Heil, Strang, Strela [17], and by Plonka [33] under the condition that the shifts of $\phi_1, \ldots, \phi_r$ be linearly independent. In [22], Jia, Riemenschneider, and Zhou gave a characterization for the accuracy without the assumption of linear independence.

In the following we give a characterization for the accuracy of ϕ in a form slightly different from that of [17]. For $m = 0, 1, 2, \ldots$, set

$$E_m := \frac{1}{m!}\sum_{\alpha \in \mathbb{Z}}(2\alpha)^m a(2\alpha) \quad \text{and} \quad O_m := \frac{1}{m!}\sum_{\alpha \in \mathbb{Z}}(2\alpha - 1)^m a(2\alpha - 1).$$

Theorem 4.1 *Let $\phi = (\phi_1, \ldots, \phi_r)^T$ be a vector of compactly supported distributions. Suppose ϕ satisfies the refinement equation (1.1) with mask a. Let k be a positive integer. If there exist $1 \times r$ vectors $c_m = (c_{m1}, \ldots, c_{mr})$ ($m = 0, 1, \ldots, k-1$) such that*

$$\sum_{m=0}^{j}(-1)^m 2^{j-m} c_{j-m} E_m = c_j \quad \text{and} \quad \sum_{m=0}^{j}(-1)^m 2^{j-m} c_{j-m} O_m = c_j \quad (4.8)$$

are true for $j = 0, 1, \ldots, k-1$, and if $c_0 \neq 0$, then ϕ has accuracy k. Moreover, under the condition $c_0 \hat{\phi}(0) = 1$, we have

$$\frac{x^j}{j!} = \sum_{\alpha \in \mathbb{Z}} \sum_{m=0}^{j} \frac{\alpha^m}{m!} c_{j-m} \phi(x - \alpha), \quad j = 0, 1, \ldots, k-1, \; x \in \mathbb{R}.$$

Conversely, if ϕ has accuracy k, and if the shifts of $\phi_1, \ldots, \phi_r$ are linearly independent, then there exist $1 \times r$ vectors c_m ($m = 0, 1, \ldots, k-1$) satisfying $c_0 \neq 0$ and the conditions in (4.8).

In fact, Theorem 4.1 remains true if, for $\xi = 0$ and $\xi = \pi$, the sequences $(\hat{\phi}_j(\xi + 2\beta\pi))_{\beta \in \mathbb{Z}}$ ($j = 1, \ldots, r$) are linearly independent.

We use the generalized Lipschitz space to measure smoothness of a given function. By $(\text{Lip}^*(\nu, L_p(\mathbb{R})))^r$ we denote the linear space of all vectors $f = (f_1, \ldots, f_r)^T$ such that $f_1, \ldots, f_r \in \text{Lip}^*(\nu, L_p(\mathbb{R}))$. The optimal smoothness of a vector $f \in (L_p(\mathbb{R}))^r$ in the L_p-norm is described by its critical exponent $\nu_p(f)$ defined by

$$\nu_p(f) := \sup\left\{ \nu : f \in \left(\text{Lip}^*(\nu, L_p(\mathbb{R}))\right)^r \right\}.$$

It is easily seen that

$$\nu_\infty(f) \geq \nu_2(f) - 1/2.$$

In [24], Jia, Riemenschneider, and Zhou gave a characterization for the smoothness order of a refinable vector of functions in terms of the p-norm joint spectral radius of two matrices associated with the mask. In particular, for the case $p = 2$, Theorem 3.4 of [24] can be restated as follows.

Theorem 4.2 *Suppose $\phi = (\phi_1, \ldots, \phi_r)^T \in (L_2(\mathbb{R}))^r$ is a compactly supported solution of the refinement equation (1.1) with mask a. Let $b := a \diamond a/2$ and let T_b be the corresponding transition operator. Then, for any positive integer k,*

$$\nu_2(\phi) \geq -\log_2 \sqrt{\rho(T_b|_W)},$$

where W is the minimal invariant subspace of T_b generated by $e_{jj}(\Delta^k \delta)$, $j = 1, \ldots, r$. Moreover, $\nu_2(\phi) = -\log_2 \sqrt{\rho(T_b|_W)}$, provided

$$k > -\log_2 \sqrt{\rho(T_b|_W)}$$

and the shifts of $\phi_1, \ldots, \phi_r$ are stable.

We are ready to provide two examples of biorthogonal double wavelets.

Example 4.3 Let ϕ_1 and ϕ_2 be two functions on $\mathbb{R}$ given by

$$\phi_1(x) := \begin{cases} 3x+2 & \text{for } x \in [-2/3, -1/3], \\ 1 & \text{for } x \in [-1/3, 1/3], \\ -3x+2 & \text{for } x \in [1/3, 2/3], \\ 0 & \text{for } x \in \mathbb{R} \setminus [-2/3, 2/3], \end{cases}$$

and

$$\phi_2(x) := \begin{cases} -3x-2 & \text{for } x \in [-2/3, -1/3], \\ 3x & \text{for } x \in [-1/3, 1/3], \\ -3x+2 & \text{for } x \in [1/3, 2/3], \\ 0 & \text{for } x \in \mathbb{R} \setminus [-2/3, 2/3]. \end{cases}$$

Then ϕ_1 is symmetric about the origin, and ϕ_2 is anti-symmetric about the origin. It can be directly verified that the shifts of ϕ_1 and ϕ_2 are linearly independent.

The vector $\phi = (\phi_1, \phi_2)^T$ is refinable:

$$\phi(x) = a(-1)\phi(2x+1) + a(0)\phi(2x) + a(1)\phi(2x-1), \qquad x \in \mathbb{R},$$

where the mask a is given by

$$a(-1) = \begin{bmatrix} 1/2 & 1/2 \\ -1/2 & -1/2 \end{bmatrix}, \quad a(0) = \begin{bmatrix} 1 & 0 \\ 0 & 1/2 \end{bmatrix}, \quad a(1) = \begin{bmatrix} 1/2 & -1/2 \\ 1/2 & -1/2 \end{bmatrix}.$$

Since the shifts of ϕ_1 and ϕ_2 are linearly independent, Theorem 3.2 tells us that there exists a dual refinable vector of compactly supported functions in $L_2(\mathbb{R})$. The corresponding mask $\tilde{a}$ must satisfy condition (a) of Theorem 3.1. We choose $\tilde{a}$ such that $\tilde{a}(\alpha) = 0$ for $\alpha \in \mathbb{Z} \setminus \{-1, 0, 1\}$, and

$$\tilde{a}(-1) = \begin{bmatrix} 1/2 & 1/2 \\ -7/8 & -7/8 \end{bmatrix}, \quad \tilde{a}(0) = \begin{bmatrix} 1 & 0 \\ 0 & 1/2 \end{bmatrix}, \quad \tilde{a}(1) = \begin{bmatrix} 1/2 & -1/2 \\ 7/8 & -7/8 \end{bmatrix}.$$

Then $\tilde{a}$ possesses the desired property. Moreover, the subdivision schemes associated with a and $\tilde{a}$ converge in the L_2-norm (see [24, Example 5.2]). Let $\tilde{\phi} = (\tilde{\phi}_1, \tilde{\phi}_2)^T$ be the solution of the refinement equation with mask $\tilde{a}$ such that $\widehat{\tilde{\phi}_1}(0) = 1$ and $\widehat{\tilde{\phi}_2}(0) = 0$. Then $\tilde{\phi} \in (L_2(\mathbb{R}))^2$ is dual to ϕ. It was proved in [24] that $\nu_\infty(\tilde{\phi}) = 0.375$ and the optimal accuracy of $\tilde{\phi}$ is 2.

In order to construct biorthogonal wavelets ψ and $\tilde{\psi}$ we need to find $c \in (\ell_0(\mathbb{Z}))^2$ and $\tilde{c} \in (\ell_0(\mathbb{Z}))^2$ such that they satisfy (4.5), (4.6), and (4.7). We choose c and $\tilde{c}$ such that $c(\alpha) = \tilde{c}(\alpha) = 0$ for $\alpha \in \mathbb{Z} \setminus \{-1, 0, 1\}$,

$$c(-1) = \begin{bmatrix} -1/2 & -1/2 \\ 1/2 & 1/2 \end{bmatrix}, \quad c(0) = \begin{bmatrix} 1 & 0 \\ 0 & 7/2 \end{bmatrix}, \quad c(1) = \begin{bmatrix} -1/2 & 1/2 \\ -1/2 & 1/2 \end{bmatrix},$$

and

$$\tilde{c}(-1) = \begin{bmatrix} -1/2 & -1/2 \\ 1/8 & 1/8 \end{bmatrix}, \quad \tilde{c}(0) = \begin{bmatrix} 1 & 0 \\ 0 & 1/2 \end{bmatrix}, \quad \tilde{c}(1) = \begin{bmatrix} -1/2 & 1/2 \\ -1/8 & 1/8 \end{bmatrix}.$$

If $\psi = (\psi_1, \psi_2)^T$ and $\tilde{\psi} = (\tilde{\psi}_1, \tilde{\psi}_2)^T$ are given by (4.4), then ψ_1, ψ_2 and $\tilde{\psi}_1, \tilde{\psi}_2$ are biorthogonal double wavelets. □

Example 4.4 Consider the following refinement equation:

$$\phi = \sum_{\alpha \in \mathbb{Z}} a(\alpha) \phi(2 \cdot - \alpha),$$

where the refinement mask a is supported on $\{-1, 0, 1\}$ and

$$a(-1) = \begin{bmatrix} 1/2 & 3/2 \\ -1/8 & -1/2 \end{bmatrix}, \quad a(0) = \begin{bmatrix} 1 & 0 \\ 0 & 1/2 \end{bmatrix}, \quad a(1) = \begin{bmatrix} 1/2 & -3/2 \\ 1/8 & -1/2 \end{bmatrix}.$$

It was proved in [24, Example 4.2] that the shifts of ϕ_1 and ϕ_2 are linearly independent. Moreover, $\nu_\infty(\phi) = 2$ and the optimal accuracy of ϕ is 3. By Theorem 3.2, there exists a dual refinable vector of compactly supported functions in $L_2(\mathbb{R})$. We choose $\tilde{a}$ to be the sequence supported on $\{-2, -1, 0, 1, 2\}$ and given by

$$\tilde{a}(-2) = \begin{bmatrix} -53/512 & -7/256 \\ 359/512 & 99/512 \end{bmatrix}, \quad \tilde{a}(-1) = \begin{bmatrix} 1/2 & 25/256 \\ -421/128 & -161/256 \end{bmatrix},$$

$$\tilde{a}(0) = \begin{bmatrix} 309/256 & 0 \\ 0 & 281/256 \end{bmatrix}, \quad \tilde{a}(1) = \begin{bmatrix} 1/2 & -25/256 \\ 421/128 & -161/256 \end{bmatrix},$$

$$\tilde{a}(2) = \begin{bmatrix} -53/512 & 7/256 \\ -359/512 & 99/512 \end{bmatrix}.$$

It is easy to verify that a and $\tilde{a}$ satisfy condition (a) of Theorem 3.1. Moreover, the subdivision schemes associated with a and $\tilde{a}$ converge in the L_2-norm. Let $\tilde{\phi} = (\tilde{\phi}_1, \tilde{\phi}_2)^T$ be the solution of the refinement equation with mask $\tilde{a}$ such that $\widehat{\tilde{\phi}_1}(0) = 1$ and $\widehat{\tilde{\phi}_2}(0) = 0$. By Theorem 4.1 we find that the optimal accuracy of $\tilde{\phi}$ is 2. The smoothness order of $\tilde{\phi}$ can be computed by using Theorem 4.2 and the result is $\nu_2(\tilde{\phi}) \approx 1.5510$. It follows that

$$\nu_\infty(\tilde{\phi}) \geq \nu_2(\tilde{\phi}) - 0.5 > 1.05.$$

So $\tilde{\phi}$ is a vector of C^1 functions.

Finally, let us construct wavelets and dual wavelets associated with ϕ

and $\tilde{\phi}$. We choose c and $\tilde{c}$ as follows:

$$c(-3) = \begin{bmatrix} 43/512 & 57/128 \\ 43/512 & 57/128 \end{bmatrix}, \qquad c(-2) = \begin{bmatrix} -7/32 & -99/128 \\ -7/32 & -99/128 \end{bmatrix},$$

$$c(-1) = \begin{bmatrix} 981/512 & 1505/128 \\ 407/512 & 703/128 \end{bmatrix}, \qquad c(0) = \begin{bmatrix} -57/16 & 0 \\ 0 & 743/64 \end{bmatrix},$$

$$c(1) = \begin{bmatrix} 981/512 & -1505/128 \\ -407/512 & 703/128 \end{bmatrix}, \qquad c(2) = \begin{bmatrix} -7/32 & 99/128 \\ 7/32 & -99/128 \end{bmatrix},$$

$$c(3) = \begin{bmatrix} 43/512 & -57/128 \\ -43/512 & 57/128 \end{bmatrix},$$

and

$$\tilde{c}(-2) = \begin{bmatrix} -1/64 & 0 \\ -1/64 & 0 \end{bmatrix}, \qquad \tilde{c}(-1) = \begin{bmatrix} 1/8 & 1/32 \\ 1/8 & 1/32 \end{bmatrix},$$

$$\tilde{c}(0) = \begin{bmatrix} -7/32 & 0 \\ 0 & 1/8 \end{bmatrix}, \qquad \tilde{c}(1) = \begin{bmatrix} 1/8 & -1/32 \\ -1/8 & 1/32 \end{bmatrix},$$

$$\tilde{c}(2) = \begin{bmatrix} -1/64 & 0 \\ 1/64 & 0 \end{bmatrix}.$$

If $\psi = (\psi_1, \psi_2)^T$ and $\tilde{\psi} = (\tilde{\psi}_1, \tilde{\psi}_2)^T$ are given by (4.4), then ψ_1, ψ_2 and $\tilde{\psi}_1, \tilde{\psi}_2$ are biorthogonal double wavelets. Note that both ψ and $\tilde{\psi}$ are supported on $[-2, 2]$. All $\phi_1, \tilde{\phi}_1, \psi_1, \tilde{\psi}_1$ are symmetric about the origin, and all $\phi_2, \tilde{\phi}_2, \psi_2, \tilde{\psi}_2$ are anti-symmetric about the origin. □

Acknowledgement. The author is grateful to Bin Han for his help in the computation done in Example 4.4.

5 References

[1] A. S. Cavaretta, W. Dahmen, and C. A. Micchelli, *Stationary Subdivision*, Memoirs of Amer. Math. Soc., Volume 93, 1991.

[2] C. K. Chui and J. A. Lian, *A study of orthonormal multi-wavelets*, J. Applied Numerical Math. **20** (1996), 273–298.

[3] C. K. Chui and J. Z. Wang, *A general framework of compactly supported spline and wavelets*, J. Approx. Theory **71** (1992), 263–304.

[4] A. Cohen and I. Daubechies, *A stability criterion for biorthogonal wavelet bases and their related subband coding scheme*, Duke Math. J. **68** (1992), 313–335.

[5] A. Cohen and I. Daubechies, *A new technique to estimate the regularity of refinable functions*, Rev. Math. Iberoamericana **12** (1996), 527–591.

[6] A. Cohen, I. Daubechies, and J.-C. Feauveau, *Biorthogonal bases of compactly supported wavelets*, Comm. Pure and Appl. Math. **45** (1992), 485–560.

[7] A. Cohen, I. Daubechies, and G. Plonka, *Regularity of refinable function vectors*, J. Fourier Anal. Appl. **3** (1997), 295–324.

[8] W. Dahmen, B. Han, R. Q. Jia, and A. Kunoth, *Biorthogonal multiwavelets on the interval: Cubic Hermite splines*, manuscript.

[9] W. Dahmen and C. A. Micchelli, *Biorthogonal wavelet expansions*, Constr. Approx. **13** (1997), 293–328.

[10] G. Deslauriers and S. Dubuc, *Symmetric iterative interpolation processes*, Constr. Approx. **5** (1989), 49–68.

[11] G. Donovan, J. S. Geronimo, D. P. Hardin, and P. R. Massopust, *Construction of orthogonal wavelets using fractal interpolation functions*, SIAM J. Math. Anal. **27**(1996), 1158–1192.

[12] G. B. Folland, *Real Analysis*, John Wiley, New York, 1984.

[13] T. N. T. Goodman, R. Q. Jia, and C. A. Micchelli, *On the spectral radius of a bi-infinite periodic and slanted matrix*, Southeast Asian Bull. Math., to appear.

[14] T. N. T. Goodman, C. A. Micchelli, and J. D. Ward, *Spectral radius formulas for subdivision operators*, in Recent Advances in Wavelet Analysis, L. L. Schumaker and G. Webb (eds.), Academic Press, 1994, pp. 335–360.

[15] B. Han and R. Q. Jia, *Multivariate refinement equations and convergence of subdivision schemes*, SIAM J. Math. Anal., **29** (1998), 1177–1199.

[16] C. Heil and D. Colella, *Matrix refinement equations: existence and uniqueness*, J. Fourier Anal. Appl. **2** (1996), 363–377.

[17] C. Heil, G. Strang, and V. Strela, *Approximation by translates of refinable functions*, Numer. Math. **73** (1996), 75–94.

[18] R. Q. Jia, *Subdivision schemes in L_p spaces*, Advances in Comp. Math. **3** (1995), 309–341.

[19] R. Q. Jia, *Shift-invariant spaces on the real line*, Proc. Amer. Math. Soc. **125** (1997), 785–793.

[20] R.-Q. Jia and C. A. Micchelli, *Using the refinement equations for the construction of pre-wavelets II: Powers of two*, in Curves and Surfaces, P. J. Laurent, A. Le Méhauté, and L. L. Schumaker (eds.), Academic Press, New York, 1991, 209–246.

[21] R. Q. Jia and C. A. Micchelli, *On linear independence of integer translates of a finite number of functions*, Proc. Edinburgh Math. Soc. **36** (1992), 69-85.

[22] R. Q. Jia, S. D. Riemenschneider, and D. X. Zhou, *Approximation by multiple refinable functions*, Canadian J. Math.,**49** (1997),944-962.

[23] R. Q. Jia, S. D. Riemenschneider, and D. X. Zhou, *Vector subdivision schemes and multiple wavelets*, Math. Comp., to appear.

[24] R. Q. Jia, S. D. Riemenschneider, and D. X. Zhou, *Smoothness of multiple refinable functions and multiple wavelets*, SIAM J. Matrix Anal. Appl., to appear.

[25] R. Q. Jia and Z. W. Shen, *Multiresolution and wavelets*, Proc. Edinburgh Math. Soc. **37** (1994), 271-300.

[26] R. Q. Jia and J. Z. Wang, *Stability and linear independence associated with wavelet decompositions*, Proc. Amer. Math. Soc. **117** (1993), 1115-1124.

[27] Q. T. Jiang, *On the regularity of matrix refinable functions*, SIAM J. Math. Anal., **29** (1998), 1157-1176.

[28] P. Lancaster and M. Tismenersky, *The Theory of Matrices*, Second Edition, Academic Press, Orlando, 1985.

[29] W. Lawton, *Necessary and sufficient conditions for constructing orthonormal wavelets*, J. Math. Phys. **32** (1991), 57-61.

[30] P. G. Lemarié-Rieusset, *On the existence of compactly supported dual wavelets*, Applied and Computational Harmonic Analysis **3** (1997), 117-118.

[31] R. L. Long and D. L. Chen, *Biorthogonal wavelet bases on $\mathbb{R}^d$*, Applied and Computational Harmonic Analysis **2** (1995), 230-242.

[32] R. L. Long, W. Chen, and S. L. Yuan, *Wavelets generated by vector multiresolution analysis*, Applied and Computational Harmonic Analysis **4** (1997), 317-350.

[33] G. Plonka, *Approximation order provided by refinable function vectors*, Constr. Approx. **13** (1997), 221-244.

[34] W. Rudin, *Functional Analysis*, McGraw-Hill Book Company, New York, 1973.

[35] Z. W. Shen, *Refinable function vectors*, SIAM J. Math. Anal., **29** (1998), 235-250.

[36] D. X. Zhou, *The p-norm joint spectral radius for even integers*, Methods and Applications of Analysis, to appear.

10
Study of Linear Independence and Accuracy of Scaling Vectors via Two-scale Factors

Jianzhong Wang [1]

ABSTRACT A scaling vector $\phi = (\phi_1, \cdots, \phi_r)^T$ is a compactly supported vector-valued distribution that satisfies a matrix refinement equation $\phi(x) = \sum \mathbf{P}_k \phi(2x - k)$, where $(\mathbf{P}_k)$ is a finite matrix sequence. We call $P(z) = \frac{1}{2}\sum \mathbf{P}_k z^k$ the symbol of ϕ. A symbol $P(z)$ is said to be two-scale similar to a polynomial matrix $Q(z)$ if there is an invertible polynomial matrix $T(z)$ that satisfies $P(z) = T(z^2)Q(z)T^{-1}(z)$. $T(z)$ is called a two-scale factor of $P(z)$. In this paper we characterize linear independence and accuracy associated with a scaling vector via the two-scale factor of its symbol. The necessary and sufficient conditions for either linear independence or accuracy are given. The relation between them is also revealed.

1 Introduction

In this paper, we study linear independence and accuracy associated with scaling vectors. A compactly supported distribution vector

$$\phi(x) = (\phi_1(x), \phi_2(x), \cdots, \phi_r(x))^T, \quad x \in \mathbb{R}, \tag{1}$$

is said to be a *scaling vector* or a *refinable vector* if it satisfies a two-scale matrix equation

$$\phi(x) = \sum_{k=m}^{n} \mathbf{P}(k)\phi(2x - k), \tag{2}$$

where the finite $r \times r$ matrix sequence $(\mathbf{P}(k))_{k=m}^{n}$ is called the *mask* of ϕ while $P(z) := \frac{1}{2}\sum \mathbf{P}(k)z^k$ is called the *symbol* of ϕ. If $r = 1$, then ϕ is reduced to a *scaling distribution*, whose mask is a sequence of complex numbers.

---

[1] Department of Mathematical and Information Sciences, Sam Houston State University, Huntsville, TX 77341, USA (e-mail: `mth_jxw@shsu.edu`)
This paper is supported by NSF of US Grant DMS-9503282 and by the Institute of Mathematical Sciences, The Chinese University of Hong Kong.

The scaling vectors in a function space such as $(L^2)^r$ are often used to generate *multiresolution analysis* (MRA) and then lead to *multiwavelet basis*, which is generated by more than one "mother" wavelets. One of the motivations to study multiwavelets is to find wavelet bases with "better" properties. As we know, in the wavelet theory, a "good" wavelet basis is often required to have short support, symmetry, certain approximation order and regularity, and orthogonality etc. However, if a basis is generated by a single wavelet, then we are in a dilemma such as a compactly supported orthonormal wavelet cannot be symmetric or anti-symmetric, a wavelet with short support cannot have high approximation order. These limitations of the single-wavelet bases push us to study scaling vectors and multiwavelets. We expect to trade-in a "better" wavelet basis (often called multiwavelet basis) by trading-off the number of scaling functions and the corresponding wavelets generating the basis.

Scaling vectors have come up in several papers for different purposes. The first multiwavelet perhaps occurred in the paper of Alpert and Rokhlin [2]. The authors of [2] constructed the multiwavelet using piecewise polynomials for developing a fast algorithm to evaluate Legendre expansions (see also [1]). Donovan, Geronimo, Hardin, and Massopust [16] first constructed continuous, compactly supported, and symmetric (or anti-symmetric) orthonormal scaling vectors and the corresponding multiwavelets based on the fractal interpolation method [17]. Their examples stimulate interest in the study of multiwavelets. A general discussion of orthonormal scaling vectors and multiwavelets, including how to construct the orthonormal multiwavelets once the scaling vectors are known, can be found in Goodman and Lee [18] and Goodman, Lee, and Tang [19]. The existence and uniqueness of the solution of a two-scale matrix equation, as well as the regularity of the solution are studied in Heil and Colella [20]. Their results were greatly improved by Cohen, Daubechies, and Plonka [13], Zhou [48], and Jiang and Shen [32]. The support of a scaling vector is also an interesting topic in multiwavelets. Massopust, Ruch, and Van Fleet in [33] revealed that the support length of a scaling vector, unlike a scaling distribution, does not always coincide with the length of its mask. [33] also gave an estimate for the support length of a scaling vector. Using an algebraic method, So and Wang [42] obtained a formula to locate the support of each component of a scaling vector. Readers also can find an interesting relation between the support length and the linear independence of a scaling vector in Ruch, So, and Wang [40]. A description of the approximation order provided by a scaling vector was given by Heil, Strang, and Strela [21]. Later, Jia, Riemenschneider, and Zhou [29] improved the result of [21] by analyzing the subdivision operator associated with a scaling vector. Chen in [8] found that the criterion for approximation order obtained in [29] can be simplified once a simple condition is added to a scaling vector. At the same time, another work on this aspect was done by Plonka [35] (see also [36]), where she introduced the notion of two-scale similarity to characterize the approximation order. Two-scale similarity is now proved to be a powerful tool in the study of properties of scaling vectors. For example, [13] and Plonka and Strela [38] used it to build new scaling vectors with higher regularity from a known one. Chui and Lian [9] used it in the construction of symmetric (or anti-symmetric) multiwavelets (see also [38]). The author of this paper in [46] applied it to characterize the stability and linear independence of scaling vectors.

In this paper, we plan to give a systematic discussion on linear independence (including stability) and approximation order of scaling vectors via their symbols. Linear independence and approximation order are two important properties of a scaling vector. It is well-kown that a scaling vector generates an MRA only if its integer translates are stable. (Stability is a "weak" form of linear independence. We shall discuss it in Section 2.) Stability of a scaling vector is also crucial for constructing its dual scaling vector. (See [19] and Chui and Wang [12].) In many cases, generators of MRA are required to be linearly independent. Hence it is desirable to give criteria for linear independence of a scaling vector. The importance of approximation order of a scaling vector is also evident. Some results on these two topics already exist in literature. Our goal is to reveal the intrinsic relation between them and contribute some new results so that the essence of these topics becomes more clear.

This paper is organized as follows. In section 2, we introduce the notions and results of linear independence and accuracy of a distribution vector, which is not necessarily refinable. In Section 3, we present necessary and sufficient conditions for linear independence in terms of two-scale factor. The characterization of accuracy of scaling vectors is established in Section 4, where we also show the relation between some existing results and ours. Finally, we put a few remarks in Section 5 to refer other relevant works which we will not discuss in detail.

2 Linear independence and accuracy of a compactly supported distribution vector

In this section we introduce the notions and results of linear independence and approximation order of a compactly supported distribution vector, of which most exist in literature.

2.1 Linear independence

The integer translates of a compactly supported distribution vector are often used to span a "shift-invariant space". Linear dependence describes some kind of redundance among these integer translates. That is, if these integer translates are required to be a basis of the space they span, they have to be linearly independent in that space. We now give the precise definition of linear independence.

Let l be the space containing all sequences of complex numbers, l^∞ be the subspace of l that contains all bounded sequences, and l_0 be the subspace of l that contains all compactly supported sequences. We define the semi-convolution of a compactly supported distribution vector ϕ with a vector sequence $\mathbf{a} := (\mathbf{a}_{j,k}) \in (l)^r$ by

$$\phi * \mathbf{a} = \sum_{j=1}^{r} \sum_{k \in \mathbb{Z}} \mathbf{a}_{j,k} \phi_j(x-k).$$

Definition 2.1 *A compactly supported distribution vector ϕ is said to have linearly independent integer translates if*

$$\phi * \mathbf{a} = 0 \Longrightarrow \mathbf{a} = 0, \quad \mathbf{a} \in (l)^r,$$

it is said to have l^∞-linearly independent integer translates if

$$\phi * \mathbf{a} = 0 \Longrightarrow \mathbf{a} = 0, \quad \mathbf{a} \in (l^\infty)^r,$$

and it is said to have finitely linearly independent integer translates if

$$\phi * \mathbf{a} = 0 \Longrightarrow \mathbf{a} = 0, \quad \mathbf{a} \in (l_0)^r.$$

Let $S(\phi) := \{\phi * \mathbf{a}; \ \mathbf{a} \in (l)^r\}$, $S^b(\phi) := \{\phi * \mathbf{a}; \ \mathbf{a} \in (l^\infty)^r\}$, and $S_0(\phi) := \{\phi * \mathbf{a}; \ \mathbf{a} \in (l_0)^r\}$. Then the semi-convolution, $\phi*$, is a mapping from $(l)^r$ to $S(\phi)$ (from $(l^\infty)^r$ to $S^b(\phi)$, and from $(l_0)^r$ to $S_0(\phi)$ respectively), which is one-to-one if and only if the integer translates of ϕ are linearly independent (l^∞-linearly independent, and finitely linearly independent respectively). Sometimes we use the kernels of the convolution mapping to describe linear independence. Let

$$K(\phi) := \{\mathbf{a} \in (l)^r; \ \phi * \mathbf{a} = 0\},$$

$$K^b(\phi) := \{\mathbf{a} \in (l^\infty)^r; \ \phi * \mathbf{a} = 0\},$$

and

$$K_0(\phi) := \{\mathbf{a} \in (l_0)^r; \ \phi * \mathbf{a} = 0\}$$

be the kernels of the convolution mapping in the spaces $(l)^r$, $(l^\infty)^r$, and $(l_0)^r$ respectively. Then ϕ has linearly independent, (l^∞-linearly independent, and finitely linearly independent) integer translates if and only if $K(\phi) = \{0\}$, ($K^b(\phi) = \{0\}$, and $K_0(\phi) = \{0\}$).

For convenience, in the rest of the paper, we will simply say that ϕ is linearly independent, (l^∞-linearly independent, and finitely linearly independent) instead of that ϕ has linearly independent, (l^∞-linearly independent, and finitely linearly independent) integer translates.

The various linear dependences describe the different degrees of redundancy of distribution vectors. Linear dependence implies a "global", or "weaker" redundancy, while finite linear dependence reflects a "local", or "stronger" redundancy.

Example 2.1 Let $B_1 = \chi_{[0,1)}$ be the character function of interval $[0,1)$ and $B_2 := B_1 * B_1$ be the hat function on $[0,2]$. Then

$$\sum_{k \in \mathbb{Z}}(B_1(x-k) - B_2(x-k)) = 0,$$

which implies that the integer translates of these two functions are linearly dependent (also l^∞-linearly dependent). However, they are finitely linearly independent, for any non-trivial finite linear combination of the integer translates of B_1 and B_2 does not vanish. In other words, the redundancy of B_1 and B_2 only occurs "globally". If we add $\phi_3(x) = xB_1(x)$ to the vector (B_1, B_2), then

$$B_1(\cdot - 1) - B_2(\cdot) + \phi_3(\cdot) - \phi_3(\cdot - 1) = 0,$$

which shows that the redundancy of B_1, B_2, and ϕ_3 exists "locally", i.e., they are finitely linearly dependent.

Jia and Micchelli in [28] pointed out that the exponential set contained in $(l)^r$ $((l^\infty)^r)$ is the testing set for linear independence (l^∞-linear independence). Let $\theta \in \mathbb{C} \setminus \{0\}$ and $a \in \mathbb{C}^r \setminus \{0\}$. The sequence vector $f_{a,\theta} := (\theta^k a)_{k \in \mathbb{Z}}$ is called an exponential sequence vector. Let $\mathcal{E}$ be the set of all exponential sequence vectors in $(l)^r$. It is obvious that $f_{a,\theta}$ is bounded if and only if $\theta \in \Gamma$, where $\Gamma = \{z \in \mathbb{C};\ |z| = 1\}$ is the unit circle on the complex plane. We now define

$$\Theta(\phi) = \{\theta \in \mathbb{C} \setminus \{0\};\ \phi * f_{a,\theta} = 0,\ f_{a,\theta} \in \mathcal{E}\}. \tag{3}$$

Theorem 2.1 ([28]) Let ϕ be a compactly supported distribution vector. Then ϕ is linearly dependent (l^∞-linearly dependent) if and only if $\Theta(\phi) \neq \emptyset$ ($\Theta(\phi) \cap \Gamma \neq \emptyset$).

The set $\Theta(\phi)$ also tests the finite linear independence. We obtained the following in [46].

Theorem 2.2 Let ϕ be a compactly supported distribution vector. Then ϕ is finitely linearly dependent if and only if $\Theta(\phi)$ is an infinite set.

Remark 2.1 Note that $\Theta(\phi)$ is an infinite set is equivalent to that $\Theta(\phi) = \mathbb{C} \setminus \{0\}$. In other words, if there exists a $\theta \notin (\Theta(\phi) \cup \{0\})$, then $\Theta(\phi)$ is finite.

From Theorem 2.1, we derive the following corollary.

Corollary 2.3 Let ϕ be a compactly supported distribution vector. Then ϕ is linearly dependent, (or l^∞-linearly dependent,) if and only if there is a vector $a \in \mathbb{C}^r \setminus \{0\}$, such that $\psi = a^T \phi$ is linearly dependent (or l^∞-linearly dependent).

Furthermore, if $\psi = a^T \phi$ is linearly dependent, (or l^∞-linearly dependent,) then there is a compactly supported distribution $\tilde{\psi} \in S(\phi)$ and a complex number $\theta \in \mathbb{C} \setminus \{0\}$ (or $\theta \in \Gamma$) such that

$$\psi = \tilde{\psi}(\cdot) - \theta \tilde{\psi}(\cdot - 1).$$

Although the function $\tilde{\psi}$ in Corollary 2.3 is compactly supported, it may not be in $S_0(\phi)$.

We have discussed linear independence in the *time domain*. We now give its characterization in the *frequency domain*. Let $\hat{f}$ denote the Fourier transform of a distribution f:

$$\hat{f}(\omega) = \int_\mathbb{R} f(x) e^{-ix\omega} dx.$$

The following theorem is a directly subsequence of Theorem 2.1 and Theorem 2.2 (see [28] and [46]).

Theorem 2.4 Let ϕ be a compactly supported distribution vector. Then

(i) ϕ is linearly independent if and only if for <u>every $\omega \in \mathbb{C}$</u> the following r sequences

$$(\hat{\phi}_l(\omega + 2k\pi))_{k \in \mathbb{Z}}, \quad l = 1, 2, \cdots, r \qquad (4)$$

are linearly independent.

(ii) ϕ is l^∞-linearly independent if and only if the sequences (4) are linearly independent for <u>every $\omega \in \mathbb{R}$</u>.

(iii) ϕ is finitely linearly independent if and only if the sequences (4) are linearly independent for <u>at least one $\omega \in \mathbb{C}$</u>.

If we set

$$\Omega(\phi) = \{\omega;\ a^T \hat{\phi}(\omega + 2k\pi) = 0,\ a \in \mathbb{C}^r \setminus \{0\},\ \forall k \in \mathbb{Z}\},$$

then Theorem 2.4 can be re-stated as following.

Corollary 2.5 ϕ is linearly independent (l^∞-linearly independent, and finitely linearly independent respectively) if and only if $\Omega(\phi) = \emptyset$ ($\Omega(\phi) \cap \mathbb{R} = \emptyset$, and $\Omega(\phi) \neq \mathbb{C}$ respectively).

The relation between $\Omega(\phi)$ and $\Theta(\phi)$ is the following.

$$\theta \in \Theta(\phi) \iff \exp(i\theta) \in \Omega(\phi).$$

Another useful corollary of Theorem 2.4 is

Corollary 2.6 *Let ϕ be a compactly supported distribution vector. Then we have the followings.*

(i) ϕ is linearly dependent if and only if there is another compactly supported distribution vector $\psi = (\psi_1, \cdots, \psi_r) \subset S_0(\phi)$ and a Laurent polynomial matrix $H(z)$, that is singular at some $z_0 \in \mathbb{C} \setminus \{0\}$, such that

$$\hat{\phi}(\omega) = H(e^{i\omega})\hat{\psi}(\omega). \tag{5}$$

(ii) ϕ is l^∞-linearly dependent if and only $H(z)$ in (5) is singular at some $z_0 \in \Gamma$.

(iii) ϕ is finitely linearly dependent if and only if $H(z)$ in (5) is singular for all $z \in \mathbb{C} \setminus \{0\}$.

Finally, we briefly introduce the notion of stability of a compactly supported distribution vector and show that it is equivalent to l^∞-linear independence. Recall that the spaces $S(\phi)$, $S^b(\phi)$ and $S_0(\phi)$ are all algebraic spaces. In application, we often consider the linear normed space generated by ϕ. Assuming $\phi \in (L^p)^r, 1 \leq p \leq \infty$, we define

$$S^p(\phi) := \{\phi * \mathbf{a}; \quad \mathbf{a} \in (l^p)^r\}.$$

where the norm of a vector sequence $\mathbf{a} \in (l^p)^r$ is defined by

$$\|\mathbf{a}\|_p = (\sum_{j=1}^r \sum_{k \in \mathbb{Z}} |\mathbf{a}_{j,k}|^p)^{1/p}, \quad 1 \leq p \leq \infty.$$

Evidently, $S^p(\phi)$ is a subspace of L^p and the convolution operator, $\phi*$, is a continuous mapping from $(l^p)^r$ to $S^p(\phi)$, $1 \leq p \leq \infty$. The stability of ϕ is defined as follows.

Definition 2.2 *A vector-valued function $\phi \in (L^p)^r$ is said to be l^p-stable, $1 \leq p \leq \infty$, if there exist two positive constants C_1 and C_2 such that, for every $\mathbf{a} \in (l^p)^r$,*

$$C_1\|\mathbf{a}\|_p \leq \|\phi * \mathbf{a}\|_p \leq C_2\|\mathbf{a}\|_p. \tag{6}$$

Thus, for any $p, 1 \leq p \leq \infty$, if ϕ is l^p-stable, then space $S^p(\phi)$ is closed, and the integer translates of ϕ form an unconditional basis of $S^p(\phi)$. It is known that (6) holds if and only if the r sequences (4) are linearly independent for every $\omega \in \mathbb{R}$ (see [28]). By Theorem 2.4, we have

Corollary 2.7 *A compactly supported vector-valued function $\phi \in (L^p)^r$ is stable if and only if it is l^∞-linearly independent.*

Recall that the concept of l^∞-linear independence makes sense not only for a vector-valued function in $(L^p)^r$, but also for any compactly supported distribution vector. Therefore, l^∞-linear independence can be considered as the generalization of stability for compactly supported distribution vectors.

2.2 Approximation order

We now turn to the discussion of approximation order of a compactly supported distribution vector. Let Π_k be the set of all polynomials of degree no more than k and Π be the set of all polynomials. Write

$$S_{(p)}\left(=S_{(p)}(\phi)\right) = S(\phi) \cap L^p,$$

and

$$S_{(p)}^h = \{g(\cdot/h), \quad g \in S_{(p)}\}.$$

Remark 2.2 *For $1 < p < \infty$, $S_{(p)}(\phi) = S^p(\phi)$. Another subspace of L^p created by ϕ is $S_p(\phi) = \text{Clos}_{L^p} S_0(\phi)$. The discussion of the relations among $S^p(\phi)$, $S_{(p)}(\phi)$, and $S_p(\phi)$ can be found in Jia [24].*

Definition 2.3 *Let $\phi \in (L^p)^r$. The space $S(\phi)$ is said to provide L^p-approximation order k if*

$$\text{dist}_p(f, S_{(p)}^h) \leq C h^k,$$

where C is a positive constant independent of h.

A distribution vector ϕ is say to have accuracy k if $\prod_{k-1} \subset S(\phi)$.

In general, these two concepts are not the same (see [6]). However, for the compactly supported vector-valued functions, Jia proved that they are equivalent.

Theorem 2.8 ([24]) *Let ϕ be a compactly supported vector-valued function in $(L^p)^r$. Then $S(\phi)$ provides L^p-approximation order k, $(1 \leq p \leq \infty,)$ if and only if ϕ has accuracy k.*

Since for the compactly supported vector-valued function in $(L^p)^r$ the concepts of accuracy and approximation order are the same, we shall not keep them distinct.

Similar to the description of linear independence, the accuracy of a compactly supported distribution vector ϕ can also be characterized through a special distribution $\psi \in S(\phi)$, which is called a *superfunction*. The following theorem is a direct consequence of the result of [24] (see also [7] and [4]).

Theorem 2.9 *Let ϕ be a compactly supported distribution vector. Then ϕ has accuracy n if and only if there is a compactly supported distribution $\psi \in S_0(\phi)$ that has the same accuracy.*

In approximation theory, the *Strang-Fix conditions* are often used to characterize the approximation order. A compactly supported distribution ψ is said to satisfy the Strang-Fix conditions of order n if the Fourier transform of ψ satisfies the following conditions

$$\hat{\psi}^{(j)}(2k\pi) = \delta_{0j}\delta_{0k}, \quad j = 0, 1, \cdots, n-1. \tag{7}$$

Note that (7) is equivalent to

$$\sum_{k\in\mathbb{Z}} k^j \psi(x-k) = x^j, \quad 0 \le j \le n-1.$$

Remark 2.3 *The Strang-Fix conditions are variated slightly by different authors. In some papers, the conditions $\hat{\phi}^{(j)} = \delta_{0j}$ are excluded from (7), or only $\hat{\phi}(0) = 1$ is reserved.*

If a superfunction ϕ has zero mean, i.e., $\hat{\phi}(0) = 0$, then it may cause an unstable approximation scheme. Hence a superfunction satisfying the Strang-Fix conditions (then it has non-zero mean) is desirable in order to construct an effective approximation scheme. The following theorem confirms the existence of such a superfunction.

Theorem 2.10 ([3]) *A compactly supported distribution ψ has accuracy n if and only if there is a compactly supported distribution $\tilde{\psi} \in S(\psi)$ that satisfies the Strang-Fix conditions of order n.*

This result can also be directly derived from the result of [24]. Note the difference between Theorem 2.9 and Theorem 2.10: in the former, $\phi \in S_0(\phi)$ but it may have zero mean while in the latter, ϕ has non-zero mean but it may not be in $S_0(\phi)$. For example, let $\phi(\cdot) = B_2(\cdot) - B_2(\cdot - 1)$, where B_2 is the hat function defined in Example 2.1. It is clear that $S(\phi)$ provides approximation order 2 and $B_2 \in S(\phi)$ satisfies the Strang-Fix conditions. But $B_2 \notin S_0(\phi)$ and there is no function in $S_0(\phi)$ satisfying the Strang-Fix conditions.

To ensure that there is a superfunction satisfying the Strang-Fix conditions in $S_0(\phi)$, Jia in [27] introduced the following condition, which was re-introduced by de Boor, DeVore, and Ron in [5].

Condition(β1): $1 \notin \Theta(\phi)$.

We can verify that Condition (β1) is equivalent to the followings.

(1) The r components of the 1-periodic distribution vector defined by $\phi_p := \sum_{k\in\mathbb{Z}} \phi(x-k)$ are linearly independent.

(2) The r sequences $\left(\hat{\phi}(2k\pi)\right)_{k\in\mathbb{Z}}$ are linearly independent.

In [27], the following result is proved (see also [5]).

Theorem 2.11 ([27]) *If the compactly supported distribution vector ϕ satisfies Condition (β1) and has accuracy n. Then there exists a unique compactly supported distribution $\psi \in S_0(\phi)$ that has the following properties:*

(i) ψ is a linear combination of $\phi_j(\cdot - i)$, $0 \le i \le n-1$ and $1 \le j \le r$. i.e., $\hat{\psi}(\omega) = \left(p(e^{-i\omega})\right)^T \hat{\phi}(\omega)$, where $p(z)$ is a polynomial vector with degree less than n.

(ii) ψ satisfies the Strang-Fix conditions of order n.

There is an interesting relation between approximation order and B-spline. The k-order cardinal B-spline $B_k(x)$ is inductively defined by

$$B_0(x) = \chi_{[0,1)}(x),$$
$$B_k(x) = \chi_{[0,1)} * B_{k-1}(x).$$

From Theorem 2.10, we have

Corollary 2.12 *Let ψ be a compactly supported distribution. Then $S(\psi)$ has approximation order n if and only if there is another compactly supported distribution γ such that*

$$\psi(x) = B_n * \gamma(x). \tag{8}$$

It is obvious that the derivative of distribution ψ in (8) is linearly dependent since $\psi'(x) = \tilde{\psi}(x) - \tilde{\psi}(x-1)$, where $\tilde{\psi}(x) = B_{n-1} * \gamma(x)$.

3 Linear independence of scaling vectors

In Section 2, we discussed the linear independence and accuracy of compactly supported distribution vectors. In this section, we study the linear independence for scaling vectors. It is obvious that all conclusions for compactly supported distribution vectors also hold for scaling vectors. We now want to obtain further results which are described via symbols of scaling vectors.

3.1 Symbol of a scaling vector

As we said at the beginning of this paper, we always assume that the symbol of a scaling vector is a polynomial matrix (in short, a P-matrix), where the term *polynomial* is also used for *Laurent polynomial*. A P-matrix is not always a symbol of a scaling vector. Daubechies and Lagarias proved the following result for scaling distribution (i.e., $r = 1$).

Theorem 3.1 ([15]) *A polynomial $p(z)$ is a symbol of a non-vanished scaling distribution ϕ if and only if $p(1) = 2^{l-1}, l \in \mathbb{N}$. Furthermore, if the symbol $p(z)$ of scaling distribution ϕ satisfies $p(1) = 2^l, l \geq 1$, then ϕ is the l-th derivative of a scaling distribution ψ, that has symbol $q(z) = 2^{-l}p(z)$ and satisfies $\hat{\psi}(0) \neq 0$.*

For convenience, the number $2^{l-1}, l \in \mathbb{N}$, is said to be of 2^n-form. Thus, Theorem 3.1 confirms that a polynomial $p(z)$ is a symbol of a scaling distribution if and only if $p(1)$ is a 2^n-form number.

Recently Jiang and Shen (see also Zhou [48]) generalized this result to scaling vectors.

Theorem 3.2 ([32]) *A P-matrix $P(z)$ is a symbol of a non-vanished scaling vector ϕ if and only if the matrix $P(1)$ has a 2^n-form eigenvalue. Furthermore, if 2^N, $N \geq 1$ is the largest 2^n-form eigenvalue of $P(1)$, then there is a scaling vector ϕ with symbol $P(z)$ which is the N-th derivative of the scaling vector ψ with symbol $2^{-N} P(z)$ and $\hat{\psi}(0) \neq 0$.*

Theorem 3.2 (Theorem 3.1) gave the criterion for a P-matrix (a polynomial) being a symbol of a scaling vector (a scaling distribution). Hence we suggest the following definition.

Definition 3.1 *A P-matrix $P(z)$ is said to be a symbol if $P(1)$ has at least one 2^n-form eigenvalue and is said to be a normal symbol if 1 is the only 2^n-form eigenvalue of $P(1)$.*

By this definition, if a P-matrix is a symbol, then there exists a scaling vector ϕ satisfying

$$\hat{\phi}(\omega) = P(e^{-i\omega/2})\hat{\phi}(\omega).$$

However it is possible that different scaling vectors have the same symbol. The scaling vectors with the same symbol can be distinguished by their "initial values".

Theorem 3.3 ([32]) *If $P(z)$ is a normal symbol, then for an arbitrary right 1-eigenvector $\vec{r}$ of $P(1)$, there is a unique scaling vector ϕ that has symbol $P(z)$ and satisfies $\hat{\phi}(0) = \vec{r}$.*

By this theorem, a scaling vector ϕ with a normal symbol can be uniquely determined by its initial value $\hat{\phi}(0)$. For a scaling vector with non-normal symbol, The following result is directly derived from Theorem 3.2.

Corollary 3.4 *If 2^N is the largest 2^n-form eigenvalue of symbol $P(z)$ and $\vec{r}$ is a 1-eigenvector of $2^{-N} P(1)$, then there is a unique scaling vector ϕ with symbol $P(z)$, which is the N-th derivative of a scaling vector ψ with $\hat{\psi}(0) = \vec{r}$.*

The initial condition for a scaling distribution is relatively simple, because two scaling distributions ϕ and ψ have the same symbol if and only if $\phi = c\psi, c \neq 0$, i.e., a scaling distribution is uniquely determined by its symbol up to constant. Hence we normalize a scaling distribution ϕ by $\lim_{\omega \to 0} \frac{\hat{\phi}(\omega)}{(-i\omega)^N} = 1$ when $p(1) = 2^N$. It is also easy to verify that a scaling distribution has a unique symbol. However, if a scaling vector is finitely linearly dependent, it has (infinitely) many symbols. A scaling vector has a unique symbol if and only if it is finitely linearly independent.

3.2 Two-scale factor of a symbol

To investigate the properties of a scaling vector via its symbol, an important tool is *two-scale factorization*.

Definition 3.2 *A P-matrix $P(z)$ is said to be* two-scale similar *to a P-matrix $Q(z)$ if there exists an invertible P-matrix $T(z)$ such that*

$$P(z) = T(z^2)Q(z)T^{-1}(z). \tag{9}$$

We call $P(z)$ the two-scale transform *of $Q(z)$, and call $T(z)$ the* two-scale transformation matrix *from $Q(z)$ to $P(z)$. $T(z)$ is also called a two-scale factor of $P(z)$ when $Q(z)$ need not be emphasized.*

For polynomials, the two-scale similar relation (9) is reduced to

$$p(z) = \frac{t(z^2)}{t(z)} q(z). \tag{10}$$

Note that if $T(z)$ is a two-scale transformation matrix from $Q(z)$ to $P(z)$, then so is $cT(z), c \neq 0$. Hence we sometimes assign a free constant factor to $T(z)$. An invertible P-matrix $T(z)$ may be singular at some $z \neq 0$. If a P-matrix is invertible for <u>all</u> $z \in \mathbb{C} \setminus \{0\}$, then we call it *fundamental*. In other words, the determinant of a fundamental P-matrix is a monomial $cz^l, c \neq 0$. Hence its inverse is also fundamental.

Our motivation of introducing the two-scale factors can be illustrated by the following theorem.

Theorem 3.5 *Let $\phi = (\phi_1, \cdots, \phi_r)^T$ and $\psi = (\psi_1, \cdots, \psi_r)^T$ be two finitely linearly independent scaling vectors with symbols $P(z)$ and $Q(z)$ respectively. If there is an invertible P-matrix $T(z) = \sum T(k)z^k$ such that*

$$\phi(x) = \sum T(k)\psi(x-k), \tag{11}$$

then $P(z)$ and $Q(z)$ satisfy (9).

Proof. By (11), we have

$$\hat{\phi}(\omega) = T(e^{-i\omega})\hat{\psi}(\omega).$$

We also have

$$\hat{\psi}(\omega) = Q(e^{-i\omega/2})\hat{\psi}(\omega/2).$$

Since $T(z)$ is invertible,

$$\begin{aligned}\hat{\phi}(\omega) &= T(e^{-i\omega})Q(e^{-i\omega/2})\hat{\psi}(\omega/2) \\ &= T(e^{-i\omega})Q(e^{-i\omega/2})T^{-1}(e^{-i\omega/2})\hat{\phi}(\omega/2).\end{aligned}$$

It follows $P(z) = T(z^2)Q(z)T^{-1}(z)$. □

If $T(z)$ in (11) is fundamental, then $P(z)$ and $Q(z)$ are two-scale similar to each other. In this case, the scaling vectors ϕ and ψ both generate the same space and both have the same accuracy, the same regularity, the same linear independence and so on. Hence a fundamental two-scale factor is said to be *trivial*.

We have seen that, for a symbol $P(z)$, the matrix $P(1)$ plays an important rule. Hence we introduce the following notations. The eigenvalue set of an operator O on a linear space is denoted by $E(O)$. For a P-matrix $P(z)$, we simply write $E(P)$ instead of $E(P(1))$. The set of all 2^n-form eigenvalues of $P(1)$ is denoted by $E_d(P)$, and the multiplicity of eigenvalue 1 of $P(1)$ is denoted by $\#^1(P)$. It is evident that, for a normal symbol $P(z)$, $E_d(P) = \{1\}$.

In order to reveal the relation of the spectrum between $P(1)$ and $Q(1)$, where $P(z)$ is two-scale similar to $Q(z)$ with two-scale transformation matrix $T(z)$, we need the following lemma.

Lemma 3.6 *Let K be a subspace of a Hilbert space H and P be a linear operator on H such that $P(K) \subset K$. Denote the orthogonal complement of K by $K^\perp$ and the identity restriction from H to a subspace $S \subset H$ by I_S. Then*

$$E(P) = E(I_K P) \cup E(I_{K^\perp} P)$$

Proof. Let $\lambda \in E(P)$ and $e \in H$ be a λ-eigenvector: $Pe = \lambda e$. If $e \in K$, then $\lambda \in E(K)$. Otherwise, we write $e = e_1 + e_2$, where $e_1 \in K$ and $e_2 \in K^\perp$. Then

$$I_{K^\perp} P e_2 = I_{K^\perp} P e - I_{K^\perp} P e_1 = \lambda I_{K^\perp}(e_1 + e_2) = \lambda e_2,$$

which implies $\lambda \in E(I_{K^\perp} P)$.

We now prove $E(I_K P) \cup E(I_{K^\perp} P) \subset E(P)$. It is obvious that if $\lambda \in E(I_K P)$, then $\lambda \in E(P)$. Note also that if $\lambda \notin E(I_K P)$ and $\lambda \in E(I_{K^\perp} P)$, then $\lambda \in E(P)$. In fact, assume that $e_2 \in K^\perp$ satisfies $I_{K^\perp} P e_2 = \lambda e_2$. By setting $a = I_K P e_2$, $e_1 = (\lambda I - I_K P)^{-1} a$, and $e = e_1 + e_2$, we have

$$Pe = I_k P e_1 + a + \lambda e_2 = \lambda e_1 + \lambda e_2 = \lambda e.$$

The proof is completed. □

From Lemma 3.6, we derive the following lemma, which describes the relation of the spectrums of $P(1)$ and $Q(1)$.

Let A and B be two finite sets in $\mathbb{R}$. If there is a constant $c \in \mathbb{R}$ and a one-to-one mapping f from A to B such that $a \leq cf(a)$ for all $a \in A$, then we write $A \leq cB$.

Lemma 3.7 *Let P-matrix $P(z)$ be two-scale similar to P-matrix $Q(z)$ with two-scale similar transformation matrix $T(z)$. Let R be the range of $T(1)$ and K be the kernel of $T(1)$. Then*

$$E\left(I_{K^\perp} Q(1)\right) = E\left(I_R P(1)\right), \qquad (12)$$

$$E\left(I_{R^\perp} Q(1)\right) \leq \frac{1}{2} E\left(I_K P(1)\right). \qquad (13)$$

Therefore, $E_d(Q) \subset E_d(P)$ and $\#^1(Q) \leq \#^1(P)$.

Proof. We decompose $T(z)$ into the form

$$T(z) = L(z)D(z)R(z), \qquad (14)$$

where $L(z)$ and $R(z)$ both are fundamental and $D(z)$ is the *Canonical Smith's form* of $T(z)$. We write

$$\tilde{Q}(z) = R(z^2)Q(z)R^{-1}(z), \qquad (15)$$

and

$$\tilde{P}(z) = L^{-1}(z^2)P(z)L(z). \qquad (16)$$

Then

$$\tilde{P}(z) = D(z^2)\tilde{Q}(z)D^{-1}(z). \qquad (17)$$

Without loss of generality, We now assume that

$$D(z) = \text{diag}\left((z-1)^{\alpha_1}d_1(z), \cdots, (z-1)^{\alpha_k}d_k(z), d_{k+1}(z), \cdots, d_r(z)\right) \qquad (18)$$

where $d_j(1) \neq 0, 1 \leq j \leq r$, $1 \leq \alpha_k \leq \ldots \leq \alpha_1$, and $k < r$. Let $i_s, 0 \leq s \leq t$, be the indices such that $i_0 = 0, i_t = k, i_j < i_{j+1}$. Set $\beta_s = \alpha_m$ for $i_{s-1} < m \leq i_s$ so that $\beta_s \geq \beta_{s+1}$. Thus, we have

$$\tilde{P}(1) = \begin{pmatrix} 2^{\beta_1}Q_{11}(1) & 0 & 0 & \cdot & \cdot & 0 \\ & 2^{\beta_2}Q_{22}(1) & 0 & \cdot & & 0 \\ \cdot & \cdot & \cdot & \cdot & & \cdot \\ * & * & 2^{\beta_t}Q_{tt}(1) & \cdot & 0 \\ * & \cdot & * & Q_{r-k,r-k}(1) & \end{pmatrix},$$

where $Q_{ss}(1)$ is the principal minor of $\tilde{Q}(1)$ containing the rows and columns from $\alpha_{i_{s-1}+1}$ to α_{i_s} and $Q_{r-k,r-k}(1)$ is the principal minor of $\tilde{Q}(1)$ containing the last $(r-k)$ rows and the last $(r-k)$ columns. Similarly,

$$\tilde{Q}(1) = \begin{pmatrix} Q_{11}(1) & * & * & \cdot & & * \\ 0 & Q_{22}(1) & * & \cdot & & * \\ \cdot & \cdot & \cdot & \cdot & & \cdot \\ 0 & 0 & 0 & Q_{tt}(1) & \cdot & * \\ 0 & 0 & \cdot & 0 & Q_{r-k,r-k}(1) & \end{pmatrix}.$$

Let R_D be the range of $D(1)$ and K_D be the kernel of $D(1)$. Then from the representations of $\tilde{P}(1)$ and $\tilde{Q}(1)$, we have

$$E\left(I_{K_D^\perp}\tilde{Q}(1)\right) = E\left(I_{R_D}\tilde{P}(1)\right)$$

and

$$E\left(I_{K_D}\tilde{Q}(1)\right) \leq \frac{1}{2}E\left(I_{R_D^\perp}\tilde{P}(1)\right).$$

It follows that

$$E\left(I_{K^\perp}Q(1)\right) = E\left(I_R P(1)\right),$$

and
$$E(I_K Q(1)) \leq \frac{1}{2} E(I_{R^\perp} P(1)).$$

By Lemma 3.6,
$$E(P) = E(I_R P(1)) \cup E(I_{R^\perp} P(1))$$
and
$$E(Q) = E(I_{K^\perp} Q(1)) \cup E(I_K Q(1)).$$

Hence, $E_d(Q) \subset E_d(P)$ and $\#^1(Q) \leq \#^1(P)$. □

Under a condition for the initial value $\hat{\phi}(0)$, we obtained the converse of Theorem 3.5 in [46]. For convenience, it is presented here with a new proof.

Theorem 3.8 *Assume the symbol $P(z)$ of a scaling vector ϕ is normal and $P(z)$ has a two-scale factor $T(z)$ that satisfies*

$$\operatorname{rank} T(1) = \operatorname{rank}(T(1), \hat{\phi}(0)). \tag{19}$$

Then there is a scaling vector $\psi \in S(\phi)$ such that (11) holds.

Proof. Let R and K be defined as in Lemma 3.7. Since (19) holds, $\hat{\phi}(0) \in R$ and $I_R P(1)\hat{\phi}(0) = \hat{\phi}(0)$, which implies that $1 \in E(I_R P(1))$. By Lemma 3.7, $1 \in E(I_{K^\perp} Q(1))$. The conditions

$$E_d(P) = \{1\} \quad \text{and} \quad E(I_K Q(1)) \leq \frac{1}{2} E(I_{K^\perp} P(1))$$

imply $E_d(I_K Q(1)) = \emptyset$. Hence, there is a vector $u \in \mathbb{C}^r$ such that $Q(1)u = u$ and $T(1)u = \hat{\phi}(0)$. Let ψ be the unique scaling vector ψ with symbol $Q(z)$ and $\hat{\psi}(0) = u$. Then (11) holds. □

Remark 3.1 *If $T(1)$ is not singular, then (19) is always true. Hence the condition (19) needs to be verified only if $T(1)$ is singular.*

3.3 Linear independence of scaling vectors

We now give the necessary and sufficient condition for the linear independence of a scaling vector. The following theorem is a directly subsequence of Theorem 3.5, Theorem 3.8, and Corollary 2.6.

Theorem 3.9 *A finitely linearly independent scaling vector ϕ with a normal symbol $P(z)$ is linearly dependent if and only if $P(z)$ has a non-trivial two-scale factor $T(z)$ that satisfies (19).*

An important case of the normal symbol $p(z)$ is that $\#^1 P(1) = 1$. For example, If a scaling vector $\phi \in (L^1)^r$ with symbol $P(z)$ satisfies Condition $(\beta 1)$, then $\#^1 P(1) = 1$ and the modulus of other eigenvalues of $P(1)$ are less than 1 (see [30]). For this kind of symbols, the condition (19) can be removed from Theorem 3.9.

Corollary 3.10 *Assume that the symbol $P(z)$ of a finitely linearly independent scaling vector ϕ is normal and 1 is the simple eigenvalue of $P(1)$. Then ϕ is linearly dependent if and only if $P(z)$ is two-scale similar to a symbol $Q(z)$ with a non-trivial two-scale transformation matrix $T(z)$.*

Proof. We only need to prove that $\#^1 P(1) = 1$ implies (19). Since $Q(z)$ is a symbol and 1 is a simple eigenvalue of $P(1)$, by Lemma 3.7, $Q(z)$ must be a normal symbol and $\#^1(Q) = 1$. By (13), $1 \notin E(I_{R^\perp} Q(1))$ and therefore $1 \in E(I_K Q(1))$. It follows from (12) that $1 \in E(I_R P(1))$. Since $\hat{\phi}(0)$ is the unique 1-eigenvector of $P(1)$ (up to a constant), we obtain (19). □

Similarly we have the criterion for l^∞-linear independence

Theorem 3.11 *A finitely linearly independent scaling vector ϕ with a normal symbol $P(z)$ is l^∞-linearly dependent if and only if $P(z)$ has a two-scale factor $T(z)$ that is singular at some $z_0 \in \Gamma$ and satisfies (19).*

We have seen that the condition (19) is necessary for ϕ being linearly dependent. A natural question arises. What property does ϕ possess if (19) fails. The following theorem gives the answer.

Theorem 3.12 *If the symbol of the scaling vector ϕ is normal and has a two-scale factor $T(z)$ with*

$$\operatorname{rank} T(1) \neq \operatorname{rank}\left(T(1), \hat{\phi}(0)\right), \tag{20}$$

then ϕ has accuracy at least 1. Furthermore, there is a vector $a \in \mathbb{C}^r$ such that the distribution $\psi = a^T \phi$ satisfies the Strang-Fix conditions of order 1.

In order to prove Theorem 3.12, we need the following lemma.

Lemma 3.13 *Assume that the symbol $P(z)$ of a scaling vector ϕ has a two-scale factor*

$$T(z) = \operatorname{diag}((z-1), 1, \cdots, 1),$$

and $\hat{\phi}_1(0) \neq 0$. Then ϕ has accuracy at least 1.

Proof. Without loss of generality, we assume $\hat{\phi}_1(0) = 1$. By (9), we have

$$\begin{aligned}\hat{\phi}_1(\omega) &= (z+1) q_{11}(z) \hat{\phi}_1(\omega/2) \\ &+ \sum_{i=2}^{r} (z^2 - 1) q_{i1}(z) \hat{\phi}_i(\omega/2), \quad z = e^{-i\omega/2}.\end{aligned} \tag{21}$$

For $k \neq 0$, let j be the odd integer that $k = 2^l j$. From (21), we have

$$\begin{aligned}\hat{\phi}_1(2k\pi) &= p_{11}(e^{-ik\pi}) \hat{\phi}_1(k\pi) \\ &= (e^{-i2^l j\pi} + 1) q_{11}(e^{-i2^l j\pi}) \hat{\phi}_1(2^l j\pi) \\ &= (2 q_{11}(1))^l (e^{-ij\pi} + 1) q_{11}(-1) \hat{\phi}_1(j\pi) \\ &= 0.\end{aligned}$$

Since $\hat{\phi}_1(0) = 1$, ϕ_1 satisfies the Strang-Fix conditions of order 1 and therefore has accuracy at least 1. □

We return to the proof of Theorem 3.12.

Proof. We assume the symbol $P(z)$ of ϕ has a two-scale factor $T(z)$ that satisfies (20). As in the proof of Lemma 3.7, we decompose $T(z)$ into the form (14), in which $D(z)$ is of the form (18). Let $\tilde{Q}(z)$ and $\tilde{P}(z)$ be defined by (15) and (16) respectively and

$$\hat{\phi}_L(\omega) = L^{-1}(e^{-i\omega})\hat{\phi}(\omega).$$

Then the symbol $\tilde{P}(z)$ of ϕ_L satisfies

$$\tilde{P}(z) = D(z^2)\tilde{Q}(z)D^{-1}(z).$$

It is obvious that

$$\operatorname{rank} D(1) \neq \operatorname{rank}\left(D(1), \hat{\phi}_L(0)\right). \tag{22}$$

By (22), at least one of $\hat{\phi}_{L,j}(0)$, $1 \leq j \leq k$ is not equal to zero. Without loss of generality, we assume $\hat{\phi}_{L,1}(0) \neq 0$. It is easy to verify that $T_1(z) := \operatorname{diag}(z-1, 1, \cdots, 1)$ is a two-scale factor of $P(z)$. By Lemma 3.13, $\phi_{L,1}$ satisfies the Strang-Fix conditions of order 1. We now set $a^T = e_1^T L^{-1}(1)$. The function $\psi := a^T \phi$ satisfies

$$\hat{\psi}(2k\pi) = \hat{\phi}_{L,1}(2k\pi) = \delta_{0k}, \quad k \in \mathbb{Z}.$$

The theorem is proved. □

Remark 3.2 *We proved Theorem 3.12 without the assumption of linear independence of the scaling vector.*

As an application of Theorem 3.12, we give a criterion for the linear dependence of scaling vectors with non-normal symbols.

Theorem 3.14 *Let the symbol $P(z)$ of the finitely linearly independent scaling vector ϕ be non-normal and 2^N be its largest 2^n-form eigenvalue. Assume that $\vec{r} := \lim_{\omega \to 0} \frac{\hat{\phi}(\omega)}{(-i\omega)^N}$ is a 1-eigenvector of $2^{-N}P(1)$. Then ϕ is linearly dependent (l^∞-linearly dependent) iff $P(z)$ has a non-trivial two-scale factor (a two-scale factor that is singular at some $z_0 \in \Gamma$).*

Proof. The proof for l^∞-linear dependence is similar to that for linear dependence. Hence we only prove the result for linear dependence. It is obvious that the linear dependence of ϕ implies the existence of the non-trivial two-scale factor of $P(z)$. We now prove the converse. Let Φ be the scaling vector with $\hat{\Phi}(0) = \vec{r}$ and the symbol of Φ be $2^{-N}P(z)$. Then $\Phi^{(N)} = \phi$. It is clear that $2^{-N}P(z)$ has the same two-scale factors as $P(z)$. Recall that $P(z)$ has a nontrivial two-scale factor $T(z)$. If (19) holds, then Φ is linearly dependent and so is ϕ. Otherwise, by Theorem 3.12, there is an $a \in \mathbb{C}^r$ such that $a^T \Phi$ satisfies the Strang-Fix conditions of order 1, i.e.,

$$\sum_{k \in Z} a^T \Phi(x-k) = 1.$$

Thus, we have

$$\sum_{k \in Z} a^T \phi(x-k) = \sum_{k \in Z} a^T \Phi^{(N)}(x-k) = 0.$$

The proof is completed. $\square$

3.4 Root criterion for linear independence

We now use Theorem 3.9 to obtain a criterion for linear independence of a scaling distribution via the roots of its symbol. We first introduce some notions of polynomial roots.

Definition 3.3 *Let $m > 1$ be an odd integer, let z_0 be a primitive m-th root of unit, and let h_m be the smallest positive integer such that $2^{h_m} \equiv 1 (\mod m)$. Then we call*

$$c_m(z) = (z_0 - z)(z_0^2 - z) \cdots (z_0^{2^{h_m-1}} - z)$$

an m-cyclic polynomial, or simply a cyclic polynomial if m need not be stressed.

Example 3.1 *The followings are examples of m-cyclic polynomials.*

1. *$m = 3, h_3 = 2$, and $c_3(z) = 1 + z + z^2$.*
2. *$m = 5, h_5 = 4$, and $c_5(z) = 1 + z + z^2 + z^3 + z^4$.*
3. *$m = 7, h_7 = 6$, and $c_7(z) = 1 + z + z^2 + z^3 + z^4 + z^5 + z^6$. However,*
4. *$m = 9, h_9 = 6$, and $c_9(z) = 1 + z^3 + z^6$.*

Since $z_0^{2^{h_m}} = z_0$, it can be verified that

$$\frac{c_m(z^2)}{c_m(z)} = c_m(-z). \tag{23}$$

We now define two special kinds of roots of a polynomial in the following.

Definition 3.4 *$p(z)$ is said to have m-cyclic zeros if $c_m(-z)$ (NOT $c_m(z)$!) is a factor of $p(z)$, where $c_m(z)$ is an m-cyclic polynomial ($m > 1$). $p(z)$ is said to have symmetric zeros if there is an $\alpha \in \mathbb{C} \setminus \{0\}$ such that $p(\alpha) = p(-\alpha) = 0$.*

It is easy to see that if $p(z)$ has the symmetric zeros $\pm\alpha$, then $z - \alpha^2$ is a two-scale factor of $p(z)$, while if $c(-z)$ provides the cyclic zeros of $p(z)$, then $c(z)$ is a two-scale factor of $p(z)$. We can now derive the following criterion from Theorem 3.9, which is first proved in [31] by Jia and the author in a different way.

Theorem 3.15 ([31]) *Assume that the symbol $p(z)$ of scaling distribution ϕ is normal. Then ϕ is linearly independent (l^∞-linearly independent) if and only if $p(z)$ has neither cyclic zeros nor symmetric zeros (nor symmetric zeros on Γ).*

Proof. The proof can be found in [31]. For reader's convenience, we give a new proof here. Since the proof for l^∞-linear independence is similar to that for linear independence. Hence we only prove the later. The proof of "ONLY IF" is trivial. we now prove "IF". Assume that ϕ is linearly dependent. Recall that $p(1) = 1$ implies $\hat{\phi}(0) \neq 0$. By Theorem 3.9, $p(z)$ has a nontrivial two-scale factor $t(z)$ that satisfies (19), i.e., $t(1) \neq 0$ and there is a symbol $q(z)$ such that $p(z) = \frac{t(z^2)}{t(z)} q(z)$. We decompose $t(z)$ into a product of linear factors $t(z) = \prod_{i=1}^m t_i(z)$, where $t_i(z) = (z_i - z)$, $z_i \neq 0$. If $t_1(z)$ is not a factor of $t(z^2)$, then it is a factor of $q(z)$. Hence, $t_1(z^2)$ is a factor of $p(z)$ which implies that $p(z)$ has symmetric zeros. If $t_1(z)$ is a factor of $t(z^2)$ then it must be a factor of $t_j(z^2)$ for some j. Since $t(1) \neq 1$, $j \neq 1$. Without loss of generality, we can assume that $t_1(z)$ is a factor of $t_2(z^2)$, which implies $z_2 = z_1^2$. We will continue this procedure until we meet an index J such that $t_J(z)$ is a factor of $q(z)$ or is a factor of $t_l(z^2)$, $1 \leq l < J$. In the first case, $t_1(z^2)$ is a factor of $p(z)$. In the second case, If $l = 1$, $\prod_{i=1}^s t_i(z)$ is a two-scale factor of $p(z)$ which implies that $p(z)$ has cyclic zeros. Otherwise, $z_1(z)$ and $\prod_{i=l}^s t_i(z)$ both are two scale factors of $p(z)$, and therefore $p(z)$ has both symmetric zeros and cyclic zeros. □

From Theorem 3.15 and Theorem 3.14 we obtain a criterion for a scaling distribution with non-normal symbol.

Corollary 3.16 *If the symbol $p(z)$ of a scaling distribution ϕ is not normal, i.e., $p(1) = 2^N, N \geq 1$, then ϕ is linearly independent (l^∞-linearly independent) if and only if $p(-1) \neq 0$ and $p(z)$ has neither cyclic zeros nor symmetric zeros (nor symmetric zeros on Γ).*

4 Accuracy of scaling vectors

We now discuss the accuracy of a scaling vector via its symbol. [21] and [35] (see also [13]) discussed it via the symbol under the assumption of linear independence. In this section, we use a new approach to the accuracy of scaling vectors without that assumption.

4.1 A criterion of accuracy of scaling vectors

A primary discussion on accuracy 1 is already in Theorem 3.12. We now develop the idea there to obtain the general results.

Theorem 4.1 *If the symbol $P(z)$ of scaling vector ϕ is normal and has a two-scale factor $T(z)$ with the Canonical Smith's form*

$$D_n(z) := diag\left((1-z)^n, 1, \cdots, 1\right), \tag{24}$$

and $T(z)$ satisfies (20), then there exist n vectors $v(k) \in \mathbb{C}^r$, $0 \le k \le n-1$, such that the distribution $\psi := \sum_{k=0}^{n-1}(v(k))^T \phi(x-k)$ satisfies the Strang-Fix conditions of order n. Therefore, ϕ has accuracy at least n.

Proof. Let $L(z)$ and $R(z)$ be two fundamental P-matrices such that $T(z) = L(z)D_n(z)R(z)$. Let ϕ_L be determined by

$$\hat{\phi}_L(\omega) := L^{-1}(e^{-i\omega})\hat{\phi}(\omega).$$

Denote the symbol of ϕ_L by $\tilde{P}(z)$. Then $\tilde{P}(z)$ is two-scale similar to the P-matrix $\tilde{Q}(z)$ with the two-scale transformation matrix $D_n(z)$. We now use mathematical induction to prove

$$\hat{\phi}_{L,1}^{(j)}(2k\pi) = 0, \quad k \in \mathbb{Z}\setminus\{0\}, j = 0, 1, \cdots, n-1. \tag{25}$$

For $j=0$, (25) is proved in Lemma 3.13. Assuming now (25) is true for $j = 0, 1, \cdots, n-2$, we prove it is also true for $j = n-1$.

At first, since $\tilde{p}_{1,l}(z) = (z^2-1)^n \tilde{q}_{1,l}(z), 2 \le l \le r$, we have

$$\left[\tilde{p}_{1,l}(e^{-i\omega/2})\right]^{(j)}_{\omega=2k\pi} = 0, \quad 2 \le l \le r, 0 \le j \le n-1.$$

Hence,

$$\hat{\phi}_{L,1}^{(n-1)}(2k\pi) = \sum_{j=0}^{n-1}\binom{n-1}{j}[(\tilde{p}_{11}(e^{-i\omega/2})]^{(j)}_{\omega=2k\pi}\frac{\hat{\phi}_{L,1}^{(n-1-j)}(k\pi)}{2^{n-1-j}}.$$

Note that

$$\tilde{p}_{11}(e^{-i\omega/2}) = (1+e^{-i\omega/2})^n \tilde{q}_{11}(e^{-i\omega/2}).$$

Hence for odd k,

$$[\tilde{p}_{11}(e^{-i\omega/2})]^{(j)}_{\omega=2k\pi} = 0, \quad j = 0, 1, \cdots, n-1$$

and for even k, by the mathematical induction, we have

$$\hat{\phi}_{L,1}^{(j)}(k\pi) = 0, \quad 0 \le j \le n-2, \quad k \neq 0.$$

Thus for any $k \neq 0$,

$$\hat{\phi}_{L,1}^{(n-1)}(2k\pi) = 2^{-n+1}\tilde{p}_{11}(e^{-ik\pi})\hat{\phi}_{L,1}^{(n-1)}(k\pi).$$

Using the same trick in Lemma 3.13, writing $k = 2^l j$ with odd j, we obtain

$$\hat{\phi}_{L,1}^{(n-1)}(2k\pi) = c\tilde{q}_{11}^l(1)\tilde{q}_{11}(-1)(e^{-ij\pi}+1)^n \hat{\phi}_{L,1}^{(n-1)}(j\pi) = 0.$$

The equation (25) is proved. From (20), we have $\hat{\phi}_{L,1}(0) = 1$. Using Leibniz' formula for the derivatives of product, we can find a polynomial $r(z) \in \Pi_{n-1}$ such that the entire function

$$\hat{\tilde{\psi}}(\omega) = r(e^{-i\omega})\hat{\phi}_{L,1}(\omega)$$

satisfies

$$\hat{\tilde{\psi}}^{(j)}(2k\pi) = \delta_{0j}\delta_{0k}, \quad k \in \mathbb{Z} \setminus \{0\}, j = 0, 1, \cdots, n-1,$$

which implies that $\tilde{\psi} \in S_0(\phi)$ satisfies the Strang-Fix conditions of order n. Let $u(z) := r(z)l_1(z)$, where $l_1(z)$ is the first row of the matrix $L^{-1}(z)$. Then

$$\hat{\tilde{\psi}}(\omega) = \left(u(e^{-i\omega})\right)^T \hat{\phi}(\omega).$$

Finally, let $v(z) = \sum_{k=0}^{n-1} v(k)z^k \in (\Pi_{n-1})^r$ be the polynomial vector that

$$\left[v(e^{-i\omega})\right]^{(j)}_{\omega=0} = \left[u(e^{-i\omega})\right]^{(j)}_{\omega=0}, \quad j = 0, 1, \cdots, n-1.$$

Then the distribution

$$\psi(x) = \sum_{k=0}^{n-1}(v(k))^T \phi(x-k)$$

satisfies the Strang-Fix conditions of order n. □

We use the following scaling vector, which is first constructed in [13], to demonstrate Theorem 4.1.

Example 4.1 *Suppose the scaling vector $\phi := (\phi_1 \ \phi_2)^T$ has the symbol*

$$P(z) = \frac{1}{4}\begin{pmatrix} z(1+z)^2 & \frac{z}{2}(1-z^2)^2 \\ 1/32 & z \end{pmatrix},$$

it can be verify that $P(z)$ has the following factorization.

$$P(z) = \begin{pmatrix} (1-z^2)^2 & 0 \\ 0 & 1 \end{pmatrix}\begin{pmatrix} \frac{z}{4} & \frac{z}{2} \\ \frac{(1-z)^2}{32} & z \end{pmatrix}\begin{pmatrix} \frac{1}{(1-z)^2} & 0 \\ 0 & 1 \end{pmatrix}.$$

Hence $T(z) = \begin{pmatrix} (1-z)^2 & 0 \\ 0 & 1 \end{pmatrix}$ *is a two-scale factor of $P(z)$. Note that* rank $T(1) = 1$ *and $\hat{\phi}(0) = (1, 1/96)^T$ imply (20). It follows that ϕ has accuracy at least 2.*

Is Theorem 4.1 convertible? In general, the answer is NO. The following is a counter example.

Example 4.2 Let $\phi(x) = \chi_{[0,2]}$. Its symbol is $p(z) = \frac{1}{2}(z^2 + 1)$. Since $\frac{1}{2}\sum \phi(x-k) = 1$, the function ϕ has accuracy 1. However, $(z^2-1)^{-1}p(z)(z-1) = \frac{z^2+1}{2(z+1)}$ is not a polynomial. Hence $z - 1$ is not a two-scale factor of $p(z)$.

In order to obtain the condition for the converse of Theorem 4.1, we introduce
Condition (θ): $-1 \notin \Theta(\phi)$.
This condition is equivalent to the following conditions:
(1) The r sequences $\left(\hat{\phi}(2k\pi + \pi)\right)_{k\in\mathbb{Z}}$ are linearly independent.
(2) The r components of $\phi_q(x) := \sum_{k\in\mathbb{Z}}(-1)^k\phi(x-k)$ are linearly independent.

Note that linear independence (l^∞-linear independence) of ϕ is equivalent to that the r sequences $\left(\hat{\phi}(\omega + 2k\pi)\right)_{k\in\mathbb{Z}}$ is linearly independent for all $\omega \in \mathbb{C}$ (for all $\omega \in \mathbb{R}$). Hence, Condition (θ) together with Condition $(\beta 1)$ is much weaker than the l^∞-linear independence of ϕ.

The following theorem shows that the converse of Theorem 4.1 is true under the conditions $(\beta 1)$ and (θ).

Theorem 4.2 If a scaling vector ϕ having accuracy n satisfies Condition $(\beta 1)$ and Condition (θ), then its symbol has a two-scale factor $T(z)$ that satisfies (20) and has Canonical Smith's form (24).

Proof. According to Theorem 2.9 and the same reason in the proof of Theorem 4.1, we only need to prove Theorem 4.2 for the case that the first component $\phi_1(x)$ of ϕ has accuracy n. Denote the symbol of ϕ by $P(z) = (p_{ij}(z))_{r\times r}$. We now prove the theorem by the mathematical induction. For $n = 1$, we have

$$\hat{\phi}_1(2k\pi) = \sum_{j=1}^r p_{1j}((-1)^k)\hat{\phi}_j(k\pi).$$

It follows that

$$\delta_{0k} = \hat{\phi}_1(4k\pi) = \sum_{j=1}^r p_{1j}(1)\hat{\phi}_j(2k\pi).$$

Since $\hat{\phi}_1(2k\pi) = \delta_{0k}$,

$$\sum_{j=1}^r (p_{1j}(1) - \delta_{0j})\hat{\phi}_j(2k\pi) = 0.$$

By Condition $(\beta 1)$, we have $p_{1j}(1) = \delta_{0j}$, $1 \le j \le r$.

Similarly,
$$0 = \hat{\phi}_1(2(2k+1)\pi) = \sum_{j=1}^{r} p_{1j}(-1)\hat{\phi}_j((2k+1)\pi).$$

By Condition (θ), we have $p_{1j}(-1) = 0$, $1 \leq j \leq r$. Therefore, $p_{11}(z)$ has factor $1+z$ and $p_{1j}(z)$ has factor $z^2 - 1$, $2 \leq j \leq r$, which implies that $P(z)$ has two-scale factor $D_1(z)$. Assuming that the theorem is true for $n-1$, we prove it is also true for n. Note that, for any polynomial $p(z)$,

$$\left[p(e^{-i\omega/2})\right]^{(n)} = (-i/2)^n p^{(n)}(e^{-i\omega/2}) e^{-in\omega/2} \qquad (26)$$
$$+ \sum_{j=1}^{n-1} c_j p^{(n-j)}(e^{-i\omega/2}) r_j(e^{-i\omega/2}),$$

where c_j is constant and $r_j(z)$ is a polynomial of degree j.

We also have

$$0 = \hat{\phi}^{(n)}(2k\pi) = \sum_{j=1}^{r} \left[p_{1j}(e^{-i\omega/2})\right]^{(n)}_{\omega=2\pi} \qquad (27)$$
$$+ \sum_{j=1}^{r}\sum_{l=0}^{n-1} \binom{n-1}{l} \left[p_{1j}(e^{-i\omega/2})\right]^{(l)}_{\omega=2\pi} \frac{\hat{\phi}^{(n-1-l)}(k\pi)}{2^{n-1-l}}.$$

For odd $k := 2s+1$, by the mathematical induction, the second sum in (27) is zero while, by (26), the first sum is reduced to

$$(-i/2)^n \sum_{j=1}^{r} p_{1j}^{(n)}(-1)\hat{\phi}_j((2s+1)\pi) = 0.$$

Using Condition(θ), we have

$$p_{1j}^{(n)}(-1) = 0, \quad j = 1, \cdots, r. \qquad (28)$$

For even $k := 2s$, the second sum in (27) is $c\delta_{0s}$, where c is a constant. Hence (27) is reduced to

$$(-i/2)^n \sum_{j=1}^{r} p_{1j}^{(n)}(1)\hat{\phi}_j(2s\pi) = 0.$$

Using Condition $(\beta 1)$, we obtain

$$p_{1j}^{(n)}(1) = 0, \quad j = 2, \cdots, r. \qquad (29)$$

Combining (28) and (29), we obtain that $(z+1)^n$ is the factor of $p_{11}(z)$ and $(z^2 - 1)^n$ is the factor of $p_{1j}(z)$, $2 \leq j \leq r$. □

Remark 4.1 As we showed in Example 4.2, Condition (θ) cannot be removed from Theorem 4.2. Note that in Theorem 4.2, we do not assume the linear independence of ϕ. Instead, we only assume $\pm 1 \notin \Theta(\phi)$.

From Theorem 4.1 and Theorem 4.2, we obtain

Theorem 4.3 Let ϕ be a scaling vector satisfying Condition ($\beta 1$) and Condition (θ). Then ϕ has accuracy at least n if and only if its symbol has a two-scale factor $T(z)$ that has the Canonical Smith's form (24) and satisfies (20).

Once the superfunction ψ in $S_0(\phi)$ that satisfies the Strang-Fix conditions is obtained, we want to find the sequences $(a_l^k) \in (l)^r$ satisfying

$$\sum_{l \in \mathbb{Z}} (a_l^k)^T \phi(x-l) = x^k, \quad 0 \le k \le n-1. \tag{30}$$

If ϕ is not linearly independent, then the sequence satisfying (30) is not unique. Hence, we try to find one of them. That can be done as follows. Let $v(z) = \sum_{j=0}^{n-1} v_j z^j$ be the polynomial vector of degree less that n (see Theorem 4.1) such that

$$\psi(x) := \sum_{j=0}^{n-1} v_j^T \phi(x+j)$$

satisfies the Strang-Fix conditions of order n. Then we have

$$\sum_{l \in \mathbb{Z}} l^k \psi(x-l) = x^k, \quad 0 \le k \le n-1,$$

and therefore

$$\sum_{l \in \mathbb{Z}} \left(\sum_{j=0}^{n-1} (l+j)^k v_j^T \right) \phi(x-l) = x^k, \quad 0 \le k \le n-1.$$

We now can select

$$a_0^k = \sum_{j=0}^{n-1} j^k v_j, \quad 0 \le k \le n-1 \tag{31}$$

and

$$a_l^k = \sum_{j=0}^{n-1} (l+j)^k v_j = \sum_{s=1}^{k} \binom{k}{s} l^{k-s} a_0^s.$$

Using the property of $v(z)$ in the proof of Theorem 4.1, we obtain

Corollary 4.4 *The sequences a_0^k in (31) satisfies the following conditions.*

$$\sum_{s=1}^{k} \binom{k}{s} (2i)^{s-k}(a_0^s)^T \left(D^{k-s}P(e^{-i\omega})\right)_{\omega=0} = 2^{-k}(a_0^s)^T,$$

$$\sum_{s=1}^{k} \binom{k}{s} (2i)^{s-k}(a_0^s)^T \left(D^{k-s}P(e^{-i\omega})\right)_{\omega=\pi} = 0^T.$$

These conditions were obtained by Plonka [35] (see also [38] and [21]) in a different way under the assumption of linear independence of the scaling vector.

4.2 Description of accuracy by two-scale linear factors

From Theorem 4.3 we can derive a criterion for the accuracy of a scaling vector by the successive two-scale linear factors of its symbol.

Theorem 4.5 *Let ϕ be a scaling vector satisfying Condition (β1) and Condition (θ). Then ϕ has accuracy at least n if and only if its symbol $P(z) := P_0(z)$ has the following decomposition:*

$$P_{i-1}(z) = A_i D_1(z^2) P_i(z) D_1^{-1}(z) A_i^{-1}, \quad i = 1, \cdots, n, \qquad (32)$$

where $P_i(z)$ is a P-matrix, A_i is a nonsingular constant matrix, the first column of each A_i, $2 \leq i \leq n$, is e_1, and the first component of vector $r := A_1^{-1}\hat{\phi}(0)$ does not vanish.

Before we prove the theorem, we remind the following fact. If $T(z)$ is a two-scaling factor of $P(z)$ and $T(z) = T_1(z)T_2(z)$, then either $T_1(z)$ or $T_2(z)$ need not be the two-scale factor of $P(z)$. For example, $p(z) := (z^2 - z + 1)(z + 2)$ has the two-scale factor $t(z) = z^2 + z + 1$, since $\frac{t(z^2)}{t(z)} = z^2 - z - 1$. We have $t(z) = (z - \omega_1)(z - \omega_1^2)$, where $\omega_1 = e^{-2\pi i/3}$. However, neither $(z - \omega_1)$ nor $(z - \omega_1^2)$ is the two-scale factor of $p(z)$.

To prove Theorem 4.5, we need the following lemma.

Lemma 4.6 *Let $V_n(z)$ be fundamental. Then the P-matrix*

$$T_n(z) = V_n(z) D_n(z)$$

has the following decomposition.

$$T_n(z) = A_1 D_1(z) A_2 D_1(z) \cdots A_n D_1(z) V_0(z), \qquad (33)$$

where all A_i, $1 \leq i \leq n$, are nonsingular, the first column of each A_i, $2 \leq i \leq n$ is e_1. On the other hand, if $T_n(z)$ has a decomposition (33) with $V_0(z) = I$, then it is the product $V_n(z)D_n(z)$, where $V_n(z)$ is fundamental.

Proof. "$\Rightarrow$". Since $V_n(z)$ is fundamental, $V_n(1)$ is nonsingular. Therefore there is a nonsingular matrix A_1 such that the first row of $A_1^{-1}V_n(1)$ is e_1^T. (The matrix of this kind is not unique. $V_n(1)$ itself is one of them.) Let

$$V_{n-1}(z) = D_1^{-1}(z)A_1^{-1}V_n(z)D_1(z). \tag{34}$$

Then $V_{n-1}(z)$ is a fundamental P-matrix and the first column of $V_{n-1}(1)$ is e_1. We now have

$$V_n(z)D_1(z) = A_1 D_1(z)V_{n-1}. \tag{35}$$

Using the mathematical induction, we reach

$$V_{n-i-1}(z) = D_1^{-1}(z)A_{i+1}^{-1}V_{n-i}D_1(z), \quad 0 \le i \le n-1,$$

where $V_{n-i-1}(z)$ is fundamental and the first column of $V_{n-i-1}(1)$ is e_1. It follows that

$$V_{n-i}(z)D_1(z) = A_{i+1}D_1(z)V_{n-1-i}, \quad 0 \le i \le n-1, \tag{36}$$

which leads to (33). We now prove that the first column of $A_i, 2 \le i \le n$ is e_1. Recall that, for $1 \le i \le n-1$, the first columns of $A_{i+1}V_{n-i}$ and $V_{n-i}(1)$ both are e_1, which implies the result.

"$\Leftarrow$". It can be proved by directly computing. $\square$

We return to the proof of Theorem 4.5.

Proof. Assume that the scaling vector ϕ satisfying Condition (β1) and Condition(θ) has accuracy n. By Theorem 4.3, the symbol $P(z)$ of ϕ has a two-scale factor $T(z)$ that satisfies (20) and has the Canonical Smith's form $D_n(z)$. Hence

$$P(z) = T(z^2)Q(z)T^{-1}(z),$$

where $Q(z)$ is a P-matrix, and there are two fundamental matrices $V_n(z)$ and $R_n(z)$ such that

$$T(z) = V_n(z)D_n(z)R_n(z).$$

Write $T_n(z) = V_n(z)D_n(z)$ and $Q_n(z) = R_n(z^2)Q(z)R_n^{-1}(z)$. We have

$$P(z) = V_n(z^2)D_n(z^2)Q_n(z)D_n^{-1}(z)V_n^{-1}(z),$$

where $Q_n(z)$ is a P-matrix. Since $V_n(z)$ is fundamental and

$$D_n(z^2)Q_n(z)D_n^{-1}(z) = V_n^{-1}(z^2)P(z)V_n(z)$$

$V_n^{-1}(z^2)P(z)V_n(z)$ is a P-matrix and so is

$$D_1^{-1}(z^2)V_n^{-1}(z^2)P(z)V_n(z)D_1(z),$$

which, by (35), turns out to be

$$V_{n-1}^{-1}(z^2)D_1^{-1}(z^2)A_1^{-1}P(z)A_1D_1(z)V_{n-1}(z).$$

Let

$$P_1(z) = D_1^{-1}(z^2)A_1^{-1}P(z)A_1D_1(z).$$

We get

$$P(z) = A_1D_1(z^2)P_1(z)D_1^{-1}(z)A_1^{-1}.$$

Using Lemma 4.6 and by mathematical induction, we obtain (32). Let $b = V_n^{-1}\hat{\phi}(0)$. Because (20) holds for $T(z)$ and $\hat{\phi}(0) = P(1)\hat{\phi}(0)$, we have $e_1^T b \neq 0$. It follows that the first component of $A_1^{-1}\hat{\phi}(0)$ does not vanish. By Lemma 4.6 and Theorem 4.3, the converse is clear. The proof is now complete. □

Decomposing the symbol of a scaling vector in this way can also be found in [35] and [38] under the assumption that the scaling vector is linearly independent. (See Theorem 1.2 in [35], Theorem 2.4 and 2.6 in [38].)

4.3 Accuracy of scaling distributions

Finally we briefly discuss the accuracy of a scaling distribution. When a scaling vector is reduced to a scaling distribution, Theorem 4.3 can be modified as

Corollary 4.7 *Let ϕ be a scaling distribution with polynomial symbol $P(z)$. Suppose $\hat{\phi}(0) = 1$ and $\sum_{k \in \mathbb{Z}}(-1)^k\phi(x-k) \neq 0$. Then ϕ has accuracy n if and only if $P(z) = (\frac{1+z}{2})^n Q(z)$, where $Q(-1) \neq 0$. Furthermore, if the symbol of ϕ is non-normal, $P(1) = 2^N, N \geq 1$, and $\sum_{k \in \mathbb{Z}}(-1)^k\phi(x-k) \neq 0$, then ϕ has accuracy n if and only if $P(z) = (z+1)^N(\frac{1+z}{2})^n Q(z)$, where $Q(-1) \neq 0$.*

Remark 4.2 *If ϕ in Corollary 4.7 is a real valued function, then Condition (θ), i.e., $\sum_{k \in \mathbb{Z}}(-1)^k\phi(x-k) \neq 0$, holds if and only if that $(1+z^{2^m}), m \in \mathbb{N}$, are not the factors of $P(z)$.*

5 Final remarks

We remark on some relative results on linear independence and accuracy of scaling vectors, in which some are not described via symbol.

(1) The criterion of finite linear dependence can be found in [46]. Note that the symbol of a finitely linearly dependent scaling vector ϕ is not unique. However they can be characterized by the following theorem.

Theorem 5.1 ([46]) *A scaling vector ϕ with the normal symbol $P(z)$ is finitely linearly dependent if and only if the following two conditions are satisfied:*

(i) $P(z) = T(z^2)Q(z)T^{-1}(z)$ with $T(z)$ being a fundamental matrix and

$$Q(z) = \begin{pmatrix} Q_s(z) & Y(z) \\ 0 & X(z) \end{pmatrix},$$

where Q_s is an $s \times s$ polynomial matrix with eigenvalue 1 and $s < r$.

(ii) There is a vector $u = (u_1, \cdots, u_s, 0, \cdots, 0)^T$ such that

$$\hat{\phi}(0) = T^{-1}(1)(u_1, \cdots, u_s, 0 \cdots, 0)^T.$$

Furthermore, if 1 is a simple eigenvalue of $P(1)$, then (ii) can be removed.

(2) The other descriptions of linear independence or stability of scaling vectors not via symbol can be found in [23], [41], and [37].

(3) There is another interesting approach to the characterization of the approximation order of scaling vectors. Strang and Strela [44], then Jia, Riemenschneider and Zhou [29], use the eigenvalues and the corresponding eigenvectors of the subdivision operator associated with ϕ to describe the approximation order of a scaling vector.

A linear mapping $S_P : (l)^r \to (l)^r$ is said to be a *subdivision operator* associated with matrix sequence $\mathbf{P} \in (l_0)^{r \times r}$ if

$$\mathbf{v} := S_P \mathbf{u}(l) = \sum_{k \in \mathbb{Z}} \mathbf{P}^T(l - 2k)\mathbf{u}(k), \quad l \in \mathbb{Z}, \mathbf{u} \in (l)^r.$$

An element $\mathbf{u} \in (l)^r$ is called a polynomial vector sequence if there exists a polynomial vector $\vec{p}(z)$ such that $\vec{p}(n) = \mathbf{u}(n)$ for all $n \in \mathbb{Z}$.

Their result is the following.

Theorem 5.2 ([29]) *Let ϕ be a scaling vector. Then ϕ has accuracy k if and only if there exists a polynomial vector sequence $\mathbf{u}$ such that*

$$\mathbf{u} \notin K(\phi) \quad \text{and} \quad S_P \mathbf{u} - \frac{1}{2^{k-1}}\mathbf{u} \in K(\phi).$$

Chen in [8] pointed out that if the scaling vector satisfies Condition (θ), then the above result can be simplified as following.

Theorem 5.3 ([8]) *If the scaling vector ϕ in the above theorem satisfies Condition (θ), then ϕ has accuracy k if and only if there exists a polynomial vector sequence $\mathbf{u} \notin K(\phi)$ such that $\mathbf{u}$ is an eigenvector of the subdivision operator S_P associated with eigenvalue $\frac{1}{2^{k-1}}$.*

6 References

[1] B. K. Alpert, *A class of bases in L^2 for the sparse Matrix*, J. Math. Anal., **24** (1993) 246–262.

[2] B. K. Alpert and V. Rokhlin, *A fast algorithm for the evaluation of Legendre expansions*, SIAM J. Sci. Stat. Comput., **12** (1991) 158–179.

[3] C. de Boor, R. DeVore, and A. Ron, *Approximation from shift-invariant subspaces of $L^2(\mathbb{R}^d)$*, Trans. Amer. Math. Soc., **341**(1994) 787–806

[4] C. de Boor, R. DeVore, and A. Ron, *The structure of finitely generated shift-invariant spaces in $L_2(\mathbb{R}^d)$*, J. Functional Analysis, **119(1)**(1994) 37–78.

[5] C. de Boor, R. DeVore, and A. Ron, *Approximation orders of FSI spaces in $L^2(\mathbb{R}^d)$*, preprint, 1996.

[6] C. de Boor and Rong-Qing Jia, *Controlled approximation and a characterization of the local approximation order*, Proc. Amer. Math. Soc., **95**(1985) 547–553.

[7] C. de Boor and A. Ron, *Fourier analysis of the approximation power of principal shift-invariant spaces*, Constr. Approx., **8**(1992) 427–462.

[8] Di-Rong Chen, *Algebraic properties of subdivision with matrix mask and their applications*, submitted to J. Approx. Theory, 1997.

[9] C. K. Chui and J. Lian, *A study of orthonormal multiwavelets*, Texas A& M University, CAT Report #351, 1995.

[10] C. K. Chui and Jianzhong Wang, *A general framework of compactly supported spline and wavelets*, J. Approx. Theory, **71**(1993) 263–304.

[11] C. K. Chui and Jianzhong Wang, *A study of compactly supported scaling functions and wavelets*, In Wavelets, Images, Surface Fitting, P. J. Laurent, A. Le Mehaute, and L. L. Schumaker (eds.), Wellesley, (1994) 121–140.

[12] C. K. Chui and Jianzhong Wang, *On compactly supported spline wavelets and a duality principle*, Trans. Amer. Math. Soc., **330**(1992) 903–916.

[13] A. Cohen, I. Daubechies, and G. Plonka, *Regularity of refinable function vectors*, preprint, 1995.

[14] I. Daubechies, *Ten Lectures on Wavelets*, SIAM, Philadelphia, **61**(1992).

[15] I. Daubechies and J. C. Lagarias, *Two-scale difference equations: I. Existence and global regularity of solutions*, SIAM J. Math. Anal., **22**(1991) 1388–1410.

[16] G. Donovan, J. S. Geronimo, D. P. Hardin, and P. R. Massopust, *Construction of orthogonal wavelets using fractal interpolation functions*, SIAM J. Math. Anal., **27**(1996) 363–377.

[17] J. S. Geronimo, D. P. Hardin and P. R. Massopust, *Fractal functions and wavelet expansions based on several scaling functions*, J. Approx. Theory, **78**(1994) 373–401.

[18] T. N. T. Goodman and S. L. Lee, *Wavelet of multiplicity r*, Trans. Amer. Math. Soc., **342**(1994) 307–324.

[19] T. N. T. Goodman, S. L. Lee and W. S. Tang, *Wavelets in wandering subspaces*, Trans. Amer. Math. Soc., **338**(1993) 639–654.

[20] C. Heil and D. Colella, *Matrix refinement equations: Existence and uniqueness*, J. Fourier Anal. Appl., **2**(1996) 363–377.

[21] C. Heil, G. Strang, and V. Strela, *Approximation by translates of refinable functions,* Numer. Math., **73**(1996) 75–94.

[22] L. Hervé, *Multiresolution analysis of multiplicity d: applications to dyadic interpolation*, Applied and Computational Harmonic Analysis, **1** (1994) 299–315.

[23] T. A. Hogan, *Stability and independence of the shifts of finitely many refinable functions*, preprint, 1996.

[24] Rong-Qing Jia, *Shift-invariant spaces on the real line*, Proc. Amer. Math. Soc., **125**(1997) 785–793.

[25] Rong-Qing Jia, *Subdivisions schemes in L_p spaces*, Advances in Comput. Math., **3**(1995) 390–341.

[26] Rong-Qing Jia, *The subdivision and transition operators associated with a refinement equation,* in Advanced Topics in Multivariate Approximation, F. Legendre, K. Jetter and P. J. Laurent (eds.) (1996) 1–16.

[27] Rong-Qing Jia, *Shift-invariant spaces and linear operator equations*, Israel J. Math., to appear, 1996.

[28] Rong-Qing Jia and C. A. Micchelli, *On linear independence for integer translates of a finite number of functions*, Proc. Edinburgh Math. Soc., **36**(1992), 69–85.

[29] Rong-Qing Jia, S. D. Riemenschneider, and Ding-Xuan Zhou, *Approximation by multiple refinable functions*, preprint, 1996.

[30] Rong-Qing Jia, S. D. Riemenschneider, and Ding-Xuan Zhou, *Vector subdivision schemes and multiple wavelets*, preprint, 1996.

[31] Rong-Qing Jia and Jianzhong Wang, *Stability and linear independence associated with wavelet decompositions*, Proc. Amer. Math. Soc., **117**(1993), 1115–1124.

[32] Q. Jiang and Z. Shen, *On existence and week stability if matrix revinable functions,* Constr. Approx., to appear, 1996.

[33] P. Massopust, D. Ruch, and P. Van Fleet, *On the support properties of scaling vectors*, Applied and Computational Harmonic Analysis, **3**(1996) 229–238.

[34] C. A. Micchelli, *Regularity of multiwavelets,* preprint, 1996.

[35] G. Plonka, *Approximation order provided by refinable function vectors*, Constr. Approx., to appear, 1996.

[36] G. Plonka, *Factorization of refinement masks for function vectors,* Universität Rostock, preprint, 1996.

[37] G. Plonka, *On Stability of scaling vectors,* Universität Rostock, preprint, 1996.

[38] G. Plonka and V. Strela, *Construction of multi-scaling functions with approximation and symmetry*, preprint, 1995.

[39] A. Ron, *A necessary and sufficient condition for the linear independence of the integer translates of a compactly supported distribution*, Constr. Approx., **5**(1989) 297–308.

[40] D. Ruch, W. So, and Jianzhong Wang, *Global support of a scaling vector*, Applied and Computational Harmonic Analysis, to appear, 1996.

[41] Z. Shen, *Refinable function vectors*, SIAM J. Math. Anal., to appear, 1996.

[42] W. So and Jianzhong Wang, *Estimating the support of a scaling vector*, SIAM J. Matrix Anal. Appl., **18**(1997) 66–73.

[43] G. Strang and V. Strela, *Short wavelets and matrix dilation equations*, J. Optical Engineering, **33**(1994) 2104–2107.

[44] G. Strang and V. Strela. *Orthogonal multiwavelets with vanishing moments,* SPIE Proceedings, Orlando; J. Optical Enginering, to appear, 1994.

[45] V. Strela, *Multiwavelets: Regularity, orthogonality and symmetry via two-scale similarity transform*, preprint, 1995.

[46] Jianzhong Wang, *Stability and linear independence associated with scaling vectors*, SIAM J. Math. Anal., to appear, 1996.

[47] Jianzhong Wang, *On solutions of two-scale difference equations*, Chinese Ann. Math. B, **15**(1994) 23–34.

[48] D. X. Zhou, *Existence of multiple refinable distributions*, Michigan Math. J., **44**(1997) to appear.

11
Self-Affine Tiles

Yang Wang[1]

ABSTRACT A *self-affine tile* in $\mathbf{R}^n$ is a set T of positive Lebesgue measure satisfying $\mathbf{A}(T) = \cup_{d \in \mathcal{D}}(T+d)$, where $\mathbf{A}$ is an expanding $n \times n$ real matrix with $|\det(\mathbf{A})| = m$ an integer, and $\mathcal{D} = \{d_1, \ldots, d_m\} \subset \mathbf{R}^n$ a set of m digits. Self-affine tiles arise in many contexts, including radix expansions, fractal geometry, and the construction of compactly supported orthonormal wavelet bases of $L^2(\mathbf{R}^n)$. They are also studied as interesting tiles. In this article we survey the fundamental properties of self-affine tiles. We examine necessary and sufficient conditions for digit sets $\mathcal{D}$ to give rise to self-affine tiles. A special class of self-affine tiles is the *integral self-affine tiles*, in which $\mathbf{A}$ is an integer matrix and $\mathcal{D} \subset \mathbf{Z}^n$. We study the tiling properties and the measures of integral self-affine tiles. We also compute the Hausdorff dimensions of the boundaries of integral self-affine tiles.

1 Introduction

Let $\mathbf{A}$ be an expanding matrix in $M_n(\mathbf{R})$, that is, one with all eigenvalues $|\lambda_i| > 1$, and suppose that $|\det(\mathbf{A})| = m$ for some integer $m > 1$. Let $\mathcal{D} = \{d_1, d_2, \ldots, d_m\} \subset \mathbf{R}^n$ be a finite set of vectors. A result of Hutchinson [27] states that there exists a unique nonempty compact set $T := T(\mathbf{A}, \mathcal{D})$ such that

$$T = \bigcup_{j=1}^{m} \mathbf{A}^{-1}(T + d_j). \tag{1.1}$$

More precisely, T is the *attractor* of the *iterated function system* $\{\phi_j(x) = \mathbf{A}^{-1}x + \mathbf{A}^{-1}d_j : 1 \leq j \leq m\}$. In fact, T is given explicitly by

$$T = \left\{ \sum_{k=1}^{\infty} \mathbf{A}^{-k} d_k \ : \ \text{each } d_k \in \mathcal{D} \right\}. \tag{1.2}$$

---

[1]School of Mathematics, Georgia Institute of Technology, Atlanta, GA 30332 Research supported in part by the National Science Foundation, grant DMS–9307601.

For most pairs $(\mathbf{A}, \mathcal{D})$ the set $T(\mathbf{A}, \mathcal{D})$ has Lebesgue measure $\mu(T) = 0$. If $T(\mathbf{A}, \mathcal{D})$ has positive Lebesgue measure we call $T(\mathbf{A}, \mathcal{D})$ a *self-affine tile*.

The name "self-affine tile" refers to the fact that

$$\mathbf{A}(T) = \bigcup_{j=1}^{m} (T + d_j) = T + \mathcal{D}; \tag{1.3}$$

geometrically it means that the affinely dilated set $\mathbf{A}(T)$ is perfectly tiled by the m translates $T + d_j$ of T. A simple example of a self-affine tile is the unit square $T = [0, 1]^2$, which satisfies $\mathbf{A}(T) = T + \mathcal{D}$ for

$$\mathbf{A} = 2I, \quad \mathcal{D} = \left\{ \begin{bmatrix} 0 \\ 0 \end{bmatrix}, \begin{bmatrix} 1 \\ 0 \end{bmatrix}, \begin{bmatrix} 0 \\ 1 \end{bmatrix}, \begin{bmatrix} 1 \\ 1 \end{bmatrix} \right\}.$$

Self-affine tiles have been studied as "exotic" tiles and as tiles giving interesting tilings of $\mathbf{R}^n$ ([1], [2] [11], [12], [13] [22], [24], [29], [33], [36], [51]). Furthermore, they arise in many other contexts, particularly in fractal geometry ([14], [15], [16], [50], compactly supported wavelet bases ([23], [32], [35]), radix expansions ([19]), and in Markov partitions ([28]). The current interests in self-affine tiles come largely from these applications.

Most of the studies on self-affine tiles employ one or both of the two approaches: algebraic and Fourier analytic. It is rather easy to see the role of algebraic methods. For example, given an expanding matrix $\mathbf{A}$ and a digit set $\mathcal{D}$, by iterating (1.3) we obtain $\mathbf{A}^k(T) = T + \mathcal{D}_{\mathbf{A},k}$ where

$$\mathcal{D}_{\mathbf{A},k} = \left\{ \sum_{j=0}^{k-1} \mathbf{A}^j d_j \, : \, \text{each } d_k \in \mathcal{D} \right\}. \tag{1.4}$$

As we shall see, many properties of $T(\mathbf{A}, \mathcal{D})$ depend fundamentally on the algebraic properties of $\mathcal{D}_{\mathbf{A},k}$. Of course, this is but one of the many instances where algebraic methods can be employed.

But harmonic analysis can be a powerful tool in the study of self-affine tiles as well. Let $T := T(\mathbf{A}, \mathcal{D})$ be a self-affine tile. The set-valued equation $T = T + \mathcal{D}$ can be written as

$$\chi_T(x) = \sum_{d \in \mathcal{D}} \chi_T(\mathbf{A}x - d). \tag{1.5}$$

Let $m_\mathcal{D}(\xi) = \frac{1}{|\mathcal{D}|} \sum_{d \in \mathcal{D}} e^{i2\pi \langle d, \xi \rangle}$. Taking the Fourier transform in (1.5) results in

$$\widehat{\chi}_T(\xi) = m_\mathcal{D}(\mathbf{B}^{-1}\xi) \widehat{\chi}_T(\mathbf{B}^{-1}\xi), \text{ where } \mathbf{B} := \mathbf{A}^T. \tag{1.6}$$

This yields

$$\widehat{\chi}_T(\xi) = c \prod_{j=1}^{\infty} m_\mathcal{D}(\mathbf{B}^{-j}\xi), \text{ where } c := \widehat{\chi}_T(0) = \mu(T). \tag{1.7}$$

By analyzing $m_\mathcal{D}(\xi)$ and the infinite product (1.7) a number of nontrivial results on the tile T and its tilings can be proved ([7], [22], [29], [34], [36]).

We shall provide a glimpse of both approaches in this overview. The fundamental question addressed by this paper is: for a given matrix $\mathbf{A}$ and digit set $\mathcal{D}$, under what conditions will $T(\mathbf{A}, \mathcal{D})$ be a tile? We derive several necessary and sufficient conditions in §2, and later in §4. In §3 we introduce integral self-affine tiles and prove some basic results concerning their measures and tilings. Some of these results are then used in §5 to study Haar-type wavelet bases. In §6 we show a method for finding the exact Hausdorff dimension of the boundaries of self-affine tiles.

Due to the restriction on the length of the paper, We have limited the discussions of this overview mostly to self-affine tiles as sets. In doing so we have made several conspicuous, and perhaps unjustified, omissions. In particular, we have left out the study on self-replicating tilings and on the topological properties of the tiles entirely. We apologize in advance for our inability to include these results and shall refer the readers to [2], [28], [30], [33], [52] for more details.

We are greatly indebt to Professor Ka-Sing Lau and the Mathematics Department of The Chinese University of Hong Kong for the kind invitation to visit. We would also like to thank Jeff Lagarias, Ka-Sing Lau, Rick Kenyon, Sze-Man Ngai and Bob Strichartz for encouraging and helpful discussions.

2 Conditions For A Tile

As mentioned in the introduction, for a given pair $(\mathbf{A}, \mathcal{D})$ where $\mathbf{A} \in M_n(\mathbf{R})$ is expanding and $\mathcal{D} \subset \mathbf{R}^n$ has cardinality $|\mathcal{D}| = |\det(\mathbf{A})|$, the corresponding attractor $T(\mathbf{A}, \mathcal{D})$ is usually not a tile. A fundamental question is thus: under what condition(s) is $T(\mathbf{A}, \mathcal{D})$ a tile? To gain some insight into this question we first look at the following example.

Example 2.1. Let $\mathbf{A} = [3]$ and $\mathcal{D} = \{0, 1, 4\}$. We show that $T = T(\mathbf{A}, \mathcal{D})$ is not a tile by showing that $\mu(T) = 0$, where μ denotes the Lebesgue measure. Note that

$$3T = T + \mathcal{D} = T + \{0, 1, 4\}.$$

Hence

$$\begin{aligned} 9T &= 3T + 3\mathcal{D} \\ &= T + \{0, 1, 4\} + \{0, 3, 12\} \\ &= T + \{0, 1, 3, 4, 7, 12, 13, 16\}. \end{aligned}$$

It follows by taking the Lebesgue measure that

$$9\mu(T) = \mu(T + \{0, 1, 3, 4, 7, 12, 13, 16\}) \leq 8\mu(T),$$

and hence $\mu(T) = 0$. □

For integral $\mathbf{A}$ and $\mathcal{D}$, the following theorem was established by Bandt [1]:

Theorem 2.1 *Let $\mathbf{A} \in M_n(\mathbf{Z})$ be an expanding matrix and let $\mathcal{D} \subset \mathbf{Z}^n$ be a set of complete coset representatives of $\mathbf{Z}^n/\mathbf{A}(\mathbf{Z}^n)$. Then $T = T(\mathbf{A}, \mathcal{D})$ has nonempty interior. Furthermore, T is the closure of its interior and $\mu(\partial T) = 0$.*

Proof. We present a new proof here. We first show that $\mu(T) > 0$. Let $T_0 = [0, 1]^n$ and

$$T_k = \bigcup_{d \in \mathcal{D}} \mathbf{A}^{-1}(T_{k-1} + d), \quad k \geq 1. \tag{2.1}$$

It is easy to check, by induction on k, that the unions in (2.1) are measure-disjoint and $\mu(T_k) = 1$ for all k. Since $T_k \longrightarrow T$ in the Haudorff metric (c.f. Hutchinson [27]), it follows that $\mu(T) \geq 1$.

Now let $\pi_n : \mathbf{R}^n \to \mathbf{T}^n$ be the canonical covering map, where $\mathbf{T}^n := \mathbf{R}^n/\mathbf{Z}^n$ is the n-torus. Then $\mathbf{A}_* := \pi_n \circ \mathbf{A} \circ \pi_n^{-1}$ is a well defined endomorphism on $\mathbf{T}^n$. Clearly,

$$\mathbf{A}_*\big(\pi_n(T)\big) = \pi_n\big(\mathbf{A}(T)\big) = \pi_n(T + \mathcal{D}) = \pi_n(T).$$

So $\pi_n(T)$ is invariant under $\mathbf{A}_*$. But $\mathbf{A}_*$ is ergodic because $\mathbf{A}$ is expanding (c.f. Walters [53]). Hence $\pi_n(T) = \mathbf{T}^n$. This means that

$$\bigcup_{\alpha \in \mathbf{Z}^n} (T + \alpha) = \mathbf{R}^n.$$

To see that $T^o \neq \emptyset$, let $\mathcal{J} \subset \mathbf{Z}^n$ be the smallest set such that $T + \mathcal{J} \supseteq (0, 1)^n$. Suppose that $T = \emptyset$. Fix an $\alpha_0 \in \mathcal{J}$. Then any $x \in (0,1)^n \cap (T+\alpha_0)$ must belong to another $T + \beta$ for some $\beta \in \mathcal{J}$. Hence $T + (\mathcal{J} \setminus \{\alpha_0\}) \supseteq (0,1)^n$, contradicting the minimality assumption of $\mathcal{J}$. So $T^o \neq \emptyset$. Now

$$\mathbf{A}(\overline{T^o}) = \overline{T^o} + \mathcal{D}.$$

By the uniqueness we must have $T = \overline{T^o}$.

Finally we prove that $\mu(\partial T) = 0$. Let $x_0 \in T^o$. For sufficiently large $k \geq 0$ the interior of $\mathbf{A}^k(T - x_0)$ will contain T. But

$$\mathbf{A}^k(T - x_0) = T + \mathcal{J}', \text{ where } \mathcal{J}' = \mathcal{D} + \mathbf{A}\mathcal{D} + \cdots + \mathbf{A}^{k-1}\mathcal{D} - \mathbf{A}^k x_0,$$

and the union $T + \mathcal{J}'$ is measure-disjoint. Since ∂T is contained in the overlapps in the union, it follows that $\mu(\partial T) = 0$. □

For any $\mathbf{A}$ and digit set $\mathcal{D}$ we denote

$$\mathcal{D}_{\mathbf{A},k} := \mathcal{D} + \mathbf{A}\mathcal{D} + \cdots + \mathbf{A}^{k-1}\mathcal{D}.$$

Note that if $0 \in \mathcal{D}$ then $\mathcal{D}_{\mathbf{A},k} \subseteq \mathcal{D}_{\mathbf{A},k+1}$. In this case we denote

$$\mathcal{D}_{\mathbf{A},\infty} := \bigcup_{k=1}^{\infty} \mathcal{D}_{\mathbf{A},k}, \quad \Delta(\mathbf{A},\mathcal{D}) := (\mathcal{D} - \mathcal{D})_{\mathbf{A},\infty}.$$

Theorem 2.1 is a special case of the following more general theorem, due to Kenyon [29] and Lagarias and Wang [33]:

Theorem 2.2 *Let* $\mathbf{A} \in M_n(\mathbf{R})$ *be an expanding matrix such that* $|\det(\mathbf{A})| = m \in \mathbf{Z}$. *Suppose that* $\mathcal{D} \subset \mathbf{R}^n$ *has cardinality* m, *with* $0 \in \mathcal{D}$. *Let* $T = T(\mathbf{A}, \mathcal{D})$. *Then the following conditions are equivalent:*

(a) T *has positive Lebesgue measure.*

(b) T *has nonempty interior.*

(c) T *is the closure of its interior, and its boundary* ∂T *has Lebesgue measure zero.*

(d) For each $k \geq 1$ *all* m^k *expansions in* $\mathcal{D}_{\mathbf{A},k}$ *are distinct, and* $\mathcal{D}_{\mathbf{A},\infty}$ *is uniformly discrete.*

Although not difficult, the proof is rather tedious. A proof can be found in Lagarias and Wang [33].

One other question we naturally ask is how does a self-affine tile $T(\mathbf{A}, \mathcal{D})$ tile $\mathbf{R}^n$. We show below that T tiles by translation.

Theorem 2.3 *Let* $\mathbf{A} \in M_n(\mathbf{R})$ *be an expanding matrix and* $\mathcal{D} \in \mathbf{R}^n$ *with* $|\mathcal{D}| = |\det(\mathbf{A})|$. *Suppose that* $T = T(\mathbf{A}, \mathcal{D})$ *has nonempty interior. Then there exists a set of translations* $\mathcal{J} \subseteq \Delta(\mathbf{A}, \mathcal{D})$ *such that* $T + \mathcal{J}$ *is a tiling of* $\mathbf{R}^n$.

Proof. The fundamental idea here is to repeatedly inflate the tile T at some interior point. Since $T^o \neq \emptyset$, by (1.2) there exists an interior point $x_0 \in T^o$ that has a finite radix expansion

$$x_0 = \sum_{j=1}^{N} \mathbf{A}^{-j} d_j^*, \text{ each } d_j^* \in \mathcal{D}.$$

Let $\tilde{T} = T - x_0$ and $\tilde{\mathcal{D}} := \mathcal{D}_{\mathbf{A},N} - \mathbf{A}^N x_0$. Then $0 \in \tilde{\mathcal{D}}$ and $\mathbf{A}^N(\tilde{T}) = \tilde{T} + \tilde{\mathcal{D}}$. Iterations yield that for all $k \geq 1$,

$$\mathbf{A}^{Nk}(\tilde{T}) = \tilde{T} + \tilde{\mathcal{D}}_{\mathbf{A}^N,k}. \tag{2.2}$$

Because 0 is in the interior of $\tilde{T}$, any ball $B_r(0)$ will be covered by $\mathbf{A}^{Nk}(\tilde{T})$ for sufficiently large k. Furthermore, $\tilde{\mathcal{D}}_{\mathbf{A}^N,k} \subseteq \tilde{\mathcal{D}}_{\mathbf{A}^N,k+1}$ because $0 \in \tilde{\mathcal{D}}$. Hence $\tilde{T}$ tiles $\mathbf{R}^n$ by translates of $\mathcal{J} := \tilde{\mathcal{D}}_{\mathbf{A}^N,\infty}$, which implies that T tiles

$\mathbf{R}^n$ by translates of $\mathcal{J}$. Now clearly we have $\mathcal{J} \subseteq \Delta(\mathbf{A}, \mathcal{D})$, proving the theorem. □

An immediate corollary of Theorem 2.3 is that if $\mathbf{A} \in M_n(\mathbf{Z})$ and $\mathcal{D} \subset \mathbf{Z}^n$, then we may find a $\mathcal{J} \subseteq \mathbf{Z}^n$ such that $T(\mathbf{A}, \mathcal{D}) + \mathcal{J}$ is a tiling of $\mathbf{R}^n$, provided that $T(\mathbf{A}, \mathcal{D})$ has nonempty interior.

3 Integral Self-Affine Tiles

A particular class of self-affine tiles is the so-called *integral self-affine tiles*, where $\mathbf{A} \in M_n(\mathbf{Z})$ and $\mathcal{D} \subset \mathbf{Z}^n$. The integrality allows us to establish many more properties about the tile $T(\mathbf{A}, \mathcal{D})$. In some applications, such as orthonormal wavelet bases, one encounters only integral self-affine tiles. Moreover a large class of self-affine tiles, including all self-affine tiles in the one dimension, are affinely equivalent to integral self-affine tiles, see Kenyon [29], and Lagarias and Wang [33].

Let $\mathbf{A}$ be an expanding matrix in $M_n(\mathbf{Z})$ and $\mathcal{D} \subset \mathbf{Z}^n$ with $|\mathcal{D}| = |\det(\mathbf{A})|$. Associated to the pair $(\mathbf{A}, \mathcal{D})$ is the smallest $\mathbf{A}$-invariant sublattice of $\mathbf{Z}^n$ containing the difference set $\mathcal{D} - \mathcal{D}$, which we denote by $\mathbf{Z}[\mathbf{A}, \mathcal{D}]$. If $0 \in \mathcal{D}$ then

$$\mathbf{Z}[\mathbf{A}, \mathcal{D}] = \mathbf{Z}[\mathcal{D}, \mathbf{A}(\mathcal{D}), \ldots, \mathbf{A}^{n-1}(\mathcal{D})]. \tag{3.1}$$

This follows from the Hamilton-Cayley Theorem that $\mathbf{A}^n \in \mathbf{Z}[\mathbf{A}^0, \mathbf{A}^1, \ldots, \mathbf{A}^{n-1}]$.

We call a digit set $\mathcal{D}$ *primitive* (with respect to $\mathbf{A}$) if $\mathbf{Z}[\mathbf{A}, \mathcal{D}] = \mathbf{Z}^n$, and we also call the associated tile $T(\mathbf{A}, \mathcal{D})$ a *primitive tile*. Most of the questions we consider here can be reduced to the case of primitive tiles.

Lemma 3.1 *Let $\mathbf{A}$ be an expanding matrix in $M_n(\mathbf{Z})$ and $\mathcal{D} \subset \mathbf{Z}^n$ with $|\mathcal{D}| = |\det(\mathbf{A})|$. Suppose that $0 \in \mathcal{D}$ and $\mathbf{Z}[\mathbf{A}, \mathcal{D}] = \mathbf{B}(\mathbf{Z}^n)$ for some $\mathbf{B} \in M_n(\mathbf{Z})$. Then there is an expanding matrix $\tilde{\mathbf{A}} \in M_n(\mathbf{Z})$ and a primitive digit set $\tilde{\mathcal{D}} \subset \mathbf{Z}^n$ with respect to $\tilde{\mathbf{A}}$, $|\tilde{\mathcal{D}}| = |\det(\tilde{\mathbf{A}})|$, such that*

$$T(\mathbf{A}, \mathcal{D}) = \mathbf{B}(T(\tilde{\mathbf{A}}, \tilde{\mathcal{D}})). \tag{3.2}$$

Proof. Since $\mathbf{Z}[\mathbf{A}, \mathcal{D}] = \mathbf{B}(\mathbf{Z}^n)$ is $\mathbf{A}$-invariant, $\mathbf{AB}(\mathbf{Z}^n) \subseteq \mathbf{B}(\mathbf{Z}^n)$. Hence $\mathbf{AB} = \mathbf{B}\tilde{\mathbf{A}}$ for some $\tilde{\mathbf{A}} \in M_n(\mathbf{Z})$. $\tilde{\mathbf{A}}$ is expanding because $\tilde{\mathbf{A}} = \mathbf{B}^{-1}\mathbf{AB}$. Now $\mathcal{D} \subseteq \mathbf{B}(\mathbf{Z}^n)$, so $\mathcal{D} = \mathbf{B}(\tilde{\mathcal{D}})$ for some $\tilde{\mathcal{D}} \subset \mathbf{Z}^n$. Let $\tilde{T} := T(\tilde{\mathbf{A}}, \tilde{\mathcal{D}})$. It satisfies $\tilde{\mathbf{A}}(\tilde{T}) = \tilde{T} + \tilde{\mathcal{D}}$, so

$$\mathbf{A}(\mathbf{B}(\tilde{T})) = \mathbf{B}\tilde{\mathbf{A}}(\tilde{T}) = \mathbf{B}(\tilde{T} + \tilde{\mathcal{D}}) = \mathbf{B}(\tilde{T}) + \mathcal{D}.$$

The uniqueness yields $\mathbf{B}(\tilde{T}) = T$. □

Theorem 3.2 *Let $\mathbf{A} \in M_n(\mathbf{Z})$ be expanding and $\mathcal{D} \subset \mathbf{Z}^n$ with $|\mathcal{D}| = |\det(\mathbf{A})|$. Then $\mu(T(\mathbf{A}, \mathcal{D})) := k \in \mathbf{Z}$. Furthermore, $T(\mathbf{A}, \mathcal{D}) + \mathbf{Z}^n$ is a perfect covering of $\mathbf{R}^n$ of multiplicity k.*

Proof. As before let $\pi_n : \mathbf{R}^n \longrightarrow \mathbf{T}^n$ be the canonical covering map. The integer matrix $\mathbf{A}$ induces an endomorphism $\mathbf{A}_* : \mathbf{T}^n \longrightarrow \mathbf{T}^n$ defined by $\mathbf{A}_* := \pi_n \circ \mathbf{A} \circ \pi_n^{-1}$. Let $\nu : \mathbf{T}^n \longrightarrow \mathbf{Z}$ denote the function $\nu(z) := |\pi_n^{-1}(z) \cap T|$ where $T := T(\mathbf{A}, \mathcal{D})$. Since T is compact, there exists a finite $k \in \mathbf{Z}$ such that
$$k = \max\{l \in \mathbf{Z} : \nu(z) \geq l \text{ for almost all } z \in \mathbf{T}^n\}.$$

It follows that there exist disjoint (up to a measure zero set) fundamental domains $F_1, F_2, \ldots, F_k$ of the lattice $\mathbf{Z}^n$ such that each $F_j \subseteq T$. Denote $F = \cup_{j=1}^k F_j$ and $\Omega = T \setminus F$. We show that $\Omega_* := \pi_n(\Omega)$ is invariant under $\mathbf{A}_*$.

To see this, note that
$$\mathbf{A}_*(\Omega_*) = \pi_n \circ \mathbf{A}(\Omega) = \pi_n\left((T + \mathcal{D}) \setminus \bigcup_{j=1}^k \mathbf{A}(F_j)\right). \tag{3.3}$$

Let $z_0 \in \mathbf{T}^n \setminus \Omega_*$. Then z_0 is covered exactly k times under $\pi_n : T \longrightarrow \mathbf{T}^n$; so it is covered exactly $k|\mathcal{D}|$ times under $\pi_n : T + \mathcal{D} \longrightarrow \mathbf{T}^n$. However, z_0 is also covered $|k \det(\mathbf{A})| = k|\mathcal{D}|$ times under $\cup_{j=1}^k \mathbf{A}(F_j)$ times because each F_j is a fundamental domain of $\mathbf{Z}^n$ and $\mathbf{A} \in M_n(\mathbf{Z})$. So $z_0 \notin \mathbf{A}_*(\Omega_*)$ by (3.3). This yields $\mathbf{A}(\Omega_*) \subseteq \Omega_*$. By the ergodicty of $\mathbf{A}_*$ the set Ω_* have measure zero or is all of $\mathbf{T}^n$. But the latter is ruled out by the definition of k. Therefore $\mu(\Omega) = 0$, and $\nu(z) = k$ for almost all $z \in \mathbf{T}^n$. This proves $\mu(T) = k \in \mathbf{Z}$ and $T + \mathbf{Z}^n$ is a perfect covering of $\mathbf{R}^n$ of multiplicity k. $\square$

As we will see in §4, a Haar-type orthonormal wavelet basis can be constructed from an integral self-affine tile $T(\mathbf{A}, \mathcal{D})$ with $\mu(T(\mathbf{A}, \mathcal{D})) = 1$. In this case $T(\mathbf{A}, \mathcal{D})$ tiles $\mathbf{R}^n$ by $\mathbf{Z}^n$-translations. The following is a necessary condition for it to hold.

Theorem 3.3 *Let $\mathbf{A} \in M_n(\mathbf{Z})$ be expanding and $\mathcal{D} \subset \mathbf{Z}^n$ with $|\mathcal{D}| = |\det(\mathbf{A})|$. Suppose that $\mu(T(\mathbf{A}, \mathcal{D})) = 1$. Then $\mathcal{D}$ is primitive and is a complete set of coset representatives of $\mathbf{Z}^n/\mathbf{A}(\mathbf{Z}^n)$.*

Proof. We project $T := T(\mathbf{A}, \mathcal{D})$ onto the n-torus $\mathbf{T}^n$ by π_n. By Theorem 2.3 there exists a $\mathcal{J} \subseteq \mathbf{Z}^n$ such that $T + \mathcal{J}$ is a tiling of $\mathbf{R}^n$. Since $\mu(T) = 1$, $\mathcal{J} = \mathbf{Z}^n$. Hence $\pi_n(T) = \mathbf{T}^n$. Now $\mathbf{A}(T) = T + \mathcal{D}$ yields
$$\mathbf{T}^n = \pi_n(T) = \bigcup_{d \in \mathcal{D}} \left(\tilde{T}_* + \pi_n(\mathbf{A}^{-1}d)\right), \text{ where } \tilde{T}_* := \pi_n(\mathbf{A}^{-1}(T)).$$

Since the measure of $\tilde{T}_*$ is at most $1/|\mathcal{D}|$, all $\pi_n(\mathbf{A}^{-1}d)$ must be distinct in $\mathbf{T}^n$. This shows that $\mathcal{D}$ must be a complet set of coset representatives of $\mathbf{Z}^n/\mathbf{A}(\mathbf{Z}^n)$.

The primitiveness of $\mathcal{D}$ follows directly from Lemma 3.1. $\square$

The converse of Theorem 3.3 is true in the one dimension (§4) but is false for $n \geq 2$. Let

$$\mathbf{A} = \begin{bmatrix} 2 & 1 \\ 0 & 2 \end{bmatrix}, \ \mathcal{D} = \left\{ \begin{bmatrix} 0 \\ 0 \end{bmatrix}, \begin{bmatrix} 3 \\ 0 \end{bmatrix}, \begin{bmatrix} 0 \\ 1 \end{bmatrix}, \begin{bmatrix} 3 \\ 1 \end{bmatrix} \right\}.$$

Then $\mathcal{D}$ is a primitive complete set of coset representatives of $\mathbf{Z}^2/\mathbf{A}(\mathbf{Z}^2)$. However, $\mu(T(\mathbf{A}, \mathcal{D}))$ has Lebesgue measure 3, see [36].

In the above example the tile $T(\mathbf{A}, \mathcal{D})$ tiles $\mathbf{R}^2$ by lattice translates, using the lattice $3\mathbf{Z} \oplus \mathbf{Z}$. In general we have:

Theorem 3.4 *Let $\mathbf{A} \in M_n(\mathbf{Z})$ be expanding and $\mathcal{D}$ be a complete set of coset representatives of $\mathbf{Z}^n/\mathbf{A}(\mathbf{Z}^n)$. Then there exists a full rank lattice $\mathcal{L} \subseteq \mathbf{Z}^n$ such that $T(\mathbf{A}, \mathcal{D})$ tiles $\mathbf{R}^n$ by $\mathcal{L}$-translations.*

For the proof of Theorem 3.4 we refer the readers to Conz, Hervè and Raugi [7] or Lagarias and Wang [36].

4 Digit Sets of Integral Self-Affine Tiles

Although Theorem 2.2 (d) provides a necessary and sufficient condition for $T(\mathbf{A}, \mathcal{D})$ to be a tile, the condition itself is rather difficult to verify. In this section we explicitly classify integral digit sets $\mathcal{D}$ that result in tiles for certain types of expanding matrices $\mathbf{A} \in M_n(\mathbf{Z})$.

Theorem 4.1 *Let p be a prime and $\mathcal{D} \subset \mathbf{Z}$ be a primitive digit set with $|\mathcal{D}| = |p|$. Then $T(\mathbf{A}, \mathcal{D})$ is a tile if and only if $\mathcal{D}$ is a complete set of residues (mod p).*

Proof. The sufficiency is already established. We prove the necessity. Without loss of generality we assume that $0 \in \mathcal{D}$ and $d \geq 0$ for all $d \in \mathcal{D}$. Let $f_\mathcal{D}(z)$ denote the characteristic polynomial $f_\mathcal{D}(z) := \frac{1}{|p|}\sum_{d \in \mathcal{D}} z^d$. We prove that there exists a $k \geq 1$ such that the cyclotomic polynomial $F_{p^k}(z) := \frac{z^{p^k}-1}{z^{p^{k-1}}-1}$ divides $f_\mathcal{D}(z)$.

Let $m_\mathcal{D}(\xi) := f_\mathcal{D}(e^{i2\pi\xi})$. Note that the characteristic function of $T := T(\mathbf{A}, \mathcal{D})$ satisfies

$$\chi_T(x) = \sum_{d \in \mathcal{D}} \chi_T(px - d). \tag{4.1}$$

Taking the Fourier transform yields $\widehat{\chi}_T = m_\mathcal{D}(p^{-1}\xi)\widehat{\chi}_T(p^{-1}\xi)$. By iteration,

$$\widehat{\chi}_T(\xi) = c\prod_{j=1}^{\infty} m_\mathcal{D}(p^{-j}\xi), \quad \text{where } c = \widehat{\chi}_T(0) = \mu(T). \tag{4.2}$$

The convergence of the infinite product (4.2) is well known. By Theorem 3.2, $T + \mathbf{Z}$ is a perfect covering of $\mathbf{R}$ of multiplicity $\mu(T) \in \mathbf{Z}$, so $\hat{\chi}_T(l) = 0$ for all nonzero integer l. In particular $\hat{\chi}_T(1) = 0$. By (4.2) there exists some integer $k \geq 1$ such that $m_\mathcal{D}(p^{-k}) = 0$. Hence $f_\mathcal{D}(e^{i2\pi p^{-k}}) = 0$, proving that $F_k(z)|f_\mathcal{D}(z)$ and hence $(z^{p^k} - 1)|f_\mathcal{D}(z)(z^{p^{k-1}} - 1)$.

Observe that if two integers satisfy $j_1 \equiv j_2 \pmod{p^k}$ then $z^{j_1} \equiv z^{j_2}$ $\pmod{(z^{p^k} - 1)}$. Because

$$f_\mathcal{D}(z)(z^{p^{k-1}} - 1) = \sum_{d \in \mathcal{D} + p^{k-1}} z^d - \sum_{d \in \mathcal{D}} z^d \equiv 0 \pmod{(z^{p^k} - 1)},$$

and because a nonzero polynomial of degree less than p^k can never be divisible by $z^{p^k} - 1$, we must have

$$\mathcal{D} + p^{k-1} \pmod{p^k} = \mathcal{D} \pmod{p^k}. \tag{4.3}$$

$p^{k-1} \in \mathcal{D} + p^{k-1}$, so $d \equiv p^{k-1} \pmod{p^k}$ for some $d \in \mathcal{D}$. Similarly now $2p^{k-1} \in \mathcal{D} + p^{k-1} \pmod{p^k}$, so $2p^{k-1} \in \mathcal{D} \pmod{p^k}$. This argument yields

$$\mathcal{D} \equiv \{0, p^{k-1}, 2p^{k-1}, \ldots, (p-1)p^{k-1}\} \pmod{p^k}.$$

But $\mathcal{D}$ is primitive, so $\gcd\{d : d \in \mathcal{D}\} = 1$. Therefore $k = 1$ and $\mathcal{D}$ is a complete set of residues $\pmod p$. □

The above theorem was due to Kenyon [29]. The same argument can be used to prove the following generalization, a proof of which can be found in Lagarias and Wang [34].

Theorem 4.2 *Let $\mathbf{A} \in M_n(\mathbf{Z})$ be expanding such that $|\det(\mathbf{A})| = p$ is a prime and $p\mathbf{Z}^n \not\subseteq \mathbf{A}^2(\mathbf{Z}^n)$. Let $\mathcal{D} \subset \mathbf{Z}^n$ with $|\mathcal{D}| = |\det(\mathbf{A})|$ be primitive. Then $T(\mathbf{A}, \mathcal{D})$ is a tile if and only if $\mathcal{D}$ is a set of complete coset representatives of $\mathbf{Z}^n/\mathbf{A}(\mathbf{Z}^n)$.*

It should be pointed out that the classification of digit sets for a given matrix $\mathbf{A}$ is in general very difficult, even in the integral case. This is evident from the fact that even for $\mathbf{A} = [6]$ in the one dimension it is not completely known what digit sets $\mathcal{D}$ result in self-affine tiles. The only other cases in which all digit sets resulting in self-affine tiles are classified are $\mathbf{A} = 2I$ for $n = 2$ ([29]) and $\mathbf{A} = [p^k]$ for $n = 1$, where p is a prime ([34]).

So far we have discussed mostly digit sets that are complete set of coset representatives of $\mathbf{Z}^n/\mathbf{A}(\mathbf{Z}^n)$. Naturally one may ask whether there are other types of digit sets $\mathcal{D}$ that also give self-affine tiles. Here is a simple example:

Example 4.1. Let $\mathbf{A} = [4]$ and $\mathcal{D} = \{0, 1, 8, 9\}$. Clearly $\mathcal{D}$ is primitive and is not a complete set of residues $\pmod 4$. But one may check directly that $T(\mathbf{A}, \mathcal{D}) = [0, 1] \cup [2, 3]$. □

Example 4.1 is an example of a class of digit sets called *product form digit sets*. Suppose that $0 \in \mathcal{E}$ is a set of complete coset representatives of $\mathbf{Z}^n/\mathbf{A}(\mathbf{Z}^n)$, and suppose that it has a factorization

$$\mathcal{E} = \mathcal{E}_1 + \mathcal{E}_2 + \cdots + \mathcal{E}_r, \quad \text{where } 0 \in \mathcal{E}_i \text{ and } |\mathcal{E}| = \prod_{i=1}^{r} |\mathcal{E}_i|. \tag{4.4}$$

A digit set $\mathcal{D}$ has the product-form if

$$\mathcal{D} := \mathbf{A}^{f_1}(\mathcal{E}_1) + \mathbf{A}^{f_2}(\mathcal{E}_2) + \cdots + \mathbf{A}^{f_r}(\mathcal{E}_r) \tag{4.5}$$

for some integers $0 \leq f_1 \leq f_2 \leq \cdots \leq f_r$.

Theorem 4.3 *Let $\mathbf{A} \in M_n(\mathbf{Z})$ be expanding and $\mathcal{D}$ is the product-form digit set defined in (4.5). Then $T(\mathbf{A}, \mathcal{D})$ is a measure-disjoint union of translates of $T(\mathbf{A}, \mathcal{E})$, and*

$$\mu\big(T(\mathbf{A}, \mathcal{D})\big) = \mu\big(T(\mathbf{A}, \mathcal{E})\big) \prod_{i=1}^{r} |\mathcal{E}_i|^{f_i}. \tag{4.6}$$

Proof. Let $\mathcal{A}_{i,k} := \{\sum_{j=0}^{k-1} \mathbf{A}^j e_{i,j} : \text{all } e_{i,j} \in \mathcal{E}_i\}$ with $\mathcal{A}_{i,0} = \{0\}$. We prove that $T(\mathbf{A}, \mathcal{D}) = T(\mathbf{A}, \mathcal{E}) + \mathcal{A}$ where

$$\mathcal{A} := \mathcal{A}_{1, f_1} + \mathcal{A}_{2, f_2} + \cdots + \mathcal{A}_{r, f_r}.$$

$T(\mathbf{A}, \mathcal{D}) = \{\sum_{j=0}^{\infty} \mathbf{A}^{-j} d_j : \text{all } d_j \in \mathcal{D}\}$ from (1.2), and by assumption $d_j = \sum_{i=0}^{r} \mathbf{A}^{f_i} e_{i,j}$ where $e_{i,j} \in \mathcal{E}_i$. So

$$\sum_{j=1}^{\infty} \mathbf{A}^{-j} d_j = \sum_{j=1}^{\infty} \mathbf{A}^{-j} \sum_{i=0}^{r} \mathbf{A}^{f_i} e_{i,j}$$

$$= \sum_{i=0}^{r} \bigg(\sum_{j=f_i}^{\infty} \mathbf{A}^{-j} \mathbf{A}^{f_i} e_{i,j} + \sum_{j=1}^{f_i} \mathbf{A}^{-j} \mathbf{A}^{f_i} e_{i,j} \bigg)$$

$$= \sum_{j=0}^{\infty} \mathbf{A}^{-j} \bigg(\sum_{i=0}^{r} e_{i, j+f_i} \bigg) \sum_{i=0}^{r} \sum_{j=1}^{f_i} \mathbf{A}^{f_i - j} e_{i,j}. \tag{4.7}$$

Since $\sum_{i=0}^{r} e_{i, j+f_i} \in \mathcal{E}$ and $\sum_{i=0}^{r} \sum_{j=1}^{f_i} \mathbf{A}^{f_i - j} e_{i,j} \in \mathcal{A}$, we have $\sum_{j=0}^{\infty} \mathbf{A}^{-j} d_j \in T(\mathbf{A}, \mathcal{E}) + \mathcal{A}$; hence $T(\mathbf{A}, \mathcal{D}) \subseteq T(\mathbf{A}, \mathcal{E}) + \mathcal{A}$.

Conversely, one verifies that any element in $T(\mathbf{A}, \mathcal{E}) + \mathcal{A}$ must be in $T(\mathbf{A}, \mathcal{D})$ by reversing (4.7) (we omit the details here), yielding $T(\mathbf{A}, \mathcal{E}) + \mathcal{A} \subseteq T(\mathbf{A}, \mathcal{D})$. Therefore, $T(\mathbf{A}, \mathcal{D}) = T(\mathbf{A}, \mathcal{E}) + \mathcal{A}$.

We still need to show that the translates of $T(\mathbf{A}, \mathcal{E})$ in $T(\mathbf{A}, \mathcal{E}) + \mathcal{A}$ are measure-disjoint. For any $m \geq 1$ we have

$$\mathbf{A}^m\big(T(\mathbf{A}, \mathcal{E})\big) = T(\mathbf{A}, \mathcal{E}) + \mathcal{E}_{\mathbf{A}, m}, \tag{4.8}$$

where $\mathcal{E}_{\mathbf{A}<m} := \{\sum_{k=0}^{m-1} \mathbf{A}^k e_k : \text{all } e_k \in \mathcal{E}\}$. Since each $\mathcal{E}_i \subseteq \mathcal{E}$ and $0 \in \mathcal{E}$, $\mathcal{A} \subseteq \mathcal{E}_{\mathbf{A},m}$ whenever $m \geq f_r$. But the translates of $T(\mathbf{A},\mathcal{E})$ in (4.8) are measure-disjoint, it follows that the translates of $T(\mathbf{A},\mathcal{E})$ in $T(\mathbf{A},\mathcal{E}) + \mathcal{A}$ must be measure-disjoint.

Finally, all expansions $\sum_{i=0}^{r} \sum_{j=0}^{f_i-1} \mathbf{A}^j e_{i,j}$ where $e_{i,j} \in \mathcal{E}_i$ in $\mathcal{A}$ are distinct because $\mathcal{E} = \mathcal{E}_1 + \cdots + \mathcal{E}_r$ is a direct sum by (4.4). The measure-disjointness of $T(\mathbf{A},\mathcal{E}) + \mathcal{A}$ yields (4.6). □

The digit set $\mathcal{D} = \{0, 1, 8, 9\}$ in Example 4.1 is a product-form digit set, with $\mathcal{E} = \{0, 1, 2, 3\} = \{0, 1\} + \{0, 2\}$ and $\mathcal{D} = \{0, 1\} + 4\{0, 2\}$. There are integral self-affine tiles whose digit sets are not product-form digit sets, see [34]. One simple such example is $\mathbf{A} = [4]$, $\mathcal{D} = \{0, 1, 8, 25\}$. Can you prove that $T(\mathbf{A},\mathcal{D})$ is a tile?

5 Haar-Type Wavelet Bases of $L^2(\mathbf{R}^n)$

Let $\psi_1(x), \ldots, \psi_r(x) \in L^2(\mathbf{R}^n)$ and $\mathbf{A} \in M_n(\mathbf{Z})$ be expanding. Suppose that
$$\{|\det(\mathbf{A})|^{\frac{m}{2}} \psi_i(\mathbf{A}^m x - \alpha) : \alpha \in \mathbf{Z}^n, 1 \leq i \leq r, m \in \mathbf{Z}\}$$
is an orthonormal basis of $L^2(\mathbf{R}^n)$. Then we call this basis a *wavelet basis* of $L^2(\mathbf{R}^n)$, and $\psi_1(x), \ldots, \psi_r(x)$ *wavelets*. The simplest wavelet is the wavelet basis of $L^2(\mathbf{R})$ constructed by A. Haar [25], which has $\mathbf{A} = [2]$ and consists of a single wavelet
$$\psi(x) = \begin{cases} 1 & 0 \leq x < 1/2 \\ -1 & 1/2 \leq x \leq 1 \\ 0 & \text{otherwise.} \end{cases}$$

A popular way to construct wavelet bases is by *multiresolution analysis*. We shall not discuss it in detail as a comprehensive discussion can be found in Daubechies [9]. Let $\mathbf{A} \in M_n(\mathbf{Z})$ be expanding. A *scaling function* (of a multiresolution analysis), from which a wavelet basis can be constructed, is a function $\phi(x) \in L^2(\mathbf{R}^n)$ such that

(i) $\phi(x)$ satisfies a dilation equation
$$\phi(x) = \sum_{\alpha \in \mathbf{Z}^n} c_\alpha \phi(\mathbf{A}x - \alpha). \tag{5.1}$$

(ii) $\{\phi(x-\alpha) : \alpha \in \mathbf{Z}^n\}$ is an orthonormal set in $L^2(\mathbf{R}^n)$, and $\int_{\mathbf{R}^n} \phi(x)\,dx \neq 0$.

A *Haar-type* wavelet basis is a one constructed from a scaling function of the form $\phi(x) = c\chi_\Omega(x)$ for some compact set $\Omega \subset \mathbf{R}^n$ and constant c. Gröchenig and Madych [23] established the following relation between Haar-type wavelet bases and self-affine tiles:

Theorem 5.1 *Let $\mathbf{A} \in M_n(\mathbf{Z})$ be expanding and $\Omega \subset \mathbf{R}^n$ be compact. Then the following are equivalent:*

(a) *$\phi(x) = c\chi_\Omega(x)$ is a scaling function with respect to $\mathbf{A}$ for some constant c.*

(b) *There exists a set of complete coset representatives of $\mathbf{Z}^n/\mathbf{A}(\mathbf{Z}^n)$ such that $\mathbf{A}(\Omega) = \Omega + \mathcal{D}$ up to a measure zero set, and $\mu(\Omega) = 1$.*

Proof. (a)$\Rightarrow$(b). By assumption $\phi(x)$ satisfies some dilation equation

$$\phi(x) = \sum_{\alpha \in \mathbf{Z}^n} c_\alpha \phi(\mathbf{A}x - \alpha). \tag{5.2}$$

The orthonormality of $\{\phi(x-\alpha) : \alpha \in \mathbf{Z}^n\}$ implies that $\Omega + \alpha$, $\alpha \in \mathbf{Z}^n$, are measure-disjoint. By letting $y = \mathbf{A}x$ and rewriting (5.2) as

$$\chi_{\mathbf{A}(\Omega)}(y) = \sum_{\alpha \in \mathbf{Z}^n} c_\alpha \chi_{\Omega+\alpha}(y), \tag{5.3}$$

it yields $c_\alpha = 0$ or $c_\alpha = 1$.

Let $\mathcal{D} = \{\alpha : c_\alpha = 1\}$. Integrating (5.3) yields $|\mathcal{D}| = |\det(\mathbf{A})|$, and the measure-disjointness of $\Omega + \alpha$ in (5.3) implies that

$$\mathbf{A}(\Omega) = \bigcup_{d \in \mathcal{D}} (\Omega + d) = \Omega + \mathcal{D} \tag{5.4}$$

up to a measure zero set.

To show that $\mu(\Omega) = 1$, let $\pi_n : \mathbf{R}^n \longrightarrow \mathbf{T}^n$ be the canonical covering map and $\mathbf{A}_* := \pi_n \circ \mathbf{A} \circ \pi_n^{-1}$. By (5.4), $\mathbf{A}_*(\pi_n(\Omega)) = \pi_n(\Omega)$ up to a measure zero set. It follows from the ergodicity of $\mathbf{A}_*$ that $\pi_n(\Omega) = \mathbf{T}^n$ up to a measure zero set. Hence $\mu(\Omega) \geq 1$. But $\mu(\Omega) \leq 1$ because $\Omega + \alpha$, $\alpha \in \mathbf{Z}^n$, are measure-disjoint. Therefore $\mu(\Omega) = 1$.

(b)$\Rightarrow$(a). By the ergodicity argument above, $\Omega + \mathbf{Z}^n$ is a covering of $\mathbf{R}^n$ up to a measure zero set. Since $\mu(\Omega) = 1$, all $\Omega + \alpha$, $\alpha \in \mathbf{Z}^n$, are measure-disjoint. Hence $\phi(x) := \chi_\Omega(x)$ satisfies $\phi(x) = \sum_{d \in \mathcal{D}} \phi(\mathbf{A}x - d)$ and $\{\phi(x - \alpha) : \alpha \in \mathbf{Z}^n\}$ is an orthonormal system in $L^2(\mathbf{R}^n)$. So $\phi(x) = \chi_\Omega(x)$ is a scaling function. $\square$

Remark. It can be shown that if a compact set Ω satisfies $A(\Omega) = \Omega + \mathcal{D}$ up to a measure zero set, then $\Omega \supseteq T(\mathbf{A}, \mathcal{D})$ and $\Omega = T(\mathbf{A}, \mathcal{D})$ up to a measure zero set. We omit the proof here.

Naturally we would like to know when will $\mu(T(\mathbf{A}, \mathcal{D})) = 1$ for any given $\mathbf{A}$ and $\mathcal{D}$. Theorem 3.3 states that $\mathcal{D}$ must be a primitive set of complete coset representatives of $\mathbf{Z}^n/\mathbf{A}(\mathbf{Z}^n)$. A counterexample is given in §3 to show that the converse of the theorem is false for $n \geq 2$. The converse, however, is true in the one dimension.

Theorem 5.2 *Let $q \in \mathbf{Z}$, $|q| > 1$ and $\mathcal{D}$ be a complete set of residues (mod q). Suppose that $\mathcal{D}$ is primitive. Then $\mu\bigl(T([q], \mathcal{D})\bigr) = 1$.*

Proof. We present the following Fourier analytic proof, due to Gröchenig and Haas [22]. The technique here is valuable for studying other type of scaling functions as well.

Without loss of generality we assume that $0 \in \mathcal{D}$. The primitiveness of $\mathcal{D}$ is equivalent to $\gcd\{d : d \in \mathcal{D}\} = 1$.

Let $m_\mathcal{D}(\xi) := \frac{1}{|q|} \sum_{d \in \mathcal{D}} e^{i 2\pi d\xi}$. Key to the proof is the following linear transition operator

$$C_\mathcal{D}(f)(\xi) = \sum_{l=0}^{|q|-1} \left| m_\mathcal{D}(q^{-1}(\xi+l)) \right|^2 f(q^{-1}(\xi+l)) \tag{5.5}$$

defined on the space of **Z**-periodic functions. Let

$$g_\mathcal{D}(\xi) := \sum_{k \in \mathbf{Z}} \mu(T \cap (T+k)) e^{i 2\pi k\xi} \tag{5.6}$$

where $T := T([q], \mathcal{D})$. One easily checks (see Gröchenig [21]), using the assumption that $\mathcal{D}$ is a complete set of residues (mod q), that

$$C_\mathcal{D}(1) = 1, \quad C_\mathcal{D}(g_\mathcal{D}) = g_\mathcal{D}. \tag{5.7}$$

Assume that $\mu(T) > 1$. Then $g_\mathcal{D}(\xi)$ is not a constant, hence the set

$$Z_\mathcal{D} := \left\{ \xi : g_\mathcal{D}(\xi) = \min_{\eta \in \mathbf{R}} g_\mathcal{D}(\eta) \right\}$$

is a nonempty discrete **Z**-periodic set, and $Z_\mathcal{D} \cap \mathbf{Z} = \emptyset$ by (5.6). Fix an $\xi_0 \in Z_\mathcal{D}$. By (5.7)

$$g_\mathcal{D}(\xi_0) = \sum_{l=0}^{|q|-1} \left| m_\mathcal{D}(q^{-1}(\xi_0+l)) \right|^2 g_\mathcal{D}(q^{-1}(\xi_0+l)). \tag{5.8}$$

But $\sum_{l=0}^{|q|-1} \left| m_\mathcal{D}(q^{-1}(\xi_0+l)) \right|^2 = 1$ by $C_\mathcal{D}(1) = 1$, so $g_\mathcal{D}(q^{-1}(\xi_0+l)) = g_\mathcal{D}(\xi_0)$ whenever $m_\mathcal{D}(q^{-1}(\xi_0+l)) \neq 0$. In particular there exists an l_1 such that $\xi_1 := q^{-1}(\xi_0 + l_1) \in Z_\mathcal{D}$. Note that $q\xi_1 \equiv \xi_0 \pmod{1}$.

Now let $\widehat{Z}_\mathcal{D} := Z_\mathcal{D} \pmod 1$. So for any $\widehat{\xi}_0 \in \widehat{Z}_\mathcal{D}$ there exists a $\widehat{\xi}_1 \in \widehat{Z}_\mathcal{D}$ such that $q\widehat{\xi}_1 = \widehat{\xi}_0$. But $\widehat{Z}_\mathcal{D}$ is finite. Hence the map $\widehat{\xi} \mapsto q\widehat{\xi}$ is a permutation on $\widehat{Z}_\mathcal{D}$.

Back to (5.8). There exists no $l_2 \neq l_1$ in the sum such that $\xi_2 := q^{-1}(\xi_0 + l_2) \in Z_\mathcal{D}$ because otherwise $q\xi_1 \equiv q\xi_2 \pmod 1$ while $\xi_1 \not\equiv \xi_2 \pmod 1$, contradicting the fact that $\widehat{\xi} \mapsto q\widehat{\xi}$ is a permutation on $\widehat{Z}_\mathcal{D}$. Hence $m_\mathcal{D}(q^{-1}(\xi_0+l)) = 0$ for all $0 \leq l \leq |q|-1$, $l \neq l_1$. This means

$$\left| m_\mathcal{D}(q^{-1}(\xi_0+l_1)) \right|^2 \left| m_\mathcal{D}(\xi_1) \right|^2 = 1. \tag{5.9}$$

Because $0 \in \mathcal{D}$, (5.9) is possible only if $e^{i2\pi d\xi_1} = 1$, and hence $d\xi_1 \in \mathbf{Z}$ for all $d \in \mathcal{D}$. But $\xi_1 \notin \mathbf{Z}$, it follows that $\gcd\{d : d \in \mathcal{D}\} > 1$, a contradiction. □

Theorem 5.2 generalizes to higher dimensions only in special cases. One such case is when $\mathbf{A} \in M_n(\mathbf{Z})$ is *irreducible*, which means that the characteristic polynomial of $\mathbf{A}$ is irreducible in $\mathbf{Q}[z]$.

Theorem 5.3 *Let* $\mathbf{A} \in M_n(\mathbf{Z})$ *be an expanding irreducible matrix, and* $\mathcal{D}$ *be a primitive set of complete coset representatives of* $\mathbf{Z}^n/\mathbf{A}(\mathbf{Z}^n)$. *Then* $\mu(T(\mathbf{A}, \mathcal{D})) = 1$.

A proof can be found in [36]. It uses a deep result of Cerveau, Conze, and Raugi [6] characterizing the set of zeros of certain trigonometric polynomials. In the case of reducible $\mathbf{A}$, $\mu(T(\mathbf{A}, \mathcal{D})) > 1$ only when the digit set has the so called *quasi-product form*, see [36].

Another interesting question is: For a given expanding $\mathbf{A} \in M_n(\mathbf{Z})$, is it always possible to construct Haar-type wavelet basis? In other words, is it always possible to find a digit set $\mathcal{D}$ such that $\mu(T(\mathbf{A}, \mathcal{D})) = 1$? The answer is clearly affirmative in the one dimension as a result of Theorem 5.2. The answer is known to be affirmative in dimensions $n = 2, 3$ ([22], [32], [35], [36]). In dimension n, Haar-type wavelet bases exist if $|\det(\mathbf{A})| > n$. But what if $|\det(\mathbf{A})| \leq n$, for example, $|\det(\mathbf{A})| = 2$? The question becomes intriguing, because in this case the digit set $\mathcal{D}$ consists of only two digits. Since we may assume that $0 \in \mathcal{D}$, we have in reality the freedom to choose for only one digit. If the dimension is large, it is not clear we can always choose this digit so that $\mathcal{D}$ is primitive, i.e. $\mathbf{Z}[\mathbf{A}, \mathcal{D}] = \mathbf{Z}^n$. Although no counterexample has been found yet, it is almost certain that they exist. This problem has a surprising connection to algebraic number theory, see Lagarias and Wang [35].

6 Boundaries of Self-Affine Tiles

An important problem in fractal geometry is to find the Hausdorff dimension of a fractal set. Since a tile by definition has positive Lebesgue measure, its Hausdorff dimension is simply the dimension of the space in which it resides. A more interesting problem is to find the Hausdorff dimension of the boundary of a self-affine tile.

Getting the exact Hausdorff dimension of a fractal set is tricky in general. This had been the case for the boundaries of self-affine tiles. Boundaries of several well known tiles, such as the Gosper Flake (Gardner [18]) or the Fractal Red Cross (Strichartz [49]), were studied and their exact Hausdorff dimension derived. We illustrate how the Hausdorff dimensions of the boundaries of some self-affine tiles can be computed by the example of the Fractal Red Cross (Figure 1) in [49], which is the self-affine tile

11. Self-Affine tiles 275

FIGURE 1. The Fractal Red Cross

$T := T(\mathbf{A}, \mathcal{D})$ with

$$\mathbf{A} = \begin{bmatrix} 2 & -1 \\ 1 & 2 \end{bmatrix}, \mathcal{D} = \left\{ \begin{bmatrix} 0 \\ 0 \end{bmatrix}, \begin{bmatrix} 1 \\ 0 \end{bmatrix}, \begin{bmatrix} -1 \\ 0 \end{bmatrix}, \begin{bmatrix} 0 \\ 1 \end{bmatrix}, \begin{bmatrix} 0 \\ -1 \end{bmatrix} \right\}.$$

By Figure 1, the boundary of $\mathbf{A}(T)$ consists of 12 pieces, each congruent to a quarter of the boundary of T. Note that there are 5×4 quarter boundaries in the five translates of T in $\mathbf{A}(T)$, but 8 of them are overlaps that become part of the interior of $\mathbf{A}(T)$. So in fact $\partial(\mathbf{A}(T)) = \mathbf{A}(\partial T)$ consists of $20-8 = 12$ quarter boundaries of T. Let $\mathcal{H}^s$ denote the s-dimensional Hausdorff measure. Then $\mathcal{H}^s(\mathbf{A}(\partial T)) = 3\mathcal{H}^s(\partial T)$. On the other hand, $\mathcal{H}^s(\mathbf{A}(\partial T)) = 5^{s/2}\mathcal{H}^s(\partial T)$ by the scaling property of $\mathcal{H}^s$. For $\mathcal{H}^s(\partial T)$ to be finite and nonzero we must have $5^{s/2} = 3$. So $s = 2\log_5 3$ is the dimension of ∂T.

The above method can be made rigorous. The drawback is that it depends fundamentally on the visualization of the tiles, making it useful only on a case by case basis. For many self-affine tiles, the method either does not work, or requires ingenuity to work.

In this section we outline a method for finding the exact Hausdorff dimension of the boundaries of integral self-affine tiles. It employs the same basic idea, but requires no visualization of the tiles and works in *all* cases where the expanding matrix $\mathbf{A} \in M_n(\mathbf{Z})$ is similar to a similarity.

Let $T := T(\mathbf{A}, \mathcal{D})$ be an integral self-affine tile with $\mu(T) = 1$. Because

T tiles $\mathbf{R}^n$ by $\mathbf{Z}^n$-translations and T is the closure of its interior,

$$\partial T = \bigcup_{\alpha \in \mathbf{Z}^n \setminus \{0\}} T \cap (T + \alpha). \tag{6.1}$$

Denote $B_\alpha := T \cap (T + \alpha)$ for all $\alpha \in \mathbf{Z}^n \setminus \{0\}$. Of course there are only finitely many nonemtpy B_α's. Let $\mathcal{K}_0 = \{\alpha \in \mathbf{Z}^n \setminus \{0\} : B_\alpha \neq \emptyset\}$. To find the Hausdorff dimension of ∂T we utilize the fact that $\{B_\alpha : \alpha \in \mathcal{K}_0\}$ form a *self-similar system*; more precisely,

$$\begin{aligned}\mathbf{A}(B_\alpha) &= \mathbf{A}(T) \cap \mathbf{A}(T + \alpha) \\ &= (T + \mathcal{D}) \cap (T + \mathcal{D} + \mathbf{A}\alpha) \\ &= \bigcup_{d,d' \in \mathcal{D}} (B_{\mathbf{A}\alpha + d' - d} + d).\end{aligned} \tag{6.2}$$

Now, label elements in $\mathcal{K}_0$ as $\mathcal{K}_0 = \{\alpha_1, \alpha_2, \ldots, \alpha_J\}$ and define

$$\mathcal{E}_{i,j} := \{d \in \mathcal{D} : \mathbf{A}\alpha_i + d' - d \in \mathcal{K}_0 \text{ for some } d' \in \mathcal{D}\}. \tag{6.3}$$

Let $B_i := B_{\alpha_i}$ for $1 \leq i \leq J$. Then we may rewrite (6.2) as

$$\mathbf{A}(B_i) = \bigcup_{j=1}^{J} (B_j + \mathcal{E}_{i,j}), \quad 1 \leq i \leq J. \tag{6.4}$$

Theorem 6.1 *Let $T := T(\mathbf{A}, \mathcal{D})$ be an integral self-affine tile with $\mu(T(\mathbf{A}, \mathcal{D})) = 1$. Suppose that $\mathbf{A}$ is similar to a similarity. Then*

$$\underline{\dim}_B(\partial T) = \overline{\dim}_B(\partial T) = \dim_H(\partial T) = \frac{n \log \rho(\mathbf{M})}{\log |\det(\mathbf{A})|}, \tag{6.5}$$

where $\mathbf{M} := [|\mathcal{E}_{i,j}|]_{J \times J}$ and $\rho(\mathbf{M})$ is its spectral radius.

We call the matrix $\mathbf{M} = [|\mathcal{E}_{i,j}|]$ the *substitution matrix of the boundary of T*.

To prove Theorem 6.1 we first observe that iterating (6.4) yields

$$\mathbf{A}^N(B_i) = \bigcup_{j=1}^{J} (B_j + \mathcal{E}_{i,j}^N), \quad 1 \leq i \leq J, \tag{6.6}$$

where for all $1 \leq i, j \leq 1$,

$$\mathcal{E}_{i,j}^N = \bigcup_{j=1}^{J} \left(\mathbf{A}(\mathcal{E}_{i,k}^{N-1}) + \mathcal{E}_{k,j} \right), \quad \mathcal{E}_{i,j}^1 := \mathcal{E}_{i,j}. \tag{6.7}$$

Lemma 6.2 $[|\mathcal{E}_{i,j}^N|] = \mathbf{M}^N$ *for all $N \geq 1$.*

Proof. We prove the lemma by induction on N. The lemma is clearly true for $N = 1$. Assume that it holds for $N - 1$; we show that it also holds for N.

Observe that $\mathcal{E}_{i,j} \subseteq \mathcal{D}$ and $\mathcal{D}$ is a complete set of coset representatives of $\mathbf{Z}^n/\mathbf{A}(\mathbf{Z}^n)$. Hence

$$\left| \mathbf{A}(\mathcal{E}_{i,k}^{N-1}) + \mathcal{E}_{k,j} \right| = |\mathcal{E}_{i,j}^{N-1}||\mathcal{E}_{i,j}|. \tag{6.8}$$

We shall establish that for all $k \neq l$,

$$\left(\mathbf{A}(\mathcal{E}_{i,k}^{N-1}) + \mathcal{E}_{k,j} \right) \cap \left(\mathbf{A}(\mathcal{E}_{i,l}^{N-1}) + \mathcal{E}_{l,j} \right) = \emptyset. \tag{6.9}$$

This is clear if we can show that $\mathcal{E}_{k,j} \cap \mathcal{E}_{l,j} = \emptyset$ because this will mean the two sets have no elements in a same coset of $\mathbf{Z}^n/\mathbf{A}(\mathbf{Z}^n)$. Assume that $d \in \mathcal{E}_{k,j} \cap \mathcal{E}_{l,j}$. Then there exist $d_1, d_2 \in \mathcal{D}$ such that

$$\mathbf{A}\alpha_k + d_1 - d = \mathbf{A}\alpha_l + d_2 - d = \alpha_j.$$

So $\mathbf{A}(\alpha_k - \alpha_l) = d_2 - d_1$, contradicting $k \neq l$. This proves (6.9).

It now follows from (6.7) and (6.8) that

$$|\mathcal{E}_{i,j}^N| = \sum_{k=1}^{J} |\mathcal{E}_{i,j}^{N-1}||\mathcal{E}_{i,j}|,$$

proving the lemma. □

Proof of Theorem 6.1. Let $\lambda := \rho(\mathbf{M})$. Since $\mathbf{M}$ is a nonnegative matrix, there is a nonnegative eignevector v associated to λ. Without loss of generality we assume that $v_1 > 0$ and that $\mathbf{A}$ is a similarity. Let $s := n \log \rho(\mathbf{M})/\log|\det(\mathbf{A})|$. We divide the proof into three parts.

(I) $\dim_B(\partial T) \geq s$.

Let $C_i(\varepsilon)$ denote the least number of ε-cubes needed to cover T_i. Observe that

$$\mathbf{A}^N(B_1) = \bigcup_{j=1}^{J} (B_j + \mathcal{E}_{1,j}^N) \supseteq B_1 + \mathcal{E}_{1,1}^N,$$

hence

$$B_1 \supseteq \mathbf{A}^{-N}(B_1) + \mathbf{A}^{-N}\left(\mathcal{E}_{1,1}^N\right). \tag{6.10}$$

This means at least $|\mathcal{E}_{1,1}^N|$ ε_N-cubes are needed to cover B_1, where $\varepsilon_N := |\det(\mathbf{A})|^{-\frac{N}{n}}$. Hence

$$C_1(\varepsilon_N) \geq |\mathcal{E}_{1,1}^N|.$$

Now for any sufficiently small $\varepsilon > 0$ there exists an $N > 0$ such that $\varepsilon_{N+1} < \varepsilon \leq \varepsilon_N$. So

$$\frac{\log C_1(\varepsilon)}{-\log \varepsilon} \geq \frac{\log C_1(\varepsilon_N)}{-\log \varepsilon_{N+1}} \geq \frac{n \log |\mathcal{E}_{1,1}^N|}{(N+1) \log |\det(\mathbf{A})|}.$$

It is well known that $\lim_{N\to\infty} \frac{\log |\mathcal{E}_{1,1}^N|}{N} = \log \lambda$. This yields

$$\underline{\dim}_B(\partial T) \geq \underline{\dim}_B(B_1) \geq s.$$

(II) $\overline{\dim}_B(\partial T) \leq s$.

Let $\delta_0 := \max\{2\mathrm{diam}(B_j) : 1 \leq j \leq J\}$ and $\delta_N := |\det(\mathbf{A})|^{-\frac{N}{n}}$. Observe that each B_j can be covered by a single δ_0-cube. The iteration

$$\mathbf{A}^N(B_i) = \bigcup_{j=1}^J (B_j + \mathcal{E}_{i,j}^N), \quad 1 \leq i \leq J$$

yields

$$C_i(\delta_N) \leq \sum_{j=1}^J |\mathcal{E}_{i,j}^N|.$$

Note that for each $1 \leq i \leq J$,

$$\limsup_{N\to\infty} \frac{\log\left(\sum_{j=1}^J |\mathcal{E}_{i,j}^N|\right)}{-\log \delta_N} \leq \frac{n \log \lambda}{\log |\det(\mathbf{A})|} = s.$$

The same techniques employed in (I) immediately gives (II).

(III) $\dim_H(\partial T) = s$.

$\dim_H(\partial T) = s$ follows easily from Falconer [16], Theorem 3.1 and 3.2. Details can be found in [50]. □

Example 6.1 One of the best known self-affine tile is the Twin Dragon, which is given by

$$\mathbf{A} = \begin{bmatrix} 1 & -1 \\ 1 & 1 \end{bmatrix}, \quad \mathcal{D} = \left\{ \begin{bmatrix} 0 \\ 0 \end{bmatrix}, \begin{bmatrix} 1 \\ 0 \end{bmatrix} \right\}.$$

In this example, one can show ([50]) that $\mathcal{K}_0 = \{e_1, -e_1, e_2, -e_2, e_1-e_2, e_2-e_1\}$ and the substitution matrix is

$$\mathbf{M} = \begin{bmatrix} 0 & 0 & 1 & 0 & 0 & 0 \\ 0 & 0 & 0 & 1 & 0 & 0 \\ 0 & 0 & 0 & 1 & 2 & 0 \\ 0 & 0 & 1 & 0 & 0 & 2 \\ 1 & 0 & 0 & 0 & 0 & 0 \\ 0 & 1 & 0 & 0 & 0 & 0 \end{bmatrix}.$$

The characteristic polynomial is $f(\lambda) = (\lambda^3 - \lambda^2 - 2)(\lambda^3 + \lambda^2 - 2)$. Hence by Theorem 6.1,

$$\dim_H(\partial T) = 2\log_2 \lambda_0$$

where λ_0 is the largest root of $\lambda^3 - \lambda^2 - 2$.

In general, the set $\mathcal{K}_0$ for any given self-affine tile can be found via a "pruning algorithm," see [50]. One can also obtain *a priori* a set $\mathcal{K}_1 \supseteq \mathcal{K}_0$ by estimating the diameter of the tile. It turns out that the substitution matrix obtained using $\mathcal{K}_1$ will have the exactly same spectral radius as the substitution matrix from $\mathcal{K}_0$.

It should be pointed out that Duvall and Keesling [14] have recently computed the exact Hausdorff dimension of the boundary of the well known Lévy Dragon, using a rather different approach. The method in [14] can handle more general self-similar tiles, although typically requires much larger matrices (in the case of the Lévy Dragon it is a 752 matrix).

7 References

[1] C. Bandt, Self-similar sets 5. integer matrices and fractal tilings of $\mathbf{R}^n$, Proc. Amer. Math. Soc. **112**, (1991) 549–562.

[2] C. Bandt and G. Gelbrich, Classification of self-affine tilings, J. London Math. Soc. **50** (1994), 581–593.

[3] M. Barnsley, *Fractals Everywhere,* Academic Press, Inc., Boston, 1988.

[4] M. Berger and Y. Wang, Multidimensional two-scale dilation equations, in: *Wavelets – A Tutorial In Theory and Applications,* C. K. Chui, Ed., Academic Press, 1992, 295–323.

[5] R. Bowen, Markov partitions are not smooth, Proc. Amer. Math. Soc. **71** (1978), 130–132.

[6] D. Cerveau, J. Conze and A. Raugi, Ensembles invariants pour un opérateur de transfert dans $\mathbf{R}^d$, preprint.

[7] J. Conze, L. Hervé and A. Raugi, Pavages auto-affines, opérateur de transfert et critères de réseau dans $\mathbf{R}^d$, preprint.

[8] L. Danzer, A family of 3-d-spacefillers not permitting any periodic or quasiperiodic tiling, preprint.

[9] I. Daubechies, *Ten Lectures on Wavelets*, SIAM: Philadelphia, 1992.

[10] I. Daubechies and J. C. Lagarias, Two scale difference equations I. global regularity of solutions, SIAM J. Math. Anal. **22** (1991), 1388–1410.

[11] C. De Boor and K. Höllig, Box spline tilings, Amer. Math. Monthly **98** (1991), 793–802.

[12] F. M. Dekking, Recurrent sets, Advances in Math. **44** (1982), 78–104.

[13] F. M. Dekking, Replicating superfigures and endomorphisms of free groups, J. Comb. Th., Series A, **32** (1982), 315–320.

[14] P. Duvall and J. Keesling, The hausdorff dimension of the boundary of the lévy dragon, preprint (1997).

[15] K. J. Falconer, *Fractal Geometry: Mathematical Foundations and Applications*, John Wiley & Sons (1990).

[16] K. J. Falconer, *Techniques in Fractal Geometry*, John Wiley & Sons (1990).

[17] T. Flaherty and Y. Wang, Haar-type multiwavelet bases and self-affine multi-tiles, preprint (1997)

[18] M. Gardner, *Penrose Tiles to Trapdoor Ciphers*, W. H. Freeman, New York, 1989.

[19] W. Gilbert, Geometry of radix representations, in: *The Geometric Vein: the Coxeter Festschrift*, 1981, 129–139.

[20] F. Girault-Beauquier and M. Nivat, Tiling the plane with one tile, in: *Topology and Category Theory in Computer Science*, Oxford U. Press, 1989, 291–333.

[21] K. Gröchenig, Orthogonality criteria for compactly supported scaling functions, Appl. Comp. Harmonic Analy. **1** (1994), 242-245.

[22] K. Gröchenig and A. Haas, Self-similar lattice tilings, J. Fourier Anal. Appl. **1** (1994), 131–170.

[23] K. Gröchenig and W. Madych, Multiresolution analysis, haar bases, and self-similar tilings, IEEE Trans. Info. Th. **IT-38**, No. 2, Part II (1992), 556–568.

[24] B. Grunbaum and G. C. Shepard, *Tilings and Patterns*, W. H. Freeman & Co., New York, 1987.

[25] A. Haar, Zur theorie der orthogonalen funktionen–systeme, Math. Ann. **69** (1910), 331–371.

[26] D. Hacon, N. Saldanha and P. Veerman, Self-similar tilings of $\mathbf{R}^n$, Experimental Math.

[27] J. E. Hutchinson, Fractals and self-similarity, Indiana U. Math. J. **30** (1981), 713–747.

[28] R. Kenyon, Self-similar tilings, Ph.D. thesis, Princeton University, 1990.

[29] R. Kenyon, Self-replicating tilings, in: *Symbolic Dynamics and Its Applications* (P. Walters, Ed.), Contemporary Math., Vol. 135, 1992, 239–264.

[30] R. Kenyon, Rigidity of planar tilings, Inventiones Math. **107** (1992), 637–651.

[31] D. Knuth, *The Art of Computer Programming: Volume 2. Seminumerical Algorithms* (Second Edition), Addison-Wesley: Reading, MA, 1981.

[32] J. C. Lagarias and Y. Wang, Haar type orthonormal wavelet basis in $\mathbf{R}^2$, J. Fourier Analysis and Appl. **2** (1995), 1-14.

[33] J. C. Lagarias and Y. Wang, Self-affine tiles in $\mathbf{R}^n$, Adv. in Math. **121** (1996), 21-49.

[34] J. C. Lagarias and Y. Wang, Integral self-affine tiles in $\mathbf{R}^n$ I. Standard and nonstandard digit sets, J. London Math. Soc. **53** (1996), 161-179.

[35] J. C. Lagarias and Y. Wang, Haar bases in $\mathbf{R}^n$ and algebraic number theory, J. Number Theory **57** (1996), 181-197.

[36] J. C. Lagarias and Y. Wang, Integral self-affine tiles in $\mathbf{R}^n$, part II: lattice tilings, J. Fourier Anal. and Appl. **3** (1997), 83-102.

[37] D. Lind, Dynamical properties of quasihyperbolic toral automorphisms, Ergod. Th. Dyn. Sys. **2** (1982), 49–68.

[38] S. Mallat, Multiresolution analysis and wavelets, Tans. Amer. Math. Soc. **315** (1989), 69-88.

[39] D. W. Matula, Basic digit sets for radix representations, J. Assoc. Comp. Mach. **4** (1982), 1131–1143.

[40] A. M. Odlyzko, Non-negative digit sets in positional number systems, Proc. London Math. Soc., 3rd Series, **37** (1978), 213–229.

[41] B. Praggastis, Markov partitions for hyperbolic toral automorphisms, Ph.D. Thesis, Univ. of Washington, 1992.

[42] C. Radin, Space tilings and substitutions, Geom. Dedicata **55** (1995), 257–264.

[43] C. Radin and M. Wolff, Space tilings and local isomorphism, Geom. Dedicata **42** (1992), 355–360.

[44] P. Schmitt, An aperiodic prototile in space, preprint, 1993.

[45] C. Stein, *Singular Integrals and Differentiability Properties of Functions,* Princeton U. Press, Princeton, NJ. 1970.

[46] R. S. Strichartz, Self-similar measures and their fourier transforms I., Indiana U. Math. J. **39** (1990), 797-817.

[47] R. S. Strichartz, Self-similar measures and their fourier transforms II., Trans. Amer. Math. Soc. **2** (1993), 335-361.

[48] R. Strichartz, Wavelets and self-affine tilings, Constructive Approx. **9** (1993), 327-346.

[49] R. Strichartz, Self-similarity in harmonic analysis, J. Fourier Anal. Appl. **1** (1994), 1-37.

[50] R. Strichartz and Y. Wang, Geometry of self-affine tiles I, preprint (1998).

[51] W. Thurston, Groups, tilings, and finite state automata, AMS Colloquium Lecture Notes, unpublished, 1989.

[52] A. Vince, Replicating tesselations, SIAM J. Discrete Math., **6** (1993), 501-521.

[53] J. Walters, *Ergodic Theory,* Springer-Verlag, 1970.